Symmetries in
Nuclear Structure

NATO Advanced Science Institutes Series

A series of edited volumes comprising multifaceted studies of contemporary scientific issues by some of the best scientific minds in the world, assembled in cooperation with NATO Scientific Affairs Division.

This series is published by an international board of publishers in conjunction with NATO Scientific Affairs Division

A Life Sciences **B Physics**	Plenum Publishing Corporation New York and London
C Mathematical and Physical Sciences	D. Reidel Publishing Company Dordrecht, Boston, and London
D Behavioral and Social Sciences **E Applied Sciences**	Martinus Nijhoff Publishers The Hague, Boston, and London
F Computer and Systems Sciences **G Ecological Sciences**	Springer Verlag Heidelberg, Berlin, and New York

Recent Volumes in Series B: Physics

Volume 87 —Relativistic Effects in Atoms, Molecules, and Solids
edited by G. L. Malli

Volume 88 —Collective Excitations in Solids
edited by Baldassare Di Bartolo

Volume 89a —Electrical Breakdown and Discharges in Gases: Fundamental Processes and Breakdown
edited by Erich E. Kunhardt and Lawrence H. Luesen

Volume 89b —Electrical Breakdown and Discharges in Gases: Macroscopic Processes and Discharges
edited by Erich E. Kunhardt and Lawrence H. Luessen

Volume 90 —Molecular Ions: Geometric and Electronic Structures
edited by Joseph Berkowitz and Karl-Ontjes Groeneveld

Volume 91 —Integrated Optics: Physics and Applications
edited by S. Martellucci and A. N. Chester

Volume 92 —The Physics of Superionic Conductors and Electrode Materials
edited by John W. Perram

Volume 93 —Symmetries in Nuclear Structure
edited by K. Abrahams, K. Allaart, and A. E. L. Dieperink

Symmetries in Nuclear Structure

Edited by

K. Abrahams

Netherlands Energy Research Foundation
Petten, The Netherlands

K. Allaart

Free University
Amsterdam, The Netherlands

and

A. E. L. Dieperink

Nuclear Accelerator Institute
Groningen, The Netherlands

Plenum Press
New York and London
Published in cooperation with NATO Scientific Affairs Division

Proceedings of a NATO Advanced Study Institute on
Symmetries in Nuclear Structure
organized by the Netherlands Physical Society (NNV),
held August 16–28, 1982,
in Dronten, The Netherlands

NNV

Library of Congress Cataloging in Publication Data

NATO Advanced Study Institute on Symmetries in Nuclear Structure (1982: Dronten, Netherlands)
Symmetries in nuclear structure.

(NATO advanced science institutes series. Series B, Physics; v. 93)
"Proceedings of a NATO Advanced Study Institute on Symmetries in Nuclear Structure organized by the Netherlands Physical Society (NNV), held August 16–28, 1982, in Dronten, the Netherlands"—T.p. verso.
Bibliography: p.
Includes index.
1. Symmetry (Physics)—Congresses. 2. Nuclear structure—Congresses. I. Abrahams, K. II. Allaart, K. III. Dieperink, A. E. L. IV. Title. V. Series.
QC793.3.S9N37 1982 539.7′2 83-2455
ISBN 0-306-41341-8

A Division of Plenum Publishing Corporation
233 Spring Street, New York, N.Y. 10013

Printed in the United States of America

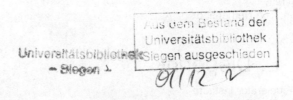

FOREWORD

The 1982 summer school on nuclear physics, organized by the Nuclear Physics Division of the Netherlands' Physical Society, was the fifth in a series that started in 1963. The number of students attending has always been about one hundred, coming from about thirty countries.

The theme of this year's school was symmetry in nuclear physics. This book covers the material presented by the enthusiastic speakers, who were invited to lecture on this subject. We think they have succeeded in presenting us with clear and thorough introductory talks at graduate or higher level.

The time schedule of the school and the location allowed the participants to make many informal contacts during many social activities, ranging from billiards to surf board sailing. We hope and expect that the combination of a relaxed atmosphere during part of the time and hard work during most of the time, has furthered the interest in, and understanding of, nuclear physics.

The organization of the summer school was made possible by substantial support from the Scientific Affairs Division of the North Atlantic Treaty Organization, the Netherlands' Ministry of Education and Science, the Foundation Physica and the Netherlands' Physical Society.

The invaluable help of the 'Bureau Congressen' of the Ministry of Education and Science and the friendly assistance of the management of the College of Agriculture in Dronten have contributed greatly to the success of the summer school. Although during the school some particular jargon developed and even unusual double entendre's came into existence, it is justified to refer to the school as 'fantastic'.

P.J. Brussaard

PREFACE

In this book it will be shown how symmetries and their breakdown play a role in the description of nuclear concepts such as electromagnetic and weak forces and their unification. Further, the current shell model as well as the interacting boson model of the atomic nucleus are described. The complementarity between theory and experiment has been exploited in several chapters on applications, such as the search for neutrino masses, and the study of electron scattering. These applications are rounded off by a representation of intermediate energy physics in a chapter on Δ dynamics and some suggestions for experimental work.

During the international summerschool in Dronten, the Netherlands, 16-28 August 1982, these topics were presented by a group of distinguished lecturers. To speed the publication of their notes, the corrected and sometimes revised manuscripts are published under reponsibility of the editors without consulting all authors.

The assistance of the secretaries and the documentalist of the Physics Department of the Netherlands Energy Research Foundation is gratefully acknowledged.

K. Abrahams
K. Allaart
A.E.L. Dieperink

CONTENTS

ELECTROMAGNETIC AND WEAK

CURRENTS IN NUCLEI*

T. W. Donnelly

Center for Theoretical Physics
Laboratory for Nuclear Science and Department of Physics
Massachusetts Institute of Technology
Cambridge, Massachusetts 02139

1. INTRODUCTION

These lectures are focussed on the basic formalism used in descriptions of electromagnetic and weak interactions with nuclei. As other speakers, whose lectures are published in these same Summer School proceedings, have emphasized the connections of this basic formalism to analyses of experimental data (particularly, I. Sick for discussions of electron scattering and E. G. Adelberger for weak interaction processes), no further attempt has been made to illustrate the ideas presented here with detailed examples. Additional illustrations can be found in the primary source material used in preparing these lectures.[1-5] Moreover, since the underlying particle physics description of the electroweak interaction is discussed in these Summer School proceedings by D. Atkinson, this will largely be taken as given in developing the formalism for application to nuclear physics (see also Ref. 5).

These lectures have been organized into nine sections: Secs. 2-6 deal with the electromagnetic interaction and Secs. 7-10 focus on the weak interaction. In Sec. 2 we begin by considering the simple problem of electron scattering from a point Dirac proton. This description is improved in Sec. 3 by allowing the nucleon to have internal structure. In Secs. 4 and 5, inclusive electron scattering from a nucleus is discussed, with Sec. 4 containing a general treatment of the (e,e') reaction and Sec. 5 specializing this to energies where discrete states in the nucleus are excited.

*This work is supported in part through funds provided by the U. S. Department of Energy (DOE) under contract DE-AC02-76ER03069.

In this latter situation, a multipole analysis of the electromagnetic current is dictated. In Sec. 6 we then conclude the discussion of the electromagnetic interaction with the general formalism involved in treating exclusive electron scattering, (e,e'x).

The treatment of the weak interaction closely parallels the approach taken for the electromagnetic interaction in Secs. 2-6. In Sec. 7 the point Dirac nucleon and dressed nucleon weak currents are discussed, followed in Sec. 8 by consideration of general inclusive reactions and their nuclear response functions (as in Secs. 4 and 5 for the electromagnetic interaction). Finally these lectures conclude by displaying specific forms for a variety of charge-changing (Sec. 9) and neutral-current (Sec. 10) weak interaction cross sections.

2. ELECTRON SCATTERING FROM A POINT DIRAC PROTON

We begin by discussing electron scattering in lowest order (one-photon-exchange) from a point Dirac proton as represented by the Feynman diagram shown in Fig. 1. Here an electron with 4-momentum K^μ is scattered to K'^μ, exchanging a photon having 4-momentum $Q^\mu = K^\mu - K'^\mu$ with the proton. The proton goes from P^μ to P'^μ, where energy-momentum conservation implies that $Q^\mu = P'^\mu - P^\mu$. A few words on conventions are in order at this point: in the present lectures the Bjorken and Drell[6] metric, etc. are employed, in contrast to much of the past work on electroweak interactions with nuclei. This is done to provide an easier bridge between a widely available standard treatment of quantum electrodynamics (namely, Bjorken and Drell[6]) and the specifics of nuclear physics. The reader is particularly directed to the appendices in Ref. 6 (pp. 281-290) and to several useful trace relationships (pp. 104-105). In the present work, 4-vectors are denoted by capital letters, $A^\mu = (A^0,\vec{A})$, $B^\mu = (B^0,\vec{B})$, etc. and the 4-vector scalar products are $A_\mu B^\mu \equiv A \cdot B = A^0 B^0 - \vec{A} \cdot \vec{B}$, where as usual summation over repeated indices is implied. The metric raising and lowering operator is $g_{\mu\nu} = g^{\mu\nu}$ = (+1 for $\mu = \nu = 0$, -1 for $\mu = \nu = 1,2,3$, 0 otherwise). Furthermore, lower case letters are used to denote the magnitudes of 3-vectors: $a = |\vec{A}|$, $b = |\vec{B}|$, etc. For the momenta involved in

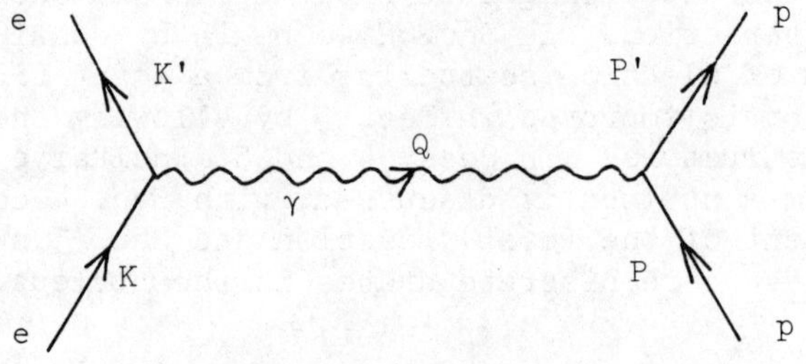

Figure 1. Electron-proton scattering.

electron scattering it is convenient to use other symbols as well:

$$K^{\mu} = (K^0, \vec{K}) \quad \text{with } \varepsilon \equiv K^0,\ k = |\vec{K}|$$
$$K'^{\mu} = (K'^0, \vec{K}') \quad \text{with } \varepsilon' \equiv K'^0,\ k' = |\vec{K}'|$$
$$P^{\mu} = (P^0, \vec{P}) \quad \text{with } E = P^0,\ p = |\vec{P}|$$
$$P'^{\mu} = (P'^0, \vec{P}') \quad \text{with } E' = P'^0,\ p' = |\vec{P}'|$$

and $Q^{\mu} = (Q^0, \vec{Q})$ with $\omega = Q^0$, the energy transfer and $q = |\vec{Q}|$, the 3-momentum transfer. Indeed we must have $K^2 = K_{\mu}K^{\mu} = m_e{}^2$, $K'^2 = m_e{}^2$ and $P^2 = P'^2 = M_p{}^2$, where m_e is the electron mass and M_p is the proton mass. These on-mass-shell conditions imply that

$$Q^2 = Q_{\mu}Q^{\mu} = \omega^2 - q^2 \leq 0 \rightarrow q \geq \omega \ ; \tag{2.1}$$

in other words, the virtual photon in electron scattering must be spacelike.

We shall only consider electromagnetic (and later, weak interaction) processes in lowest order, relying on the fact that the basic coupling at each vertex is proportional to $\sqrt{\alpha}$, where α is the fine-structure constant, and so, for example, two-photon-exchange amplitudes are generally much suppressed with respect to exchange of a single virtual photon. It is this weakness of the basic process which makes electron scattering such a powerful, relatively easily interpreted tool for studying nuclear (and particle) physics.

The application of the Feynman rules to e-p scattering proceeds straightforwardly (see Ref. 6, p. 108). In Fig. 2 the appropriate

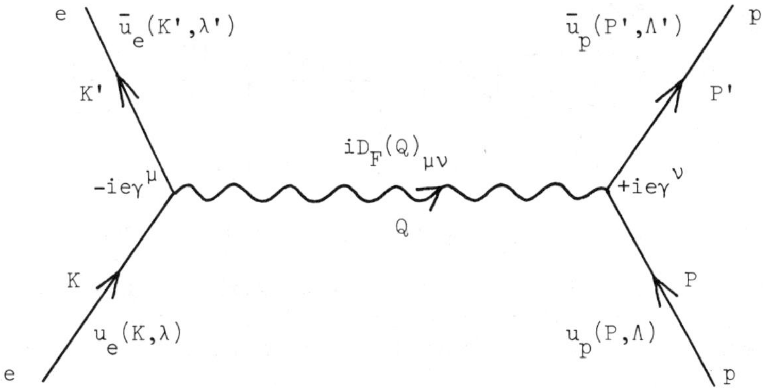

Figure 2. Application of the Feynman rules to lowest-order electron scattering from a point Dirac proton.

factors to be associated with the lines and vertices in the diagram are indicated. Here we take $e > 0$ to be the proton charge (and hence minus the electron charge). We have electron (proton) spinors u_e, $\bar{u}_e$ (u_p, $\bar{u}_p$), labelled with momentum and helicity; for example, for the incident electron of momentum K and helicity λ, we have $u_e(K,\lambda)$. The virtual photon exchanged between the electron and proton is represented by the photon propagator, $D_F(Q)_{\mu\nu} = g_{\mu\nu}/(-Q^2)$. Assembling all of these factors leads to the relevant invariant matrix element:

$$\mathcal{M}_{fi} = [\bar{u}_e(K',\lambda')(-ie\gamma^\mu)u_e(K,\lambda)] \times \left[\frac{-ig_{\mu\nu}}{Q^2}\right] \times \left[\bar{u}_p(P',\Lambda')(+ie\gamma^\nu)u_p(P,\Lambda)\right] \tag{2.2}$$

$$= \frac{i}{Q^2}\left(\frac{\varepsilon\varepsilon' EE'}{m_e{}^2 M_p{}^2}\right)^{\frac{1}{2}} j_e(K',\lambda';K,\lambda)_\mu \, j_p(P',\Lambda';P,\Lambda)^\mu , \tag{2.3}$$

when the electromagnetic currents of the electron and proton are given respectively by

$$j_e(K',\lambda';K,\lambda)_\mu = -e\left(\frac{m_e{}^2}{\varepsilon\varepsilon'}\right)^{\frac{1}{2}} \bar{u}_e(K',\lambda')\gamma_\mu u_e(K,\lambda) \tag{2.4a}$$

$$j_p(P',\Lambda';P,\Lambda)^\mu = +e\left(\frac{M_p{}^2}{EE'}\right)^{\frac{1}{2}} \bar{u}_p(P',\Lambda')\gamma^\mu u_p(P,\Lambda) . \tag{2.4b}$$

Then following Bjorken and Drell (Ref. 6, Appendix B) we obtain the differential cross section in the laboratory system, where $|\vec{P}| = p = 0$, $P^0 = E = M_p$:

$$d\sigma^{lab} = \frac{1}{\beta}\left(\frac{m_e}{\varepsilon}\right) \overline{\sum_{if}} |\mathcal{M}_{fi}|^2 \left[\frac{m_e}{\varepsilon'}\frac{d\vec{K}'}{(2\pi)^3}\right] \left[\frac{M_p}{E'}\frac{d\vec{P}'}{(2\pi)^3}\right] \times (2\pi)^4\delta^{(4)}\big((K + P) - (K' + P')\big) , \tag{2.5}$$

with $\beta \equiv k/\varepsilon$ and with $\overline{\sum_{if}}$ indicating the average over initial and sum over final states (see below). Now we wish to discuss the process p(e,e')p, that is, where the scattered proton is not presumed to be detected. Thus the $\int d\vec{P}'$ must be performed, leaving

$$d\sigma^{lab} = \frac{m_e{}^2 M_p}{(2\pi)^2 k\varepsilon' E'} \overline{\sum_{if}} |\mathcal{M}_{fi}|^2 \delta^{(1)}\big((\varepsilon + M_p) - (\varepsilon' + E')\big) d\vec{K}', \tag{2.6}$$

with $p'=|\vec{P}'|=|\vec{K}-\vec{K}'|=|\vec{Q}|=q$. Furthermore, we have

$$d\vec{K}' = (k')^2 dk' d\Omega_e, \tag{2.7}$$

where, as usual, $k'=|\vec{K}'|$ and here we take the scattering angle of the electron to be θ_e (solid angle $d\Omega_e$). Thus we have

$$\left(\frac{d\sigma}{d\Omega_e}\right)^{lab}_{(e,e')} = \frac{m_e^2 M_p}{(2\pi)^2 k} \int d\varepsilon' \frac{\partial k'}{\partial \varepsilon'} \frac{(k')^2}{\varepsilon' E'} \times \delta^{(1)}((\varepsilon + M_p) - (\varepsilon' + E')) \overline{\sum_{if}} |\mathcal{M}_{fi}|^2. \tag{2.8}$$

Since $k'=((\varepsilon')^2-m_e^2)^{1/2}$, we have $\partial k'/\partial\varepsilon'=\varepsilon'/k'$. Also, if we let $s\equiv\varepsilon+E=\varepsilon+M_p$ and $s'\equiv\varepsilon'+E'$, then converting to $\int ds'\delta^{(1)}(s'-s)x....$ using

$$\frac{\partial s'}{\partial \varepsilon'} = 1 + \frac{\varepsilon'}{k'}\left(\frac{k'-k\cos\theta_e}{E'}\right) \equiv \left(\frac{M_p}{E'}\right) f_{rec}, \tag{2.9}$$

where f_{rec} is a recoil correction factor, we obtain

$$\left(\frac{d\sigma}{d\Omega_e}\right)^{lab}_{(e,e')} = \frac{m_e^2}{(2\pi)^2} \frac{k'}{k} f_{rec}^{-1} \overline{\sum_{if}} |\mathcal{M}_{fi}|^2. \tag{2.10}$$

Now it remains to consider the invariant matrix element given in (2.2) and (2.3). We have upon substituting

$$\overline{\sum_{if}} |\mathcal{M}_{fi}|^2 = \overline{\sum_{if}} \frac{e^4}{(Q^2)^2} \Big| [\bar{u}_e(K',\lambda')\gamma_\mu u_e(K,\lambda)] \times [\bar{u}_p(P',\Lambda')\gamma^\mu u_p(P,\Lambda)] \Big|^2 \tag{2.11a}$$

$$= \frac{(4\pi\alpha)^2}{(Q^2)^2} \frac{1}{2} \sum_{\substack{\lambda=\pm\frac{1}{2} \\ \lambda'=\pm\frac{1}{2}}} (\bar{u}_e(K',\lambda')\gamma_\mu u_e(K,\lambda))^*(\bar{u}_e(K',\lambda')\gamma_\nu u_e(K,\lambda))$$

$$\times \frac{1}{2} \sum_{\substack{\Lambda=\pm\frac{1}{2} \\ \Lambda'=\pm\frac{1}{2}}} (\bar{u}_p(P',\Lambda')\gamma^\mu u_p(P,\Lambda))^*(\bar{u}_p(P',\Lambda')\gamma^\nu u_p(P,\Lambda)), \tag{2.11b}$$

where we have presumed that the desired cross section is to represent unpolarized electron scattering and hence have averaged over the initial spin projections $(\frac{1}{2}\sum_\lambda \cdot \frac{1}{2}\sum_\Lambda)$ and summed over the final spin projections $(\sum_{\lambda'} \cdot \sum_{\Lambda'})$. For convenience we define an electron tensor $\eta_e(K',K)_{\mu\nu}$ (and an analogous quantity for the proton), where

$$\eta_e(K';K)_{\mu\nu} \equiv \tfrac{1}{2} \sum_{\substack{\lambda=\pm\frac{1}{2}\\ \lambda'=\pm\frac{1}{2}}} \bar{u}_e(K,\lambda)\gamma_\mu u_e(K',\lambda')\bar{u}_e(K',\lambda')\gamma_\nu u_e(K,\lambda), \quad (2.12)$$

using properties of the Dirac γ-matrices. Then, using standard techniques, we insert positive energy projection operators, $\Lambda_{e,+}(K')$ just after the γ_μ and $\Lambda_{e,+}(K)$ just after the γ_ν and so may sum over all four components of the spinors. Explicitly these projection operators are given by[6]

$$\Lambda_{e,+}(K) = \frac{1}{2m_e}(\not{K} + m_e), \text{ where } \not{K} \equiv K^\alpha\gamma_\alpha . \quad (2.13)$$

So we obtain the electron tensor

$$\eta_e(K';K)_{\mu\nu} = \tfrac{1}{2}\mathrm{Tr}\left\{\gamma_\mu \frac{\not{K}'+m_e}{2m_e} \gamma_\nu \frac{\not{K}+m_e}{2m_e}\right\} , \quad (2.14)$$

where Tr{...} indicates that the spinor trace be taken. Using well-known trace identities (see Ref.6, pp.104-105) we have

$$\eta_e(K';K)_{\mu\nu} = \frac{1}{2m_e^2} [K_\mu K'_\nu + K'_\mu K_\nu - g_{\mu\nu}(K\cdot K' - m_e^2)] . \quad (2.15)$$

Several properties of $\eta_e(K';K)_{\mu\nu}$ should be noted for future reference: (1) It is a second-rank Lorentz tensor; (2) It is symmetric under the interchange ($\mu\leftrightarrow\nu$); (3) It is constructed from the product of one polar vector (j_μ) with another polar vector (j_ν); (4) Since specifically those polar vectors are the electromagnetic current matrix elements and we have current conservation, $Q^\mu j_\mu = 0$, we must have

$$Q^\mu \eta_e(K';K)_{\mu\nu} = Q^\nu \eta_e(K';K)_{\mu\nu} = 0 , \quad (2.16)$$

where as usual $Q^\mu = K^\mu - K'^\mu$. Substituting

$$\overline{\sum_{if}} |\mathcal{M}_{fi}|^2 = \frac{(4\pi\alpha)^2}{(Q^2)^2} \eta_e(K';K)_{\mu\nu}\, \eta_p(P';P)^{\mu\nu} \quad (2.17)$$

into (2.10) with appropriate values for the momenta we obtain the desired cross section. In fact, rather than writing this general result we make two simplifying assumptions:

(1) We assume that we are in the extreme-relativistic limit for the electron (ERL_e), in which case $\varepsilon, \varepsilon' >> m_e$ and so $k \approx \varepsilon$, $k' \approx \varepsilon'$. We then find for example that the recoil factor takes on its more familiar form:

$$f_{rec}^{ERL_e} = 1 + \frac{2\varepsilon \sin^2 \Theta_e}{M_p} ; \tag{2.18}$$

(2) We assume that we are in the non-relativistic limit for the proton (NRL_p), that is, where $p' << M_p$.

We then obtain the standard result for electron scattering from a point Dirac proton in lowest order:

$$\left(\frac{d\sigma}{d\Omega_e}\right)_{(e,e')}^{lab,ERL_e,NRL_p} = f_{rec}^{-1}\, \sigma_M , \tag{2.19}$$

where $\sigma_M = \left(\frac{\alpha \cos \Theta_e/2}{2\varepsilon \sin^2 \Theta_e/2}\right)^2$ is the Mott cross section which may be recognized as having the characteristic $1/\sin^4 \Theta_e/2$ behavior of the Rutherford cross section modified by the numerator's $\cos^2 \Theta_e/2$ behavior which reflects the relativistic nature of the (spin ½) electron.

While using nothing but techniques which are entirely standard in discussions of quantum electrodynamics, this section has been displayed in some detail as it provides the basis for all of the remaining sections in which less detail is given but where the approach is completely analogous.

3. ELECTRON SCATTERING FROM PHYSICAL NUCLEONS

Let us now improve this picture in several steps. Recall that for a point Dirac proton we had (2.4b)

$$j_p(P'\Lambda';P\Lambda)^\mu \Big|_{\text{point Dirac proton}} = e\left(\frac{M_p^2}{EE'}\right)^{\frac{1}{2}} \bar{u}_p(P',\Lambda')\gamma^\mu\, u_p(P,\Lambda) .$$

Now, in fact, in addition to this term (see Fig. 3(a)) there may be contributions from other diagrams (such as Figs. 3(b),(c),...) which altogether yield the electromagnetic current of the dressed or physical proton. Alternatively, in the language of quarks and gluons (QCD), once again the proton must have structure and not merely be treated as a point Dirac particle (which presumably is the case for a quark). What we do know is that the physical electromagnetic current is a Lorentz 4-vector, is a polar vector under spatial

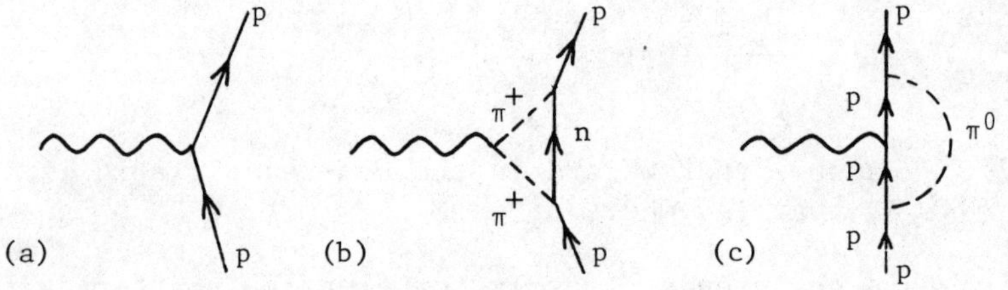

Figure 3. Diagrams contributing to the electromagnetic current of the physical proton.

inversion and is conserved. That it is a polar Lorentz 4-vector (called a vector current) implies that in general we must replace $\bar{u}_p(P',\Lambda')\gamma^\mu u_p(P,\Lambda)$ above by

$$\bar{u}_p(P',\Lambda')[F_1(Q^2)\gamma^\mu + \tilde{F}_2(Q^2)\sigma^{\mu\nu}Q_\nu + F_S(Q^2)Q^\mu]u_p(P,\Lambda) \ ,$$

where $\sigma^{\mu\nu} = \frac{i}{2}[\gamma^\mu,\gamma^\nu]$ (see Ref. 6, Appendix A), and where the proton form factors $F_1(Q^2)$, $\tilde{F}_2(Q^2)$ and $F_S(Q^2)$ reflect the finite spatial extent of the proton. Furthermore, since this vector current is conserved, $Q_\mu j_p(P'\Lambda';P\Lambda)^\mu = 0$ with $Q_\mu = P'_\mu - P_\mu$, we must have

$$\bar{u}_p(P',\Lambda')[F_1(Q^2)\not{Q} + \tilde{F}_2(Q^2)\sigma^{\mu\nu}Q_\mu Q_\nu + F_S(Q^2)Q^2]u_p(P,\Lambda) = 0 \tag{3.1}$$

The second term is automatically zero, being a product of an antisymmetric quantity ($\sigma^{\mu\nu}$) with a symmetric quantity ($Q_\mu Q_\nu$). In the first term we have $\not{Q} = \not{P}' - \not{P} = (\not{P}' - M_p) - (\not{P} - M_p)$ and we may use the Dirac equation to the right and to the left,

$$\begin{aligned} (\not{P} - M_p)u_p(P,\Lambda) &= 0 \\ \bar{u}(P',\Lambda')(\not{P}' - M_p) &= 0 \end{aligned} \tag{3.2}$$

to show that it also vanishes. Thus we conclude that

$$Q^2F_S(Q^2) \quad \bar{u}_p(P',\Lambda')u_p(P,\Lambda) = 0 \ , \tag{3.3}$$

or, since the product $\bar{u}_p u_p$ is in general not zero, that $F_S(Q^2) = 0$. Rewriting this in a more standard form we have seen that the physical proton electromagnetic current should be

$$j_p(P'\Lambda';P\Lambda)^\mu\Big|_{\text{physical proton}}$$

$$= e\left(\frac{M_p}{EE'}\right)^{1/2} \bar{u}_p(P',\Lambda')[\gamma^\mu F_1^{\,p}(Q^2) + \frac{i}{2M_p}\sigma^{\mu\nu}Q_\nu \kappa^p F_2^{\,p}(Q^2)]u_p(P,\Lambda) \ . \tag{3.4}$$

Thus, in addition to the usual Dirac piece (now modified by a form factor, $F_1{}^p(Q^2)$), we have a contribution from the anomalous magnetic moment of the proton, $\kappa^p = 1.79$, also multiplied by a form factor, $F_2{}^p(Q^2)$. The normalization has been chosen so that $F_1{}^p(0) = F_2{}^p(0) = 1$.

Of course the same ideas apply to the neutron, for, although it does not have a total charge, it does have a charge distribution, and so $F_1{}^n(Q^2)$ is not equal to zero for all values of Q^2. Moreover the neutron has an anomalous magnetic moment, $\kappa^n = -1.91$ and an attendant form factor $F_2{}^n(Q^2)$. So (3.4) may be used for neutrons as well as protons after making the following replacements:

$$\begin{aligned} \text{proton} &\rightarrow \text{neutron} \\ M_p &\rightarrow M_n \\ \bar{u}_p, u_p &\rightarrow \bar{u}_n, u_n \\ \kappa^p &\rightarrow \kappa^n \\ F_1{}^p, F_2{}^p &\rightarrow F_1{}^n, F_2{}^n \; . \end{aligned}$$

The normalization is such that $F_1{}^n(0) = 0$, $F_2{}^n(0) = 1$. Finally it is useful to extend the proton and neutron spinors to nucleon spinors, that is, spinors in the usual sense (4 component) as well as in isospin space, $u_N(P,\Lambda,\xi)$, with $\xi = +\frac{1}{2} \leftrightarrow$ proton. Then letting

$$\begin{aligned} F_1(Q^2) &\equiv F_1{}^s(Q^2) + \tau_3 F_1{}^v(Q^2) \\ F_2(Q^2) &\equiv F_2{}^s(Q^2) + \tau_3 F_2{}^v(Q^2) \; , \end{aligned} \tag{3.5}$$

that is, having isoscalar (s) and isovector (v) parts, where

$$\begin{aligned} F_1{}^s &\equiv \frac{1}{2}(F_1{}^p + F_1{}^n) \; , \quad F_2{}^s \equiv \frac{1}{2}(\kappa^p F_2{}^p + \kappa^n F_2{}^n) \\ F_1{}^v &\equiv \frac{1}{2}(F_1{}^p - F_1{}^n) \; , \quad F_2{}^v \equiv \frac{1}{2}(\kappa^p F_2{}^p - \kappa^n F_2{}^n) \; , \end{aligned} \tag{3.6}$$

we have (taking $M_p \cong M_n \equiv M_N$, the nucleon mass)

$$j_N(P'\Lambda'\xi';P\Lambda\xi)^\mu\big|_{\text{physical nucleon}}$$

$$= e\left(\frac{M_N}{EE'}\right)^{1/2} \bar{u}_N(P',\Lambda',\xi')[\gamma^\mu F_1(Q^2) + \frac{i}{2M_N}\sigma^{\mu\nu}Q_\nu F_2(Q^2)]u_N(P,\Lambda,\xi) \; . \tag{3.7}$$

Making this replacement of $j^{\mu}_{\text{point Dirac proton}}$ with $j^{\mu}_{\text{physical nucleon}}$ in the developments of Sec. 2 will lead to a physical nucleon electromagnetic tensor, that is,

$$\eta_p(P';P)^{\mu\nu}\Big|_{\text{point Dirac proton}} \qquad \text{(compare Eq. (2.12))}$$

$\rightarrow \eta_N(P'\xi';P\xi)^{\mu\nu}\Big|_{\text{physical nucleon}}$. Of course traces now must be performed where $\sigma^{\mu\nu}$ is involved. The result will be the Rosenbluth cross section for electron scattering from the nucleon. We shall take as given comparisons with experiment for e-p scattering (and with e - ^{2}H scattering to extract information on e-n form factors) and so shall assume that the nucleon form factors $F_1{}^{p,n}(Q^2)$ and $F_2{}^{p,n}(Q^2)$ are known. In fact this continues to be an interesting subject in its own right: for example, the charge form factor of the neutron is rather poorly known. For our subsequent discussion of nuclear physics it will usually be sufficient to approximate the Q^2-dependence of all four form factors by

$$f_N(Q^2) = [1 - Q^2/\alpha_N{}^2]^{-2}\ , \text{ where } \alpha_N = 855 \text{ MeV}. \qquad (3.8)$$

We now turn in the next section to the problem of electron scattering from a nucleus, again building on the development to this point.

4. GENERAL DISCUSSION OF INCLUSIVE ELECTRON SCATTERING, (e,e')

We now turn to a discussion of the problem of electron scattering from the entire nucleus. In Secs. 4 and 5 we shall presume that only the scattered electron is detected, that is, we shall focus on inclusive electron scattering (in Sec. 6 we return to consider exclusive scattering as well). The process under consideration is shown diagrammatically in Fig. 4. Here we take the initial nuclear state to be $|i\rangle$ with 4-momentum P_i and the final nuclear state to be $|f\rangle$ with P_f, where as before the momentum transfer (now from the electron to the entire nucleus) is $Q^{\mu} = K^{\mu} - K'^{\mu} = P_f{}^{\mu} - P_i{}^{\mu}$. We may have elastic (i = f) or inelastic (i ≠ f) scattering depending on the kinematics of the reaction. With $P_i{}^{\mu}P_{i\mu} = P_i{}^2 = M_i$ and $P_f{}^2 = M_f{}^2$, where M_i and M_f are the initial and final state masses respectively

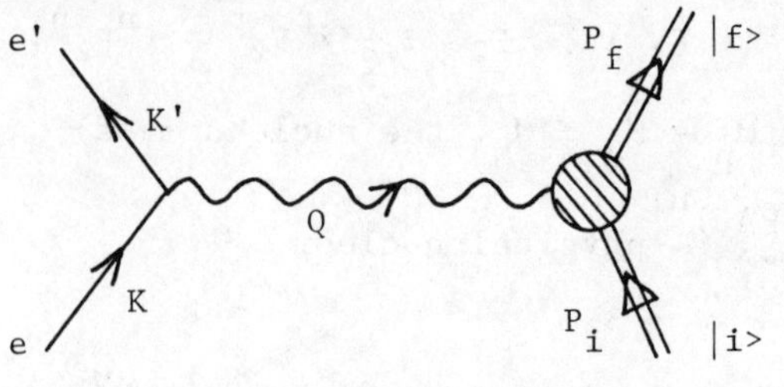

Figure 4. Electron scattering from the nucleus lowest order.

(note that M_f includes the internal excitation energy of the nucleus if the state $|f>$ is an excited state), we again find that the momentum transfer must be spacelike: $Q^2 \leq 0 \rightarrow \omega \leq q$.

We now build on the material in Secs. 2 and 3 to obtain cross sections for electron scattering from nuclei. In (2.3) we had the invariant matrix element for electron scattering from a point Dirac proton. The proton electromagnetic current in this case was given by (2.4b). When the dressed nucleon was discussed (Sec. 3), we saw that this expression should be replaced by the physical nucleon electromagnetic current (3.7), with the corresponding substitution in the equation for the invariant matrix element (2.3). Now we go further and replace the nucleon electromagnetic current by current matrix elements of the nuclear current operator*, $\hat{J}^\mu(X^\alpha)$, taken between the nuclear states $|i>$ and $|f>$:

$$J^\mu(X^\alpha)_{fi} = <f|\hat{J}^\mu(X^\alpha)|i> , \qquad (4.1)$$

with the corresponding Fourier transforms for use in momentum space. The invariant matrix element in (2.3) has this same substitution (with, of course, the electron current and photon propagator unchanged) and, in leading to an expression for $\bar{\sum}_{if}|\mathcal{M}_{fi}|^2$ in analogy to the previous developments, requires a new nuclear tensor to replace the earlier nucleon electromagnetic tensor $\eta_N{}^{\mu\nu}$. Following standard nomenclature we shall call this nuclear electromagnetic tensor $W^{\mu\nu}$. In terms of matrix elements of the electromagnetic current operators (4.1) it is given by

$$W^{\mu\nu} = (2\pi)^3 \bar{\sum_{if}} P^0 \delta^{(4)}(Q - (P_f - P_i))J^\mu(0)_{fi}J^\nu(0)_{if} . \qquad (4.2)$$

For the interaction of a single photon with a nucleus where no polarizations are measured, as we have here, there are three momenta in the problem ($Q^\mu, P_i{}^\mu$ and $P_f{}^\mu$) and one momentum conservation condition ($P_f{}^\mu = P_i{}^\mu + Q^\mu$); hence there are two independent 4-momenta insofar as the nuclear response is concerned, taken here to be Q^μ and $P_i{}^\mu$. From these we can form three scalar quantities, $P_i{}^2$, Q^2 and $Q \cdot P_i$, where $P_i{}^2 = M_i{}^2$ is taken as given. Thus, for a given target mass M_i, we have two independent scalar quantities, Q^2 and $Q \cdot P_i$. Note that in the laboratory frame, $Q \cdot P_i = \omega M_i$ and so Q^2 and $Q \cdot P_i$ can readily be expressed in terms of q and ω. Now $W^{\mu\nu}$ by construction must be a second rank Lorentz tensor, so we must have

$$W^{\mu\nu} = X_1 g^{\mu\nu} + X_2 Q^\mu Q^\nu + X_3 P_i{}^\mu P_i{}^\nu$$
$$+ X_4 Q^\mu P_i{}^\nu + X_5 P_i{}^\mu Q^\nu + X_6 \varepsilon^{\mu\nu\rho\sigma} Q_\rho P_{i\sigma} , \qquad (4.3)$$

where $X_i = X_i(Q^2, Q \cdot P_i)$ are functions of the two independent scalar

*Here the caret "^" is used to denote a second quantized operator, operating in the nuclear Hilbert space.

quantities in the problem. This is the most general form consistent with Lorentz covariance. In addition, from (4.2) we see that $W^{\mu\nu}$ behaves like a product of two polar vectors ($\sim J^\mu J^\nu$); hence a term proportional to $\varepsilon^{\mu\nu\rho\sigma}Q_\rho P_{i\sigma}$ cannot be present as it has the wrong parity properties and we have $X_6 = 0$. Furthermore, $W^{\mu\nu}$ involves the nuclear electromagnetic currents J^μ and J^ν (see 4.2), which are conserved ($Q_\mu J^\mu = 0$) and so we have

$$Q_\mu W^{\mu\nu} = Q_\nu W^{\mu\nu} = 0 \; . \tag{4.4}$$

From these constraints it is straightforward to show that $W^{\mu\nu}$ is a symmetric tensor ($W^{\mu\nu} = W^{\nu\mu}$, that is, $X_5 = X_4$) and that there are two further relationships amongst the X_i's. It is convenient to define a new 4-vector

$$S^\mu \equiv \frac{1}{M_i}[P_i{}^\mu - \left(\frac{Q \cdot P_i}{Q^2}\right) Q^\mu] \; , \tag{4.5}$$

where we have $Q_\mu S^\mu = 0$ and in the laboratory frame, $S^{\mu,\mathrm{lab}} = \delta_{\mu 0} - (\omega/Q^2)Q^\mu$. Then it can be shown that, using the four relationships involving the X_i's, the nuclear tensor may be written in terms of only two remaining nuclear electromagnetic response functions, $W_1(Q^2, Q \cdot P_i)$ and $W_2(Q^2, Q \cdot P_i)$:

$$W^{\mu\nu} = -W_1\left(g^{\mu\nu} - \frac{Q^\mu Q^\nu}{Q^2}\right) + W_2 S^\mu S^\nu \; , \tag{4.6}$$

a theorem due to Van Gehlen, Gourdin and Bjorken.[7] Substituting into the developments in Sec. 2 we arrive at an expression for unpolarized inclusive electron scattering from a nucleus in lowest order:

$$\left(\frac{d^2\sigma}{d\Omega_e d\varepsilon'}\right)^{\mathrm{lab,ERL}_e}_{(e,e')} = \frac{1}{M_i}\,\sigma_M\left\{W_2(Q^2, Q \cdot P_i) + 2W_1(Q^2, Q \cdot P_i)\tan^2\frac{\theta e}{2}\right\} , \tag{4.7}$$

where the extreme relativistic limit for the elctron has been assumed and where a (generally small) recoil correction (see 2.9) has been dropped. It is important to note that there are three independent electron kinematic variables which can be fixed by performing experiments under specified conditions, namely ε, ε' and θ_e. Equivalently we have the set Q^2, $Q \cdot P_i$ and θ_e or the set q, ω and θ_e. Indeed in the ERL_e for example we have

$$Q^2 = -4\varepsilon\varepsilon'\sin^2\frac{\theta e}{2} = \omega^2 - q^2$$

$$Q \cdot P_i = \omega M_i \; , \quad \omega = \varepsilon - \varepsilon' \tag{4.8}$$

Thus, it is possible to fix Q^2 and $Q \cdot P_i$ and, by varying $\tan^2(\theta e/2)$, to separate the two response functions W_1 and W_2. It is useful to rewrite (4.7) in the form

$$\left(\frac{d^2\sigma}{d\Omega_e d\varepsilon'}\right)^{\mathrm{lab,ERL}_e}_{(e,e')} = \frac{1}{M_i}\,\sigma_M\left\{v_L W_L(q,\omega) + v_T W_T(q,\omega)\right\} , \qquad (4.9)$$

where $W_L \equiv (q^2/Q^2)W_1 + (q^2/Q^2)^2 W_2$ is called the longitudinal response function and $W_T \equiv 2W_1$ is called the transverse response function and where the electron kinematic functions are defined by

$$v_L \equiv \left(\frac{Q^2}{q^2}\right)^2 \qquad (4.10a)$$

$$v_T \equiv -\frac{1}{2}\left(\frac{Q^2}{q^2}\right) + \tan^2\frac{\theta_e}{2} . \qquad (4.10b)$$

Here the terms longitudinal and transverse refer to polarizations in a coordinate system whose z-axis is along the momentum transfer $\vec{Q}$. In fact the term longitudinal is really a misnomer, since for W_L contributions both from a longitudinal polarization and from the time component of J^μ (that is, $J^0 = \rho$ the charge density) are involved (see also Sec. 5). Again we see that the two independent response functions (here W_L and W_T) may be separated, in this case by fixing q and ω and varying θ_e. If we examine either of the response functions in the accessible area of the (q,ω)-plane, that is, for $q \geq \omega$ we see reflected the response of the nuclear system to excitation by the virtual photon. As ω is varied, we are varying the excitation energy of the nucleus. Thus we progress from ω = 0 (except for recoil) where elastic scattering is the only possible process, to ω ∿ few MeV where low-lying (discrete) excitations of the nucleus are being explored, to ω ∿ 15-30 MeV where giant resonance modes in the nucleus are excited, to beyond where nucleons are ejected from the nucleus (giving rise to the quasi-elastic cross section which peaks at roughly $\omega = q^2/2M_N$) and where pions are produced. At some fixed value of ω (and hence fixed nuclear excitation energy) it is possible to vary the three-momentum transfer over all values $q \geq \omega$. In effect by doing this we are fixing the nature of the nuclear excitation and mapping out the Fourier transforms of the spatial distributions of the transition charge, convection current and magnetization current. For further qualitative descriptions of the nuclear response surfaces the reader is directed to Refs. 1-3.

We turn now in Sec. 5 to discussions of these general response functions in the particular kinematic regime where discrete nuclear excitations are involved.

5. EXCITATION OF DISCRETE STATES

Let us denote the Fourier transforms of the matrix elements of the electromagnetic current operators by

$$J^{\mu}(\vec{Q})_{fi} = \int d\vec{X}\, e^{i\vec{Q}\cdot\vec{X}} J^{\mu}(\vec{X})_{fi} \quad . \tag{5.1}$$

Then proceeding to a differential cross section in the standard way (that is, inserting into the general developments of Sec. 4) we obtain

$$d\sigma = \frac{2\alpha^2}{(Q^2)^2}\frac{1}{k}\,\delta^{(1)}\big((\varepsilon - \varepsilon') - (E_f - E_i)\big)\frac{d\vec{K}'}{(2\pi)^3}$$

$$\times\left[\,\left|(K_{\mu} + K_{\mu}')J^{\mu}(\vec{Q})_{fi}\right|^2 + Q^2 J_{\mu}{}^*(\vec{Q})_{fi} J^{\mu}(\vec{Q})_{fi}\right] , \tag{5.2}$$

where current conservation, $Q_{\mu}J^{\mu}(\vec{Q})_{fi} = 0$, has been used.

Now, examining the different pieces of $J^{\mu}(\vec{Q})_{fi}$, we see that we have

$\mu = 0$: $J^0(\vec{Q})_{fi} = \rho(\vec{Q})_{fi}$ = Fourier transform of the transition charge density

$\mu = 1,2,3$: $\vec{J}(\vec{Q})_{fi}$ = Fourier transform of the transition (convection and magnetization) current distributions (3-vectors).

Let us define unit spherical vectors, $\vec{e}(\vec{Q};1m)$:

$$\vec{e}(\vec{Q};10) = \vec{e}_z = \vec{Q}/q$$

$$\vec{e}(\vec{Q};1 \pm 1) = \mp\frac{1}{\sqrt{2}}\,(\vec{e}_x \pm i\,\vec{e}_y) \ , \tag{5.3}$$

choosing the z-axis along $\vec{Q}$ and the x- and y-axes perpendicular to $\vec{Q}$ (say the x-axis in the electron scattering plane and the y-axis normal to this plane, see Fig. 5). Then expanding the 3-vector current matrix elements we have

$$\vec{J}(\vec{Q})_{fi} = \sum_{m=0,\pm1} J(\vec{Q};m)_{fi}\ \vec{e}^*(\vec{Q};1m) \tag{5.4a}$$

$$= J(\vec{Q};+1)_{fi}\ \vec{e}^*(\vec{Q};1+1) + J(\vec{Q};-1)_{fi}\ \vec{e}^*(Q;1-1)$$

$$+ J(\vec{Q};0)_{fi}\ \vec{e}^*(\vec{Q};10) \ , \tag{5.4b}$$

that is, two transverse components and one longitudinal component, where

$$J(\vec{Q};m)_{fi} = \vec{e}(\vec{Q};1m) \cdot \vec{J}(\vec{Q})_{fi} . \tag{5.5}$$

With this separation into longitudinal and transverse, the current conservation condition reads

$$Q_\mu J^\mu(\vec{Q})_{fi} = 0 = Q^0 J^0(\vec{Q})_{fi} - \vec{Q} \cdot \vec{J}(\vec{Q})_{fi} \tag{5.6a}$$

$$= \omega\rho(\vec{Q})_{fi} - qJ(\vec{Q};0)_{fi} , \tag{5.6b}$$

so we have

$$J(\vec{Q};0)_{fi} = \left(\frac{\omega}{q}\right) \rho(\vec{Q})_{fi} . \tag{5.7}$$

Thus, we have eliminated the longitudinal projection of the current in favor of the time projection (the charge density) and in (5.2) need to deal only with three independent pieces of the current: $J(\vec{Q};\pm1)_{fi}$, the transverse projections, and $\rho(\vec{Q})_{fi}$. We begin with the latter.

From the original Fourier transform (5.1) we have

$$\rho(\vec{Q})_{fi} = \int d\vec{X} e^{i\vec{Q}\cdot\vec{X}} \langle f|\hat{\rho}(\vec{X})|i\rangle . \tag{5.8a}$$

$$= \int d\vec{X}[4\pi \sum_{J\geq0} \sum_M i^J j_J(qx) Y_J^M(\Omega_X) Y_J^M(\Omega_Q)^*] \langle f|\hat{\rho}(\vec{X})|i\rangle , \tag{5.8b}$$

using the expansion of a plane wave.[8] Thus, collecting all of the behavior involving X, we are led to define Coulomb operators:[1,2]

$$\hat{M}_{JM}(q) \equiv \int d\vec{X}\, j_J(qx) Y_J^M(\Omega_X)\hat{\rho}(\vec{X}) , \tag{5.9}$$

where we have $J = 0,1,2, \ldots$ with corresponding Coulomb multipoles, C0,C1,C2, . . . with parity $\pi = (-)^J$, that is natural parity. Then it may be shown for the terms in (5.2) involving $\hat{\rho}$ that

$$\frac{1}{2J_i + 1} \sum_{M_i} \sum_{M_f} |\rho(\vec{Q})_{fi}|^2$$

$$= \frac{4\pi}{2J_i + 1} \sum_{J \geq 0} |\langle J_f||\hat{M}_J(q)||J_i\rangle|^2 , \tag{5.10}$$

where (see also (4.2)) we have performed the average over initial and sum over final magnetic substates for nuclear states labelled by angular momentum quantum numbers $|J_i,M_i\rangle$ and $|J_f,M_f\rangle$ and expressed the results in terms of reduced matrix elements (denoted $||$, see Ref. 8) of the Coulomb operators. We proceed likewise for the transverse projections, except that we need the vector plane wave expansion[8] (see (5.4) and (5.5)):

$$\vec{e}(\vec{Q};1m)e^{i\vec{Q}\cdot\vec{X}} = -\sqrt{2\pi}\sum_{J\geq 1} i^J\sqrt{2J+1}$$

$$\times\ [mj_J(qx)\vec{Y}_{JJ1}{}^m(\Omega_X) + \frac{1}{q}\vec{\nabla}\times\left(j_J(qx)\vec{Y}_{JJ1}{}^m(\Omega_X)\right)]\ , \quad \text{for } m = \pm 1. \tag{5.11}$$

We are then led to define the transverse electric and magnetic multipole operators[1,2]:

$$\hat{T}^{el}_{JM}(q) \equiv \int d\vec{X}\frac{1}{q}\vec{\nabla}\times\left(j_J(qx)\vec{Y}_{JJ1}{}^M(\Omega_X)\right)\cdot\hat{\vec{J}}(\vec{X}) \tag{5.12a}$$

$$\hat{T}^{mag}_{JM}(q) \equiv \int d\vec{X}\left(j_J(qx)\vec{Y}_{JJ1}{}^M(\Omega_X)\right)\cdot\hat{\vec{J}}(\vec{X})\ , \tag{5.12b}$$

where we have J = 1,2,3, . . . with corresponding electric and magnetic multipoles, E1,E2,E3, . . . and M1,M2,M3, . . . with parity $\pi = (-)^J$ (natural) and $\pi = (-)^{J+1}$ (non-natural) respectively. Again performing the average over initial and sum over final projections of angular momentum it may be shown that

$$\frac{1}{2J_i+1}\sum_{M_i}\sum_{M_f} J(\vec{Q};m)^*_{fi}J(\vec{Q};m')_{fi}$$

$$= \delta_{m'm}\frac{2\pi}{2J_i+1}\sum_{J\geq 1}\left\{|\langle J_f||T^{e\ell}_J(q)||J_i\rangle|^2\right.$$

$$\left. + |\langle J_f||\hat{T}_J{}^{mag}(q)||J_i\rangle|^2\right\}\ , \quad m = m' = \pm 1. \tag{5.13}$$

Note that no $T^{e\ell}/T^{mag}$ interference terms occur since we assume the nuclear states to be eigenstates of parity and for a given J the electric and magnetic operators have different parity, thus connecting a given initial state to different final states. If the states are not eigenstates of parity (see E. G. Adelberger's lectures) then interferences will appear. Finally it may be shown (using the same properties of angular momentum, namely, the Wigner-Eckart theorem and orthogonality of 3-j symbols[8]) that no interferences between the charge and transverse projections occur; that is, there are no interference terms involving Coulomb multipoles and electric or magnetic multipoles. It should be remarked that these comments apply only to the <u>inclusive</u> (e,e') cross sections (see Sec. 6).

Putting all of these together using (5.2) we obtain the inclusive electron scattering differential cross section for exciting a discrete state at energy $\omega = E_f - E_i$:

$$\left(\frac{d\sigma}{d\Omega_e}\right)^{\text{lab,ERL}_e}_{(e,e')} = 4\pi\sigma_M f_{rec}^{-1} F^2(q,\theta) \ , \qquad (5.14a)$$

where f_{rec} is as defined in (2.18) except with M_i replacing M_p and where F (q,θ) is the form factor given by

$$F^2(q,\theta) = v_L F_L^{\ 2}(q) + v_T F_T^{\ 2}(q) \ , \qquad (5.14b)$$

with v_L and v_T given in (4.10). The longitudinal and transverse form factors are given by

$$F_L^{\ 2}(q) = \frac{1}{2J_i + 1} \sum_{J\geq 0} |<J_f||\hat{M}_J(q)||J_i>|^2 \qquad (5.15a)$$

$$F_T^{\ 2}(q) = \frac{1}{2J_i + 1} \sum_{J\geq 1} \left\{ |<J_f||\hat{T}_J^{e\ell}(q)||J_i>|^2 + |<J_f||\hat{T}_J^{mag}||J_i>|^2 \right\} \ , \qquad (5.15b)$$

respectively and, once again, we see that they may be separated by fixing q and ω and varying θ_e (Rosenbluth decomposition). Both elastic (f = i) and inelastic (f $\neq$ i) scattering is contained here. We see that in F_L^2 only natural parity transitions can occur, while in F_T both natural and unnatural parities enter. Two cases are somewhat special: (1) $0^+ \leftrightarrow 0^+$ and $0^- \leftrightarrow 0^-$ transitions contribute only for F_L^2 (they are C0 transitions; whereas for the transverse form factor no monopole is possible, $J \geq 1$) and (2) $0^+ \leftrightarrow 0^-$ cannot occur in lowest order (one-photon-exchange) at all.

Finally, we note that real-photon processes such as γ-decay or photoexcitation may be discussed using the same language.[1-3] However as the photon is now real ($Q^2 = 0$) instead of virtual as in electron scattering ($Q^2 \leq 0$), one now has a fixed relationship between the photon energy ω and momentum q, namely, $q = \omega$. Furthermore, a real photon is transverse and so all real-photon processes involve only $F_T^{\ 2}(q = \omega)$:

$$\gamma\text{-decay rate, } \omega_\gamma = [8\pi\alpha q F_T^{\ 2}(q)]_{q=\omega} \qquad (5.16a)$$

photoabsorption cross section integrated over an

$$\text{absorption line, } \Sigma_\gamma = \left[\frac{(2\pi)^3\alpha}{q} F_T^{\ 2}(q)\right]_{q=\omega} \qquad (5.16b)$$

Note that here, in addition to no $0^+ \leftrightarrow 0^-$ transitions, there are no $0^+ \leftrightarrow 0^+$ or $0^- \leftrightarrow 0^-$ one-photon transitions.

6. GENERAL DISCUSSION OF EXCLUSIVE ELECTRON SCATTERING, (e,e'x)

To end this discussion of electromagnetic interactions with nuclei let us turn briefly to general exclusive reactions where in addition to presuming that the scattered electron is detected we now assume also that some particle x from the nucleus is detected in coincidence with the electron. This particle x may be an ejected nucleon (e,e'p), (e,e'n) , a pion $(e,e'\pi^{0,\pm})$, a light fragment (e,e'd), $(e,e'\alpha)$, etc. , a heavy fragment (e,e'f), etc. , etc. We shall set up the problem following the procedures outlined above in Sec. 4.

The kinematics for the (e,e'x) reaction are shown in Fig. 5. We assume that the electron scattering defines the xz-plane with the virtual photon and momentum transfer lying along the z-axis. The detected particle from the nucleus is assumed to have a 4-momentum P , so that the entire nuclear vertex involves the absorption of the virtual photon of momentum Q on a nucleus with momentum P_i proceeding to a total final momentum P_f composed of particle x with momentum P and daughter system (undetected fragment or fragments) with momentum $P_f - P$. As indicated in the figure, the momenta of particle x and the daughter system do not in general lie in the electron scattering plane, but lie in a plane inclined at an angle ϕ_x with respect to the xz-plane (here we have assumed that we are in the laboratory system so that $P_i = 0$, $E_i = M_i$). As before we have momentum conservation at each vertex:

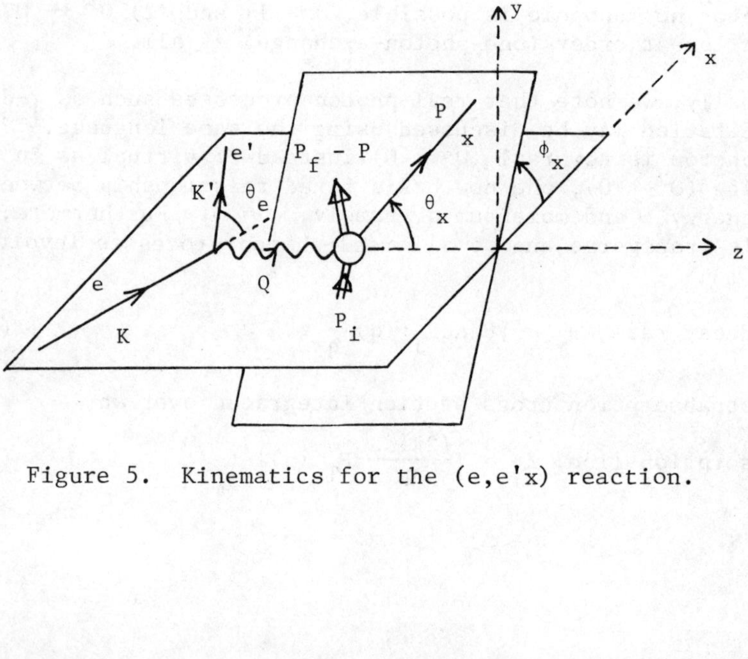

Figure 5. Kinematics for the (e,e'x) reaction.

$$Q^\mu = K^\mu - K'^\mu = P_f^{\ \mu} - P_i^{\ \mu} \ . \tag{6.1}$$

Now let us reapply the arguments of Sec. 4. Here we have three independent momenta at the nuclear vertex (Q^2, P_i^μ and P^μ) instead of two. We may form six independent scalar quantities from these: Q^2, P_i^2, P^2, $Q \cdot P_i$, $Q \cdot P$, $P \cdot P_i$. Of these we take two as given, $P_i^2 = M_i^2$ and $P^2 = M_x^2$, since the masses M_i and M_x are presumed to be known. Thus we have four independent scalars remaining: Q^2, $Q \cdot P_i$, $Q \cdot P$ and $P \cdot P_i$. In the laboratory system we have $Q \cdot P_i = \omega M_i$, $P \cdot P_i = E_x M_i$, where $E_x = \sqrt{p^2 + M_x^2}$ is the total energy of particle x and $Q \cdot P = Q^0 P^0 - \vec{Q} \cdot \vec{P} = \omega E_x - qp \cos\theta_x$, where θ_x is the angle between $\vec{Q}$ and $\vec{P}$ (see Fig. 5). Therefore the four scalar quantities may be rewritten as functions of four other variables: q, ω, E_x and θ_x. Once again we have a general nuclear electromagnetic tensor, called $\tilde{W}^{\mu\nu}$ here, and in analogy with (4.3) we may expand in the most general form consistent with Lorentz covariance:

$$\begin{aligned}\tilde{W}^{\mu\nu} &= \tilde{X}_1 g^{\mu\nu} + \tilde{X}_2 Q^\mu Q^\nu + \tilde{X}_3 P_i^{\ \mu} P_i^{\ \nu} + \tilde{X}_4 P^\mu P^\nu \\ &+ \tilde{X}_5 Q^\mu P_i^{\ \nu} + \tilde{X}_6 P_i^{\ \mu} Q^\nu + \tilde{X}_7 Q^\mu P^\nu + \tilde{X}_8 P^\mu Q^\nu \\ &+ \tilde{X}_9 P^\mu P_i^{\ \nu} + \tilde{X}_{10} P_i^{\ \mu} P^\nu + (\text{terms like } \varepsilon^{\mu\nu\rho\sigma} Q_\rho P_{i\sigma}, \text{ etc.}) \ ,\end{aligned} \tag{6.2}$$

where $\tilde{X}_i = \tilde{X}_i(Q^2, Q \cdot P_i, Q \cdot P, P \cdot P_i)$ are functions of the four independent scalar quantities in the problem. Once again the terms involving $\varepsilon^{\mu\nu\rho\sigma}$ are eliminated on parity grounds, since the tensor $\tilde{W}^{\mu\nu}$ must behave like the product of two polar vectors ($\sim J^\mu J^\nu$). Furthermore we again have current conservation, $Q_\mu \tilde{W}^{\mu\nu} = Q_\nu \tilde{W}^{\mu\nu} = 0$, which yields six equations in the 10 unknowns, $\tilde{X}_i, i = 1, \ldots, 10$. In other words we are left, after applying the general symmetry principles, with four general response functions. Using the definition of S^μ in (4.5) and defining a new 4-vector

$$T^\mu \equiv \frac{1}{M_x}[P^\mu - \left(\frac{Q \cdot P}{Q \cdot P_i}\right) P_i^{\ \mu}] \ , \tag{6.3}$$

where we have $Q_\mu T^\mu = 0$, we may write

$$\begin{aligned}\tilde{W}^{\mu\nu} &= -\tilde{W}_1 \left(g^{\mu\nu} - \frac{Q^\mu Q^\nu}{Q^2}\right) + \tilde{W}_2 S^\mu S^\nu \\ &+ \tilde{W}_3 T^\mu T^\nu + \tilde{W}_4 (S^\mu T^\nu + T^\mu S^\nu) \ ,\end{aligned} \tag{6.4}$$

to be compared with (4.6).

Combining the nuclear tensor with the electron tensor (2.15), we will find that the cross sections depend on

$$\eta_e(K';K)_{\mu\nu}\tilde{W}^{\mu\nu} \sim \tilde{W} \equiv v_L\tilde{W}_L + v_T\tilde{W}_T$$

$$+ v_{TT}\tilde{W}_{TT}\cos 2\phi_x + v_{TL}\tilde{W}_{TL}\cos\phi_x \ , \tag{6.5}$$

where in addition to v_L and v_T defined in (4.10) we have

$$v_{TT} \equiv \frac{1}{2}\left(\frac{Q^2}{q^2}\right) \tag{6.6a}$$

$$v_{TL} \equiv \frac{1}{\sqrt{2}}\left(\frac{Q^2}{q^2}\right)\sqrt{-\left(\frac{Q^2}{q^2}\right) + \tan^2\frac{\theta e}{2}} \ . \tag{6.6b}$$

The four terms correspond to longitudinal (L), transverse (T), transverse-transverse interference (TT) and transverse-longitudinal interference (TL), where as above "longitudinal" is taken to mean longitudinal and time (or charge) components. Finally we may write the multiple-differential (unpolarized) coincidence electron scattering cross section in terms of $\tilde{W}$:

$$\left(\frac{d^3\sigma}{d\Omega_e d\Omega_x dE_x}\right)^{\text{lab,ERL}_e}_{(e,e'x)} = \frac{1}{M_i}\sigma_M\tilde{W} \ , \tag{6.7}$$

again ignoring a (generally small)recoil correction and assuming the extreme relativistic limit for the electron for simplicity. Likewise the real-photon cross section may be obtained:

$$\left(\frac{d\sigma}{d\Omega_x}\right)^{\text{lab}}_{(\gamma,x)} = \left[\frac{2\pi^2\alpha}{M_i q}\tilde{W}_T\right]_{q=\omega} \tag{6.8}$$

In the expression for $\tilde{W}$ given in (6.5) the four response functions are functions of the scalar quantities in the problem, Q^2, $Q \cdot P_i$, $Q \cdot P$ and $P \cdot P_i$, which as noted above, may be re-expressed in terms of q, ω, E_x and θ_x. Note, however, that the entire dependence on the azimuthal angle ϕ_x (see Fig. 5) is manifested in the two factors $\cos\phi_x$ and $\cos 2\phi_x$ in (6.5). Thus, by varying the angle between the electron scattering plane and the plane containing the momentum transfer and the momentum of particle x, the interference response functions, $\tilde{W}_{TT}$ and $\tilde{W}_{TL}$, may be separated from the terms which do not depend on ϕ_x, namely $\tilde{W}_L$ and $\tilde{W}_T$, and from each other. Alternatively, one has the different behavior of the various electron kinematic factors, v_L, v_T, v_{TT} and v_{TL}, to help in separating the four response functions.

Finally, to recover the inclusive response functions discussed in Sec. 4 we must convert from $d^3\sigma/d\Omega_e d\Omega_x dE_x$ to $d^3\sigma/d\Omega_e d\Omega_x d\varepsilon'$, which is straightforward, and then perform the $\int d\Omega_x$. This will yield the contribution to the inclusive cross section made by the process where particle x is to be found in the final state (of course, there may be other final states). In particular, the $\int d\Omega_x$ involves the integral $\int_0^{2\pi} d\phi_x$ and, since

$$\int_0^{2\pi} d\phi_x \cos\phi_x = \int_0^{2\pi} d\phi_x \cos 2\phi_x = 0 , \qquad (6.9)$$

we see that no TT or TL interference terms survive and that the integrated cross section has only two response functions (L and T) as in (4.9). For further discussion the reader is directed to Ref. 9 with the caution that the four response functions defined there are linear combinations of the present set of response functions.

7. SEMI-LEPTONIC CHARGE-CHANGING WEAK INTERACTIONS WITH THE NUCLEON

Let us now turn to discussion of semi-leptonic weak interactions with nuclei. The development will very closely parallel the approaches taken in Secs. 2-6 and presented in various publications (see, for example, Refs. 2, 4, 5 and 10). No attempt will be made to motivate the underlying description of the electroweak interaction between leptons and hadrons as this is the subject of the lectures by D. Atkinson and given in the proceedings of this Summer School (see also Ref. 5). Let us begin, as in Sec. 2, with processes involving a point Dirac nucleon. In fact, in strict analogy with Figs. 1 and 2, let us now consider the process displayed in Fig. 6. In Sec. 2 we considered the elementary process $e^- + p \to e^- + p$; now, to illustrate the application of Feynman rules to the weak interaction, we choose to consider the process $e^- + p \to \nu_e + n$. Both are semi-leptonic reactions in that leptons interact with hadrons. It is of interest of course to discuss purely leptonic and purely hadronic weak interactions as well, but we shall not do so here. We shall assume that the interaction is of the vector (V) plus axial-vector (A) type (as motivated by the underlying gauge theory) and so, in analogy to the application of the Feynman rules to electron scattering summarized in Fig. 2, we now have the various factors from Fig. 6 to assemble.

Of course for other weak-interaction processes it is necessary only to replace the u's and $\bar{u}$'s with the appropriate spinors[4] (including spinors v and $\bar{v}$ when anti-particles are involved). Note that, as we are now exchanging a massive vector boson (the $W^{\pm}$) instead of a photon (Fig. 2), the photon propagator $D_F(Q)_{\mu\nu} = g_{\mu\nu}/(-Q^2)$ must be replaced by the W propagator (see Ref. 6,

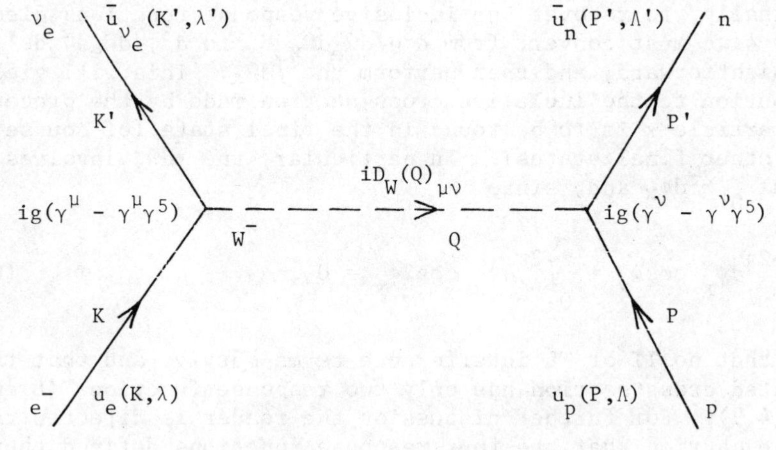

Figure 6. Application of the Feynman rules to the reaction $e^- p \to \nu_e n$ in lowest-order for point Dirac nucleons.

Appendix B),

$$D_W(Q)_{\mu\nu} = \frac{g_{\mu\nu} - \frac{Q_\mu Q_\nu}{M_W^2}}{-(Q^2 - M_W^2)} , \tag{7.1}$$

where M_W is the mass of the W. Furthermore, here we take the elementary weak interaction coupling strength to be g, replacing e for the electromagnetic interaction.

Proceeding as in Sec. 2 we now assemble the factors from the Feynman diagram and write the invariant matrix element for the process:

$$\begin{aligned} \mathcal{M}_{fi} = & [\bar{u}_{\nu_e}(K',\lambda')\{ig(\gamma^\mu - \gamma^\mu\gamma^5)\}u_e(K,\lambda)] \\ & \times [-i\left(g_{\mu\nu} - \frac{Q_\mu Q_\nu}{M_W^2}\right)/(Q^2 - M_W^2)] \\ & \times [\bar{u}_n(P',\Lambda')\{ig(\gamma^\nu - \gamma^\nu\gamma^5)\}u_p(P,\Lambda)] , \end{aligned} \tag{7.2}$$

where the various contributions have been grouped into three major factors corresponding to a leptonic current ($e^- \to \nu_e$), a W propagator and a hadronic current ($p \to n$). This is rewritten in (7.3):

$$\mathcal{M}_{fi} = -i\left\{g^2\left(g_{\mu\nu} - \frac{Q_\mu Q_\nu}{M_W{}^2}\right)/(-Q^2 + M_W{}^2)\right\}$$

$$\times [\bar{u}_{\nu_e}(\gamma^\mu - \gamma^\mu\gamma^5)u_e][\bar{u}_n(\gamma^\nu - \gamma^\nu\gamma^5)u_p] . \qquad (7.3)$$

This is to be compared with (2.2) or (2.3) for the electromagnetic interaction. There the photon propagator gave rise to a factor $\sim 1/Q^2$, which led to the characteristic $1/\sin^4(\theta_e/2)$ behavior of the Rutherford or Mott cross sections. Here the situation is very different: the mass of the W is presumed to be quite large ($M_W{}^2 \sim (80\ \text{GeV})^2$), whereas the scale of momentum transfers appropriate for nuclear physics is very much smaller, ranging from $-Q^2 \sim$ (few $\times$ MeV)^2 to $-Q^2 \sim$ (few $\times$ 100 MeV)^2. Thus in (7.3) we may generally drop all terms involving a momentum transfer compared to a W mass and for nuclear physics purposes we have for the curly bracket in (7.3):

$$\{\ \} \longrightarrow \left(\frac{g^2}{M_W{}^2}\right) g_{\mu\nu} . \qquad (7.4)$$

So in momentum space we have a momentum-independent interaction between the leptonic and hadronic currents which implies that in coordinate space we have a contact current-current interaction. In fact from a unified description of the electroweak interaction the coupling g is of order e and so the elementary weak interaction strength is comparable to the electromagnetic coupling. However, $M_W{}^2$ is very large and so $g^2/M_W{}^2$ is very small. Indeed it is conventional to replace this combination by the Fermi weak interaction coupling G where $GM_p{}^2 = 1.023 \times 10^{-5}$ (for a careful discussion of these ideas and the relationships between the couplings with appropriate factors of $\sqrt{2}$, etc., see the lectures of D. Atkinson from the Summer School, particularly his Eq. (3.40)).

Now we proceed to cross sections in the standard manner: we calculate $\overline{\sum_{if}}|\mathcal{M}_{fi}|^2$ and then include the appropriate kinematic factors and constants using the Feynman rules.[6] Just as in the developments of the electron scattering cross section in Sec. 2, we will obtain leptonic and nucleonic tensors analogous to (2.15). Specifically here we have for the leptonic weak interaction tensor (we take the neutrino mass to be zero, $m_\nu = 0$):

$$\eta'_{e\to\nu_e}(K';K)_{\mu\nu} =$$

$$[K_\mu K'_\nu + K'_\mu K_\nu - g_{\mu\nu}(K \cdot K')] - i\varepsilon_{\mu\nu\rho\sigma}K^\rho K'^\sigma . \qquad (7.5)$$

In particular, note the presence of a term proportional to $\varepsilon_{\mu\nu\rho\sigma}$ which did not occur in the electromagnetic case (2.15): this reflects the V-A nature of the weak interaction and the non-conservation of parity.

In a similar manner a tensor $\eta'_{p\to n}(P';P)^{\mu\nu}$ may be constructed for the point Dirac nucleon. This will be built from bilinear products of the nucleon current

$$j_{p\to n}(P',\Lambda';P'\Lambda)^{\mu}\Big|_{\text{point Dirac nucleon}} \sim \bar{u}_n(P',\Lambda')[\gamma^{\mu} - \gamma^{\mu}\gamma^5]u_p(P,\Lambda) . \tag{7.6}$$

Again, as in the discussions in Sec. 3, the nucleon should be dressed, so that instead we should use the physical nucleon current

$$\begin{aligned} j_{p\to n}(P',\Lambda';P,\Lambda)^{\mu}\Big|_{\text{physical nucleon}} &\sim \bar{u}_n(P',\Lambda')[F_1^{wk}(Q^2)\gamma^{\mu} + \frac{i}{2M_N}F_2^{wk}(Q^2)\sigma^{\mu\nu}Q_{\nu} + F_S^{wk}(Q^2)Q^{\mu}]u_p(P,\Lambda) \\ &+ \bar{u}_n(P',\Lambda')[G_A^{wk}(Q^2)\gamma^5\gamma^{\mu} + \frac{1}{2M_N}G_P^{wk}(Q^2)\gamma^5Q^{\mu} + iG_T^{wk}(Q^2)\gamma^5\sigma^{\mu\nu}Q_{\nu}] \\ &\times u_p(P,\Lambda) , \end{aligned} \tag{7.7}$$

which should be compared with the developments leading to (3.4). We see that the physical nucleon current has vector and axial-vector pieces, as does the bare current (7.6), but with relative strengths which are not in general 1:1 (as they are for leptons). This is the most general V-A form consistent with Lorentz covariance and has, in addition to the expected vector and axial-vector contributions which go as γ^{μ} and $\gamma^5\gamma^{\mu}$ (multiplied by form factors F_1^{wk} and G_A^{wk}) respectively, four new contributions (analogous to the anomalous magnetic moment contribution for electromagnetic interactions (3.4)). These are called weak magnetism ($\sim F_2^{wk}$), induced scalar ($\sim F_S^{wk}$), induced pseudoscalar ($\sim G_P^{wk}$) and induced tensor ($\sim G_T^{wk}$). Of the six pieces of the current four are called first class (F_1^{wk}, F_2^{wk}, G_A^{wk} and G_P^{wk} contributions) while the other two are second class (F_S^{wk} and G_T^{wk} contributions), where the nomenclature is due to Weinberg[11] (see also Ref. 4). These second-class currents are felt to be absent and thus we shall take $F_S^{wk}(Q^2) = F_T^{wk}(Q^2) = 0$. It is useful to extend the notation to include isospin (as we did in Sec. 3, see Eq. (3.7)), in which case the expression in (7.7) has the form

$$\bar{u}_N(P',\Lambda',\xi')[\ .\ .\ .\ ]\ \tau_{-}u_N(P,\Lambda,\xi) , \tag{7.8}$$

where the square brackets contain the form factors, γ-matrices, etc. as in (7.7). Thus we have a pure isovector current with projection -1 (charge lowering). In a completely analogous fashion we may obtain a $n \to p$ current with projection +1 (charge raising). Remember that for electromagnetic interactions we had both isoscalar and isovector currents and had only projection 0 (neutral current), as summarized in (3.5). We now make the conserved vector current (CVC) hypothesis which states that the vector part of the weak interaction current is the same as the isovector part of the electromagnetic current (up to rotations in isospin space):

$$F_1^{wk}(Q^2) = F_1^{em}(Q^2)_{isovector}$$

$$F_2^{wk}(Q^2) = F_2^{em}(Q^2)_{isovector} \quad . \tag{7.9}$$

That is, we have for this part of the electroweak current a single isovector with projections 0 (electromagnetic current, isovector part) and ±1 (weak current, vector part). Finally, using weak interaction processes such as neutron β-decay, μ^--capture in hydrogen, ν-reactions involving nucleons, it is possible to determine the remaining form factors, $G_A^{wk}(Q^2)$ and $G_P^{wk}(Q^2)$. For example, from neutron β-decay we find that $G_A^{wk}(0) = 1.23$.

With this brief discussion of charge-changing weak interactions with the nucleon we turn in the next sections to a brief summary of some of the formalism for semi-leptonic weak interactions with nuclei.

8. GENERAL DISCUSSION OF INCLUSIVE CHARGE-CHANGING SEMI-LEPTONIC WEAK INTERACTIONS WITH NUCLEI

The types of processes of interest here are shown diagrammatically in Fig. 7. The analysis closely parallels the treatment in Sec. 4 and more detail may be found in Ref. 4 as well. Instead of the purely polar vector electromagnetic current matrix elements we had in (4.1), we now have vector and axial-vector currents, where the latter are denoted with a "5" to indicate the extra γ^5 which occurs in the nucleon currents from which they are constructed (see Eq. (7.7)):

$$\mathcal{J}_\mu^{(\pm)}(X^\alpha)_{fi} = J_\mu^{(\pm)}(X^\alpha)_{fi} + J_\mu^{5(\pm)}(X^\alpha)_{fi} \quad . \tag{8.1}$$

The (±) here is used to indicate where the process is charge raising or lowering and will be suppressed in what follows. Following (4.2) we may construct a weak interaction nuclear tensor $W_{\mu\nu}^{wk}$ which behaves like the product $\mathcal{J}_\mu \mathcal{J}_\nu \sim (J_\mu + J_\mu^5)(J_\nu + J_\nu^5)$ and so has

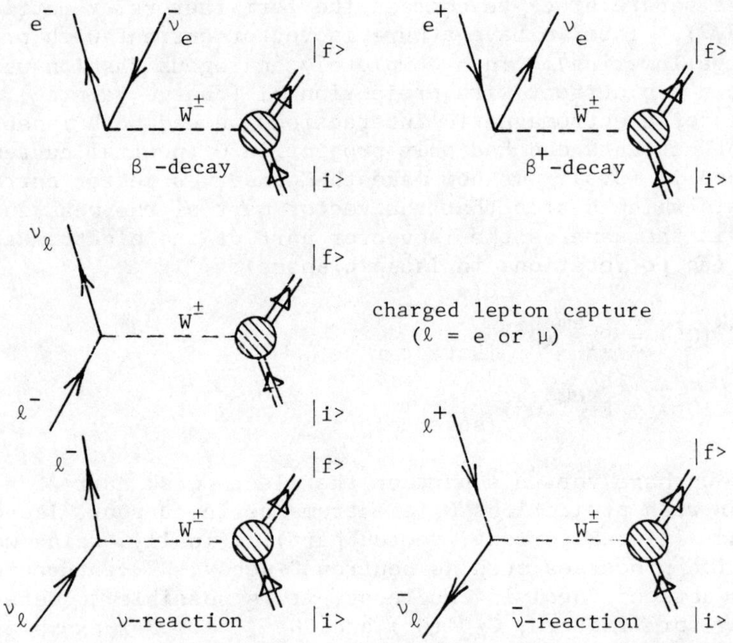

Figure 7. Semi-leptonic charge-changing weak interaction processes.

pieces which behave as the product of two polar vectors (VV), two axial-vectors (AA) and a parity-violating piece involving the interference between a polar vector and an axial-vector (VA):

$$W_{\mu\nu}^{wk} = W_{\mu\nu}^{wk,VV} + W_{\mu\nu}^{wk,AA} + W_{\mu\nu}^{wk,VA} \qquad (8.2)$$

Proceeding just as in Sec. 4 we may use Lorentz covariance, parity transformation properties, and, in the case of the VV contribution, current conservation to obtain the most general form of the nuclear tensor. Note that the axial-vector current is <u>not</u> conserved ($Q^{\mu}J_{\mu}{}^{5} \neq 0$) and so, as a consequence, constraints such as (4.4) cannot be applied where the axial-vector current is involved. The general forms are found to be:[4]

$$W_{\mu\nu}^{wk,VV} = -\overline{W}_1\left(g_{\mu\nu} - \frac{Q_{\mu}Q_{\nu}}{Q^2}\right) + \overline{W}_2 S_{\mu}S_{\nu} , \qquad (8.3a)$$

to be compared with (4.6) where the arguments are identical (in fact as noted in Sec. 7, $\overline{W}_{1,2}$ are directly related to $(W_{1,2})_{isovector}$);

$$W_{\mu\nu}^{wk,AA} = \overline{W}_3 g_{\mu\nu} + \frac{1}{M_i^2}\left\{\overline{W}_4 Q_\mu Q_\nu + \overline{W}_5 P_{i\mu} P_{i\nu}\right.$$

$$\left. + \overline{W}_6 (P_{i\mu} Q_\nu + Q_\mu P_{i\nu}) + \overline{W}_7 (P_{i\mu} Q_\nu - Q_\mu P_{i\nu})\right\} , \qquad (8.3b)$$

which looks like (4.3) without the $\varepsilon_{\mu\nu\rho\sigma}$ term (eliminated by parity considerations), that is, like the electromagnetic case before current conservation was employed; and

$$W_{\mu\nu}^{wk,VA} = \frac{1}{M_i^2} \overline{W}_8 \, \varepsilon_{\mu\nu\rho\sigma} P_i^\rho Q^\sigma , \qquad (8.3c)$$

that is, for this parity-violating VA interference term only the $\varepsilon_{\mu\nu\rho\sigma}$ term from (4.3) survives. As before, all of the response functions are functions of the two scalar quantities in the problem, $\overline{W}_k = \overline{W}_k(Q^2, Q \cdot P_i)$, $k = 1, \ldots, 8$. Cross sections may then be expressed in terms of these eight response functions (just as inclusive electron scattering was expressible in terms of W_1 and W_2, see Eq. (4.7)). For example, consider the neutrino reaction (ν_ℓ, ℓ^-), where $\ell = e$ or μ. The inclusive double-differential cross section may be written[4]

$$\left(\frac{d^2\sigma}{d\Omega d\varepsilon_\ell}\right)^{ERL_\ell}_{(\nu_\ell,\ell^-)} = \frac{1}{M_i} (G\varepsilon_\ell)^2 \left\{ S_2 \cos^2\frac{\theta}{2} + 2S_1 \sin^2\frac{\theta}{2} \right.$$

$$\left. - 2S_3 \sin\frac{\theta}{2} \sqrt{-Q^2 \cos^2\frac{\theta}{2} + q^2 \sin^2\frac{\theta}{2}} \Big/ M_i \right\} , \qquad (8.4)$$

where S_1, S_2 and S_3 are particular combinations of the $\overline{W}_k$ and where θ is the angle between the incident neutrino and the exiting charged lepton of energy ε_ℓ.

Note in comparing (4.7) with (8.4) that the dramatic behavior of the Mott cross section ($\sim 1/\sin^4(\theta_e/2)$) has been replaced in the weak interaction case with a very mild dependence on scattering angle.

Of course we might continue to a discussion of exclusive (coincidence) weak interaction processes as we did for the electromagnetic interaction in Sec. 6. However, in the limited space allowed here let us instead conclude with two further sections, one on excitation of discrete states by the charge-changing weak interaction and a final section sketching the properties of the neutral current weak interaction.

9. EXCITATION OF DESCRETE STATES BY CHARGE-CHANGING SEMI-LEPTONIC WEAK INTERACTIONS

We now specialize the discussion of the previous section to excitation of discrete states, requiring in the process an extension of the multipole analysis presented in Sec. 5 to include the axial-vector as well as vector current. Recall that the treatment of the electromagnetic (vector) current in Sec. 5 led to three types of multipole operators, $\hat{M}_{JM}$, $\hat{T}_{JM}^{e\ell}$ and $\hat{T}_{JM}^{mag}$ (see (5.9) and (5.12)). So here for the vector part of the weak current we have the same multipoles. For the axial-vector pieces we have analogous expressions defining $\hat{M}_{JM}^{5}$, $\hat{T}_{JM}^{e\ell 5}$ and $\hat{T}_{JM}^{mag5}$ in precisely the same way, but with the axial-vector current replacing the vector current in (5.9) and (5.12). However now, because the axial-vector current is not conserved, we must explicitly introduce a longitudinal projection:[4,10]

$$\hat{L}^{5}_{JM}(q) \equiv \int d\vec{x}\, \frac{i}{q}\vec{\nabla}\left(j_J(qx)\vec{Y}^{M}_{JJ1}(\Omega_x)\right) \cdot \hat{\vec{J}}^{5}(\vec{x}) \ , \tag{9.1}$$

where a similar definition can be made for the vector current, yielding a multipole operator $\hat{L}_{JM}$ (which may be eliminated using current conservation in favor of $\hat{M}_{JM}$). In turn we may denote the multipole operators for the entire V-A current with script letters:

$$\begin{aligned}
\hat{\mathcal{M}}_{JM} &= \hat{M}_{JM} + \hat{M}^{5}_{JM} \ , \quad J \geq 0 \\
\hat{\mathcal{L}}_{JM} &= \hat{L}_{JM} + \hat{L}^{5}_{JM} \ , \quad J \geq 0 \\
\hat{\mathcal{T}}^{e\ell}_{JM} &= \hat{T}^{e\ell}_{JM} + \hat{T}^{e\ell 5}_{JM} \ , \quad J \geq 1 \\
\hat{\mathcal{T}}^{mag}_{JM} &= \hat{T}^{mag}_{JM} + \hat{T}^{mag5}_{JM} \ , \quad J \geq 1 \ .
\end{aligned} \tag{9.2}$$

These are divided into operators with natural parity ($\pi = (-)^{J}$), $\hat{M}_{JM}$, $\hat{L}_{JM}$, $\hat{T}_{JM}^{e\ell}$ and $\hat{T}_{JM}^{mag5}$, and operators with non-natural parity ($\pi = (-)^{J+1}$), $\hat{M}_{JM}^{5}$, $\hat{L}_{JM}^{5}$, $\hat{T}_{JM}^{e\ell 5}$ and $\hat{T}_{JM}^{mag}$.

Recall the expressions given in Sec. 5 for electron scattering, (5.14). Similar expressions may be obtained for weak interaction processes (see Ref. 4 and 10 for details); here we merely restate the answers:

β-decay

$$d\omega_{\beta^{\mp}} = \frac{G^2}{2\pi^3} k\varepsilon(W_0 - \varepsilon)^2 d\varepsilon \frac{d\Omega_e}{4\pi}\frac{d\Omega_\nu}{4\pi}\frac{4\pi}{2J_i + 1}$$
$$\times \Big\{ \sum_{J\geq 0} [v_1 |\langle J_f||\hat{\mathcal{M}}_J||J_i\rangle|^2 + v_2 |\langle J_f||\hat{\mathcal{L}}_J||J_i\rangle|^2$$
$$+ v_3 \mathrm{Re}\langle J_f||\hat{\mathcal{M}}_J||J_i\rangle^* \langle J_f||\hat{\mathcal{L}}_J||J_i\rangle]$$
$$+ \sum_{J\geq 1} [v_4 (|\langle J_f||\hat{\mathcal{T}}_J^{e\ell}||J_i\rangle|^2 + |\langle J_f||\hat{\mathcal{T}}_J^{mag}||J_i\rangle|^2)$$
$$\pm v_5 \mathrm{Re}\langle J_f||\hat{\mathcal{T}}_J^{e\ell}||J_i\rangle^* \langle J_f||\hat{\mathcal{T}}_J^{mag}||J_i\rangle] \Big\} , \tag{9.3}$$

where the electron (positron) has energy ε, maximum energy W_0 and 3-momentum $k = |\vec{K}|$. The quantities v_i, $i = 1, . . . ,5$ are specific lepton kinematic factors[4,10] analogous to v_L and v_T defined in (4.10).

Charged-lepton capture

$$\omega_{fi} = \frac{G^2 E_\nu^2}{2\pi}\frac{4\pi}{2J_i + 1} \Big\{ \sum_{J\geq 0} |\langle J_f||(\hat{\mathcal{M}}_J' - \hat{\mathcal{L}}_J')||J_i\rangle|^2$$
$$+ \sum_{J\geq 1} |\langle J_f||(\hat{\mathcal{T}}_J^{e\ell\prime} - \hat{\mathcal{T}}_J^{mag\prime})||J_i\rangle|^2 \Big\} , \tag{9.4}$$

where the neutrino energy $E_\nu = q = m_\ell - (E_f - E_i) - \varepsilon_b$, with m_ℓ the lepton mass ($\ell = e$ or μ), E_i and E_f the initial and final nuclear energies and ε_b the atomic binding energy of the lepton. Here we assume that the capture takes place from the 1S atomic level of the lepton and the primes indicate that the radial wave function for that orbit is to be included in the radial integrals which result in calculating the matrix elements in (9.4).

Neutrino reactions

$$\left(\frac{d\sigma}{d\Omega}\right)_{\substack{(\nu_\ell,\ell^-)\\(\bar{\nu}_\ell,\ell^+)}} = \frac{G^2}{4\pi^2} k\varepsilon \frac{4\pi}{2J_i + 1} \Big\{ \sum_{J\geq 0} [v_1' |\langle J_f||\hat{\mathcal{M}}_J||J_i\rangle|^2$$
$$+ v_2' |\langle J_f||\hat{\mathcal{L}}_J||J_i\rangle|^2$$
$$+ v_3' \mathrm{Re}\langle J_f||\hat{\mathcal{M}}_J||J_i\rangle^* \langle J_f||\hat{\mathcal{L}}_J||J_i\rangle]$$

$$+ \sum_{J\geq 1} \Big[v_4' \left(|\langle J_f || \hat{\mathcal{T}}_J^{e\ell} || J_i \rangle|^2 + |\langle J_f || \hat{\mathcal{T}}_J^{mag} || J_i \rangle|^2 \right)$$

$$\pm v_5' \mathrm{Re} \langle J_f || \hat{\mathcal{T}}_J^{e\ell} || J_i \rangle^* \langle J_f || \hat{\mathcal{T}}_J^{mag} || J_i \rangle \Big] \Big\} \; , \qquad (9.5)$$

where again the charged lepton has energy ε and 3-momentum k and where v_i', i = 1, . . . ,5 are specific lepton kinematic factors.[4,10]

The main purpose of displaying the formulas (9.3-9.5) here is to show the essential unity of approach taken in discussing the electromagnetic interaction (and leading to (5.14)) and the charge-changing weak interaction.

10. NEUTRAL-CURRENT SEMI-LEPTONIC WEAK INTERACTIONS WITH NUCLEI

In addition to the charge-changing weak interaction processes shown in Fig. 7 and discussed in the previous sections, we may also have neutral-current weak interaction processes such as those in Fig. 8. The upper diagrams for neutrino and anti-neutrino scattering are pure neutral current processes, while the lower graph provides a weak interaction amplitude which interferes with the (much larger) electromagnetic amplitude from Fig. 4, which is also a neutral current process.

The reader is directed to Ref. 5 for a more complete discussion

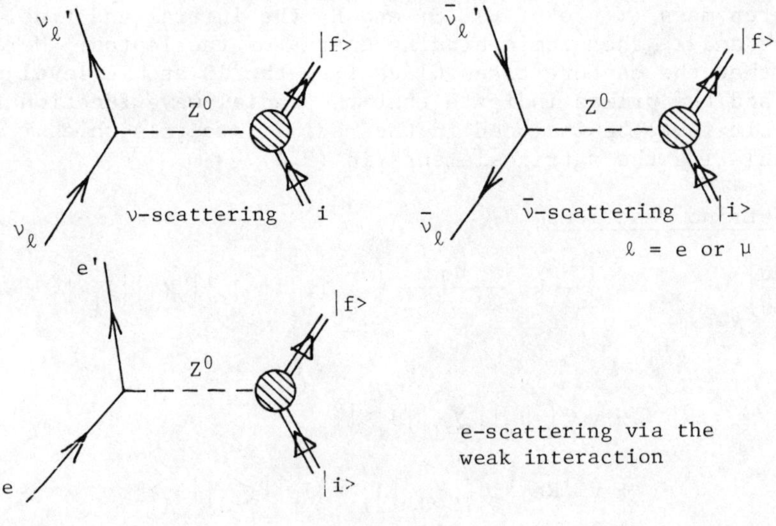

Figure 8. Neutral current weak interaction processes.

of neutral current effects in nuclei; here we provide only a brief outline of the approach taken to illustrate once again the essential unity in studying all aspects of the electroweak interaction with nuclei. Let us first reiterate the isospin structure of the single-nucleon currents studied thus far:

Electromagnetic current (Cf. (3.5-3.7))

$$J_\mu^{em} = J_\mu^{(0)} I_0^{\ 0} + J_\mu^{(1)} I_1^{\ 0} , \tag{10.1}$$

where[5]

$$I_T^{\ M_T} \equiv \frac{1}{2} \times \begin{cases} 1 & T = 0, \quad M_T = 0 \\ \tau_0 = \tau_3 & T = 1, \quad M_T = 0 \\ \tau_{\pm 1} = \mp \frac{1}{\sqrt{2}} (\tau_1 \pm i\tau_2) & T = 1, \quad M_T = \pm 1 , \end{cases} \tag{10.2}$$

that is, it contains isoscalar (T = 0) and isovector (T = 1) pieces and is a neutral current (M_T = 0).

Charge-changing weak current (Cf. (7.8))

$$\mathcal{J}_\mu^{(\pm)} = (J_\mu^{(1)} + J_\mu^{5(1)}) I_1^{\ \pm 1} , \tag{10.3}$$

that is, it contains only isovector contributions (T = 1) with charge raising or lowering operators (M_T = ±1), and has both vector and axial-vector pieces.

Neutral-current[4]

Within a wide class of gauge theory models of the electroweak interaction (see Ref. 4 and also the lectures of D. Atkinson) we have the structure

$$\begin{aligned} \mathcal{J}_\mu^{(0)} &= \alpha_V^{(0)} J_\mu^{(0)} I_0^{\ 0} + \alpha_V^{(1)} J_\mu^{(1)} I_1^{\ 0} \\ &+ \alpha_A^{(0)} J_\mu^{5(0)} I_0^{\ 0} + \alpha_A^{(1)} J_\mu^{5(1)} I_1^{\ 0} \\ &+ \alpha_{em} J_\mu^{em} , \end{aligned} \tag{10.4}$$

that is, we have isoscalar (T = 0) and isovector (T = 1) contributions both with (M_T = 0) and have in general both vector and axial-vector currents. The quantities $\alpha_V^{(T)}$, $\alpha_A^{(T)}$, T = 0,1 and α_{em} are gauge model couplings; for example, the standard model (W - S - GIM, see Ref. 4) yields

$$\alpha_V^{(0)} = \alpha_A^{(0)} = 0$$

$$\alpha_V^{(1)} = \alpha_A^{(1)} = 1$$

$$\alpha_{em} = -2 \sin^2\theta_W \ . \qquad (10.5)$$

Using (10.1) let us write (10.4) in the form

$$\mathcal{J}_\mu^{(0)} = \beta_V^{(0)} J_\mu^{(0)} I_0^{\ 0} + \beta_V^{(1)} J_\mu^{(1)} I_1^{\ 0} + \beta_A^{(0)} J_\mu^{5(0)} I_0^{\ 0} + \beta_A^{(1)} J_\mu^{5(1)} I_1^{\ 0} \ , \qquad (10.6)$$

with $\beta_V^{(T)} \equiv \alpha_V^{(T)} + \alpha_{em}$, $\beta_A^{(T)} \equiv \alpha_A^{(T)}$, $T = 0,1$. Then we can extend all of our previous developments for the electromagnetic interaction (Secs. 2-6) and the charge-changing weak interaction (Secs. 7-9) to encompass now the neutral-current weak interaction. For example, we obtain multipole operators $\hat{M}_{JM}$, $\hat{L}_{JM}$, $\hat{T}_{JM}^{e\ell}$ and $\hat{T}_{JM}^{mag}$ as before, but now multiplied by the appropriate $\beta_V^{(T)}$ (depending on the isospin nature required) and $\hat{M}_{JM}^5$, $\hat{L}_{JM}^5$, $\hat{T}_{JM}^{e\ell 5}$ and $\hat{T}_{JM}^{mag 5}$ now multipled by the appropriate $\beta_A^{(T)}$. We are immediately led to expressions for neutral current weak interaction cross sections which are analogous to those already displayed in Secs. 8 and 9. Take as just one specific example neutrino or anti-neutrino scattering: for $(d\sigma/d\Omega)_{(\nu_\ell,\nu_\ell')}$ and $(d\sigma/d\Omega)_{(\bar{\nu}_\ell,\bar{\nu}_\ell')}$ we will find an expression very similar to (9.5), but having the above-mentioned modifications to the currents (i. e., multiplication by appropriate factors $\beta_V^{(T)}$ and $\beta_A^{(T)}$), having different structure in isospin space and having modified kinematic factors (we take $m_\nu = m_{\bar{\nu}} = 0$). The main point to be made here once again is that the procedures leading to any of this (now quite large)

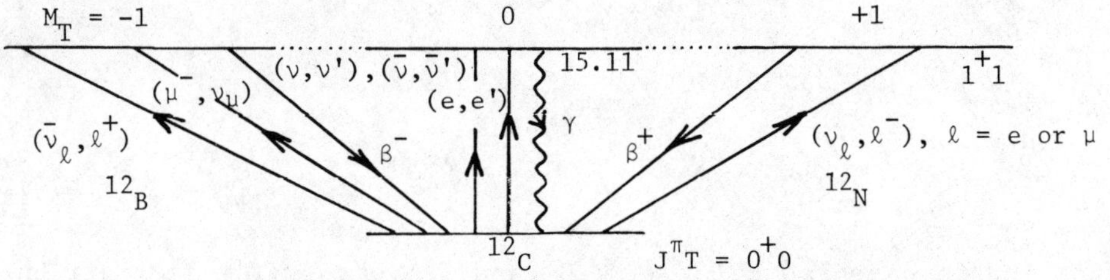

Figure 9. Electroweak processes in the A = 12 system.

class of electroweak interaction cross sections are uniformly approached in the same way. To see further illustration of this main theme in these lectures the reader is directed especially to Refs. 1, 2, 4, 5 and 10.

Let us conclude with a classic example of how the various processes are inter-related in Fig. 9.

REFERENCES

1. T. deForest, Jr. and J. D. Walecka, Adv. in Phys. 15 (1966) 1.
2. T. W. Donnelly and J. D. Walecka, Ann. Rev. Nucl. Sci. 25 (1975) 329.
3. H. Überall, Electron Scattering from Complex Nuclei, Parts A and B, (Academic, 1971).
4. J. D. Walecka, in Muon Physics, Vol. II, eds. V. W. Hughes and C. S. Wu, (Academic, 1975) p. 113.
5. T. W. Donnelly and R. D. Peccei, Phys. Reports 50 (1979) 1.
6. J. D. Bjorken and S. D. Drell, Relativistic Quantum Mechanics, (McGraw-Hill, 1964).
7. R. Von Gehlen, Phys. Rev. 118 (1960) 1455; M. Gourdin, Nuov. Cim. 21 (1961) 1094; J. D. Bjorken (1960, unpublished).
8. A. R. Edmonds, Angular Momentum in Quantum Mechanics, (Princeton University, 1960).
9. T. deForest, Jr., Ann. Phys. (N. Y.) 45 (1967) 365.
10. J. S. O'Connell, T. W. Donnelly and J. D. Walecka, Phys. Rev. C6 (1972) 719.
11. S. Weinberg, Phys. Rev. 112 (1958) 1375.

UNIFICATION OF WEAK AND ELECTROMAGNETIC FORCES

D. Atkinson

Institute for Theoretical Physics

Groningen, The Netherlands

1. INTRODUCTION

The purpose of these lectures is to introduce the nuclear physicist to the modern theory of the electroweak interaction, and to apply it to β-decay.

First quantum electrodynamics is reviewed, in a fully relativistic notation, and the theory is illustrated by a calculation of Bhabha scattering in lowest order. A treatment of the weak interaction , in which the photon is replaced by the charged W-boson, is then applied to electron-neutrino scattering. The important lesson is that the part of a langrangian which is quadratic in the fields yields the particle propagators, while cubic and quartic terms give respectively 3- and 4-line vertices in Feynman diagrams.

Next the leptonic sector of the $SU(2)_L \times U(1)$ electroweak theory is introduced(1); and the spontaneous symmetry-breaking phenomenon is explained in detail. Electron-neutrino scattering is briefly reconsidered.

The semileptonic electroweak interactions, in which quarks as well as leptons interact, are treated in the manner of Glashow, Iliopolous and Maiani(2), in which the charmed quark is used to suppress unwanted strangeness-changing neutral current effects. The application here is to nuclear β^+-decay, in which a distinction is made between the pure Fermi and pure Gamow-Teller cases.

Finally, an extension of the above theory to a left-right symmetrical $SU(2)_L \times SU(2)_R \times U(1)$ model(3), in which low-energy parity violation is a consequence of spontaneous symmetry breakdown, is explained; and β^+-decay is reconsidered in this light. The result of a recent Groningen experiment that tests this theory is finally given(4).

In these lectures, we consider only the electron and muon,

with their neutrinos, and the four lightest quarks. The theory may be extended to include the tau lepton and its neutrino, together with a top and bottom quark. However, the top quark has not been detected, so this extension is still speculative.

2. ELECTROMAGNETIC AND WEAK LEPTONIC INTERACTIONS

The free Dirac equation for an electron of mass m_e is

$$(i\gamma^\rho \partial_\rho - m_e)\psi(x) = 0, \tag{2.1}$$

where ψ is a four-component spinor field, and γ^ρ are four-dimensional complex matrices satisfying the anticommutation rules

$$\{\gamma^\rho, \gamma^\sigma\} = 2g^{\rho\sigma}. \tag{2.2}$$

The minimal coupling prescription

$$\partial_\rho \to D_\rho = \partial_\rho - ieA_\rho, \tag{2.3}$$

where A_ρ is the electromagnetic four-potential, yields the Dirac equation with electromagnetic interaction:

$$(i\gamma^\rho D_\rho - m_e)\psi(x) = 0 \tag{2.4}$$

which can be obtained as the Euler-Lagrange equation of motion for ψ from the lagrangian density

$$\overline{\psi}(x)(i\gamma^\rho D_\rho - m_e)\psi(x). \tag{2.5}$$

In order to have a non-trivial equation of motion for the photon field, we add the term

$$-\tfrac{1}{4}A_{\rho\sigma}(x)A^{\rho\sigma}(x), \tag{2.6}$$

where the field-tensor is

$$A_{\rho\sigma}(x) = \partial_\rho A_\sigma(x) - \partial_\sigma A_\rho(x). \tag{2.7}$$

The gauge transformation $\psi(x) \to e^{-i\Theta(x)}\psi(x)$, $A_\rho(x) \to A_\rho(x) - \frac{1}{e}\partial_\rho\Theta(x)$, leaves the lagrangian invariant. This is trivial so far as (2.6) is concerned; but in (2.5) etc. the key observation is that

$$\begin{aligned} i\gamma^\rho D_\rho\psi &= (i\gamma^\rho\partial_\rho + e\gamma^\rho A_\rho)\psi \\ &\to (i\gamma^\rho\partial_\rho + e\gamma^\rho A_\rho - \gamma^\rho\partial_\rho\Theta)e^{-i\Theta}\psi \\ &= e^{-i\Theta}(i\gamma^\rho\partial_\rho + e\gamma^\rho A_\rho)\psi = e^{-i\Theta}i\gamma^\rho D_\rho\psi. \end{aligned} \tag{2.8}$$

The sum of (2.5) and (2.6) can be written

$$\mathcal{L} = \mathcal{L}_f + \mathcal{L}_i \tag{2.9}$$

where the quadratic part

$$\mathcal{L}_f = \bar{\psi}(i\gamma^\rho \partial_\rho - m_e)\psi - \tfrac{1}{4}A_{\rho\sigma}A^{\rho\sigma}, \tag{2.10}$$

is called the free part of the lagrangian. In a perturbation expansion of the scattering amplitude, it gives rise to the electron propagator, $i(\gamma^\rho p_\rho - m_e)^{-1}$, and the photon propagator, $-ig^{\rho\sigma}/k^2$. The cubic part,

$$\mathcal{L}_i = e\bar{\psi}\gamma^\rho A_\rho \psi, \tag{2.11}$$

is called the interaction term: it gives rise to vertices, in which an electron line emits a photon. For example, in Bhabha scattering, one has in second order in e the two graphs shown in Fig. 1. The exchange graph (1a) gives

$$-e^2\bar{u}(p_e')\gamma^\rho u(p_e)\frac{-ig_{\rho\sigma}}{k^2}\,\bar{v}(p_p)\gamma^\sigma v(p_p') \tag{2.12}$$

and the annihilation graph (1b) yields

$$+e^2\bar{v}(p_p)\gamma^\rho u(p_e)\frac{-ig_{\rho\sigma}}{q^2}\,\bar{u}(p_e')\gamma^\sigma v(p_p'). \tag{2.13}$$

For simplicity, I have left out normalization factors.

The important thing is that an electron (or positron) line enters and leaves each vertex, and one photon line is attached. Further, each vertex contributes a factor e and a matrix γ, which has to be placed correctly between the spinors. All these things can be read off from (2.11). Further, each vertex contributes a factor i, because the perturbation series effectively comes from the series expansion of $\exp[i\int\mathcal{L}_i]$: this gives the initial minus sign in (2.12). In (2.13), this minus sign is cancelled because of the anticommutativity of fermi fields. The square of the absolute value of the sum of (2.12) and (2.13), suitably normalized, is the Bhabha differential cross-section, to second order in e.

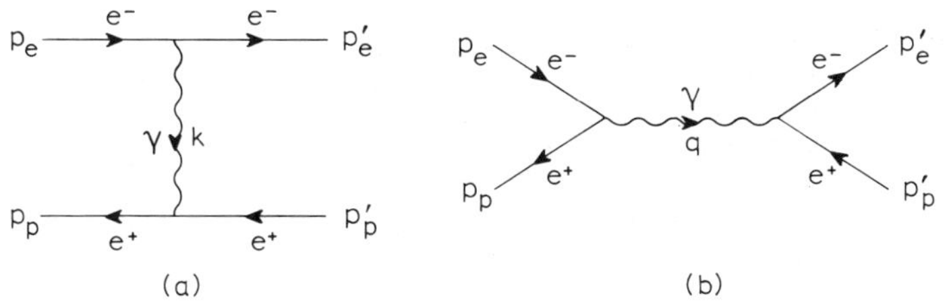

Figure 1. Feynman graphs for Bhabha scattering.

Quantum electrodynamics can be extended to include the muon by adding to the above lagrangian the term

$$\bar{\psi}(\mu)(i\gamma^{\rho}D_{\rho} - m_{\mu})\psi(\mu), \qquad (2.14)$$

Where the label, μ, refers to the muon, and where the *covariant derivative*, D_{ρ}, is the same as in (3), with the same *gauge field*, A_{ρ}, the photon. In muon-positron scattering, one obtains the exchange term (2.12), but where the u-spinors now refer to the muon. There is no annihilation graph, because electron and muon number are separately conserved.

Muons and electrons have also weak interactions. Indeed, the decay

$$\mu^- \to e^- + \bar{\nu}_e + \nu_{\mu}, \qquad (2.15)$$

with a half-life of 2×10^{-6} seconds, is a weak effect. It is mathematically closely related to the scattering process

$$\nu_{\mu} + e^- \to \nu_e + \mu^-, \qquad (2.16)$$

which is depicted in Fig. 2. This is like the Bhabha exchange graph, except that the gauge vector particle, W, must be electrically charged, and it must be massive, since the weak interactions are of short range.

A further complication is that the neutrinos are massless (probably!) and they have no right hands. More precisely, the helicity of a neutrino is always negative. Mathematically, we can define right- and left-handed helicity projection operators for massless spin-½ particles,

$$P_R = \tfrac{1}{2}(1 + \gamma^5), \quad P_L = \tfrac{1}{2}(1 - \gamma^5), \qquad (2.17)$$

where $\gamma^5 = i\gamma^0\gamma^1\gamma^2\gamma^3$. Note that $\gamma^{5\dagger} = \gamma^5$ and $(\gamma^5)^2 = 1$, also that γ^5 anticommutes with the other γ's. These projection operators satisfy

$$P_R P_L = 0 = P_L P_R, \; P_R + P_L = 1, \; P_R^2 = P_R, \; P_L^2 = P_L. \qquad (2.18)$$

The neutrino field is such that $P_R\psi(\nu_e) = 0$. Hence

$$\psi(\nu_e) = (P_R + P_L)\psi(\nu_e) = P_L\psi(\nu_e). \qquad (2.19)$$

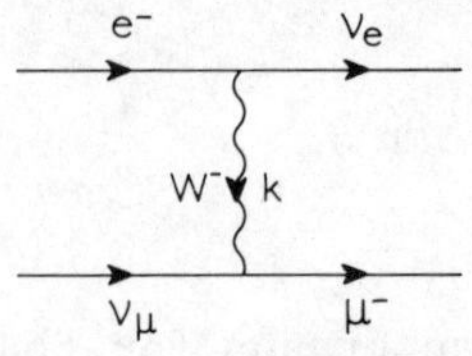

Figure 2. Feynman graph for electron-μ-neutrino scattering.

By analogy with the electromagnetic interaction lagrangian (2.11), we might guess that the electron-W vector-neutrino vertex has the form

$$e\bar{\psi}(e)\gamma^\rho W_\rho \psi(\nu_e) = e\bar{\psi}(e)\gamma^\rho W_\rho P_L \psi(\nu_e) \quad (2.20)$$

where I have tentatively put the coupling constant, e, equal to that in (2.11). We treat the muon and its neutrino similarly. The diagram gives a contribution looking like (2.12), except that the v's are replaced by u's, which are associated with the four fermions in question, and that the photon propagator, $-ig_{\rho\sigma}/k^2$, is replaced by

$$D_{\rho\sigma} = -i\,\frac{g_{\rho\sigma} - \dfrac{k_\rho k_\sigma}{m_W^2}}{k^2 - m_W^2}, \quad (2.21)$$

where m_W is the W-mass. If m_W is large and $k^2 << m_W^2$, then (2.21) can be approximated by $ig_{\rho\sigma}/m_W^2$ and the effective coupling constant at low energies, e^2/m_W^2, will be much smaller than the electromagnetic fine-structure constant.

In the elastic scattering process

$$\nu_e + e^- \rightarrow \nu_e + e^- \quad (2.22)$$

there is also a contribution from a neutral weak gauge vector, Z (see Fig. 3). The second graph is missing for reaction (2.16), because of separate electron-and muon-number conservation. At the $\bar{\nu}$ - Z - ν vertex, we expect an interaction

$$e\bar{\psi}(\nu_e)\gamma^\rho Z_\rho \psi(\nu_e) = e\bar{\psi}(\nu_e)\gamma^\rho Z_\rho P_L \psi(\nu_e), \quad (2.23)$$

where the mass of the Z might be different from that of the W. At the $\bar{e}$ - Z - e vertex, one might naïvely expect

$$e\bar{\psi}(e)\gamma^\rho Z_\rho \psi(e) = e\bar{\psi}(e)\gamma^\rho Z_\rho (P_R + P_L)\psi(e), \quad (2.24)$$

where now P_R does not disappear, since the electron is ambidextrous. In fact, since parity is not conserved in the weak interactions, the RHS of (2.24) could be generalized by making the coefficients of P_R and P_L unequal. In the Weinberg-Salam model, this is what happens, as we shall see in the next section.

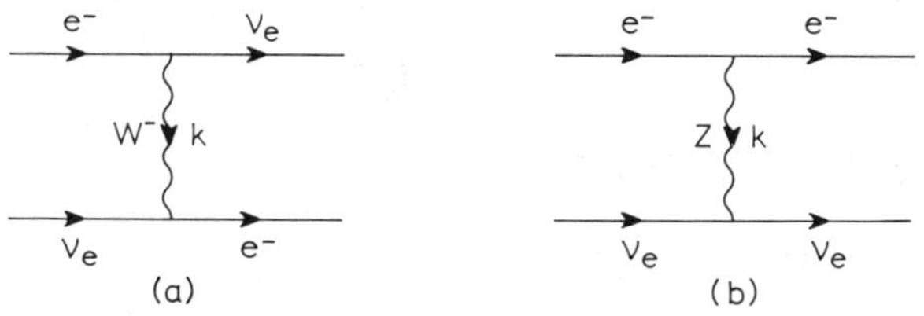

Figure 3. Feynman graphs for electron-neutrino scattering.

3. ELECTROWEAK LEPTONIC INTERACTIONS

In the Weinberg-Salam theory, the form of interaction between the leptons and the gauge vector bosons, the γ, W and Z, is strictly circumscribed by the requirement of the gauge invariance of the lagrangian. For the moment we consider only the electron and its neutrino. The free part of the lagrangian has the form

$$\overline{\psi}(e)(i\gamma^\rho\partial_\rho - m_e)\psi(e) + \overline{\psi}(\nu_e)(i\gamma^\rho\partial_\rho)\psi(\nu_e) \tag{3.1}$$

and this is gauge invariant under a phase-factor change in ψ, together with a minimal substitution, as in the previous section. This is called a U(1) gauge invariance, because a phase-factor is a one-dimensional unitary matrix.

The neutrino is left-handed,

$$\psi(\nu_e) = P_L\psi(\nu_e), \tag{3.2}$$

whereas the electron is not:

$$\psi(e) = (P_R + P_L)\psi(e). \tag{3.3}$$

Weinberg puts the neutrino and the electron's left hand into a doublet,

$$L(e) = \begin{bmatrix} P_L\psi(\nu_e) \\ P_L\psi(e) \end{bmatrix}, \tag{3.4}$$

and the electron's right hand into a singlet,

$$R(e) = P_R\psi(e). \tag{3.5}$$

In this notation, (3.1) can be rewritten

$$\overline{L}(e)i\gamma^\rho\partial_\rho L(e) + \overline{R}(e)i\gamma^\rho\partial_\rho R(e) - m_e\overline{\psi}(e)\psi(e). \tag{3.6}$$

If we interpret L as an SU(2) isospinor and R as an SU(2) scalar, then the first two terms in (3.6) are SU(2) invariant, but the mass term spoils this invariance. Following Weinberg, we throw it away: SU(2) invariance does not permit one to put in a mass term by hand. We shall see how to generate a mass for the electron when the time comes!

Without the mass-term, (3.6) is invariant under a global SU(2) mixing and the phase-factor U(1) multiplication, and we can make this local by generalizing the minimal coupling prescription.:

$$\mathcal{L}(e) = \overline{L}(e)i\gamma^\rho D_\rho L(e) + \overline{R}(e)i\gamma^\rho D_\rho R(e) \tag{3.7}$$

where the covariant derivative is

$$D_\rho = \partial_\rho - ig\vec{T}\cdot\vec{A}_\rho - i\tilde{g}YB_\rho \tag{3.8}$$

Here $\vec{T}$ are the generators for SU(2): acting on L(e), they are simply $\frac{1}{2}\vec{\tau}$, where $\vec{\tau}$ are the Pauli matrices; acting on R(e), they vanish, since R(e) is an SU(2) invariant.

Y is the generator of the abelian group U(1): it need not have the same value for L(e) and for R(e). For each of the four generators, there is a gauge field: the three $\vec{A}_\rho$ for SU(2) and B_ρ for U(1). The group SU(2) is associated with a coupling constant, g, and U(1) with another, $\tilde{g}$. Under a U(1) phase-change of L(e) and R(e), (3.7) remains invariant if B_ρ simultaneously changes as in the electromagnetic case. Under an SU(2) mixing,

$$L(e) \to \exp\left[\frac{i\vec{\tau}\cdot\vec{\Theta}}{2}\right] L(e), \tag{3.9}$$

where the $\vec{\Theta}$ depend on space and time, (3.7) remains invariant if the $\vec{A}_\rho$ change suitably. The rule is quite complicated; we do not need the details, it is sufficient to know it exists.

The eigenvalues of $\vec{T}$ are called weak left-handed isospin, the eigenvalues of Y weak hypercharge, and the isospin-hypercharge group is usually written $SU(2)_L \times U(1)$. The analogy with strong isospin and hypercharge is not accidental: the mathematics is the same, and the freedom to allow L(e) and R(e) to have different Y-eigenvalues can be exploited to ensure satisfaction of the weak version of the equality

$$Q = T_3 + \tfrac{1}{2}Y, \tag{3.10}$$

where Q is the electric charge (see Table 1).
Using the values of Y (-1 for L and -2 for R), we can rewrite (3.7) as

$$\begin{aligned} \mathcal{L}(e) = {} & i\bar{L}(e)\gamma^\rho(\partial_\rho - ig\,\frac{\vec{\tau}}{2}\cdot\vec{A}_\rho + i\tilde{g}B_\rho)L(e) + \\ & + i\bar{R}(e)\gamma^\rho(\partial_\rho + 2i\tilde{g}B_\rho)R(e). \end{aligned} \tag{3.11}$$

The charged W fields are defined by

$$W_\rho = 2^{-\frac{1}{2}}(A_{1\rho} + iA_{2\rho}), \tag{3.12}$$

and the neutral Z field and the photon field are

Table 1. Weak isospin and hypercharge assignments.

	$P_L\psi(\nu_e)$	$P_L\psi(e)$	$P_R\psi(e)$
T	$\frac{1}{2}$	$\frac{1}{2}$	0
T_3	$\frac{1}{2}$	$-\frac{1}{2}$	0
Q	0	-1	-1
Y	-1	-1	-2

$$Z_\rho = A_{3\rho}\cos\Theta_W - B_\rho \sin\Theta_W, \tag{3.13}$$

$$A_\rho = A_{3\rho}\sin\Theta_W + B_\rho \cos\Theta_W, \tag{3.14}$$

where Θ_W is called Weinberg's angle, defined by

$$\tan\Theta_W = \frac{2\tilde{g}}{g}. \tag{3.15}$$

With these definitions, (3.11) gives the coupling between the lepton and the gauge boson field.

To the electron lagrangian must be added a free gauge vector part, to allow the γ, Z and W to propagate. The massless abelian case was treated in eq. (2.6) et seq. For the massive abelian B_ρ, we would be inclined to add

$$-\tfrac{1}{4}B_{\rho\sigma}B^{\rho\sigma} + \tfrac{1}{2}m^2 B_\rho B^\rho, \tag{3.16}$$

where

$$B_{\rho\sigma} = \partial_\rho B_\sigma - \partial_\sigma B_\rho, \tag{3.17}$$

and where m is a mass. However, the mass-term spoils the gauge-invariance of the lagrangian, and must be rejected. The same applies to the $\vec{A}_\rho$ fields. The correct form of the gauge part of the lagrangian is

$$\mathcal{L}(g) = -\tfrac{1}{4}\vec{F}_{\rho\sigma}\cdot\vec{F}^{\rho\sigma} - \tfrac{1}{4}B_{\rho\sigma}B^{\rho\sigma} \tag{3.18}$$

where $B_{\rho\sigma}$ is given by (3.17) and where

$$F^a_{\rho\sigma} = \partial_\rho A^a_\sigma - \partial_\sigma A^a_\rho + g\varepsilon_{abc}A^b_\rho A^c_\sigma. \tag{3.19}$$

The last term in (3.19) is needed to ensure the invariance of $\mathcal{L}(g)$ under the $SU(2)_L$ gauge transformations. Note that (3.18) contains terms quadratic in the fields, the free part of the gauge lagrangian, and terms cubic and quartic, corresponding to vertices at which three or four gauge fields interact.

If we rewrite (3.18) in terms of the physical gauge fields (3.12) - (3.14) we find

$$\mathcal{L}(g) = -\tfrac{1}{2}W_{\rho\sigma}^{\dagger}W^{\rho\sigma} - \tfrac{1}{4}Z_{\rho\sigma}Z^{\rho\sigma} - \tfrac{1}{4}A_{\rho\sigma}A^{\rho\sigma} \tag{3.20}$$

+ three- and four-vector vertex interactions.

where

$$\begin{aligned} W_{\rho\sigma} &= \partial_\rho W_\sigma - \partial_\sigma W_\rho \\ Z_{\rho\sigma} &= \partial_\rho Z_\sigma - \partial_\sigma Z_\rho \\ A_{\rho\sigma} &= \partial_\rho A_\sigma - \partial_\sigma A_\rho . \end{aligned} \tag{3.21}$$

No explicit mass terms are allowed for the W- and Z-fields: they would spoil the $SU(2)_L \times U(1)$ gauge symmetry. There is another reason why explicit mass terms are ruinous. For large k, the massive vector propagator, Eq. (2.21), behaves like

$$\frac{i}{m_W^2}\frac{k_\rho k_\sigma}{k^2},$$

whereas the massless propagator, $-ig_{\rho\sigma}/k^2$, tends to zero. The consequence is that QED is renormalizable, so that higher-order corrections can be calculated, whereas a theory with massive vectors is in general not renormalizable.

The theory described by a lagrangian formed from the sum of (3.11) and (3.20) is theoretically satisfactory; the difficulty is a practical one: the electron, the W and Z vectors would be massless. The solution, the famous Higgs mechanism with Goldstone cannibalism, is very beautiful and works as follows.

Weinberg adds an $SU(2)_L$ complex isospinor of Lorentz scalar fields,

$$\phi = \begin{bmatrix}\phi^+ \\ \phi^0\end{bmatrix} = \frac{1}{2^{\frac{1}{2}}}\begin{bmatrix}\phi_1^+ + i\phi_2^+ \\ \phi_1^0 + i\phi_2^0\end{bmatrix} \tag{3.22}$$

and the following lagrangian,

$$\partial_\rho\phi^\dagger\partial^\rho\phi - m^2\phi^\dagger\phi - \lambda(\phi^\dagger\phi)^2, \tag{3.23}$$

where the last term is a self-interaction. The scalar fields, ϕ, are coupled to the gauge fields, $\vec{A}_\rho$ and B_ρ, by replacing the ordinary by the covariant derivative, (3.8). The weak isospin and hypercharge assignments of ϕ are given in Table 2.
So the scalar lagrangian with minimal coupling is

$$\mathcal{L}(s) = \left[(\partial_\rho - ig\,\frac{\vec{\tau}}{2}\cdot\vec{A}_\rho - i\tilde{g}B_\rho)\phi\right]^\dagger\left[(\partial^\rho - ig\,\frac{\vec{\tau}}{2}\cdot\vec{A}^\rho - i\tilde{g}B^\rho)\phi\right] - m^2\phi^\dagger\phi - \lambda(\phi^\dagger\phi)^2 \tag{3.24}$$

The vacuum state, the state of lowest energy, is obtained by considering the hamiltonian as a functional of all the fields, and minimizing it. It can be shown that this minimum corresponds to

Table 2. Weak quantum numbers of the scalar fields.

	ϕ_+	ϕ_0
T	$\frac{1}{2}$	$\frac{1}{2}$
T_3	$\frac{1}{2}$	$-\frac{1}{2}$
Q	1	0
Y	1	1

setting the lepton and gauge fields equal to zero, and the scalar field, ϕ, to a constant value (i.e. space- and time-independent). This constant has to be chosen such that the surviving part of the hamiltonian density,

$$\mathcal{H}_{min} = -\mathcal{L}_{min} = m^2\phi^\dagger\phi + \lambda(\phi^\dagger\phi)^2, \tag{3.25}$$

is minimum; and this can be done by looking at the first two derivatives of $\mathcal{H}_{min}$ with respect to the components of ϕ. If λ and m^2 are both positive, the "normal" case, this minimum occurs when ϕ, like the other fields, vanishes. However, if $m^2 < 0$ and $\lambda > 0$, the minimum is reached when

$$2\phi^\dagger\phi \equiv (\phi_1^+) + (\phi_2^+)^2 + (\phi_1^0)^2 + (\phi_2^0)^2 = -\frac{m^2}{\lambda} \equiv v^2. \tag{3.26}$$

Any value of the fields on this hypersphere yields an absolute minimum of the hamiltonian. Evidently m, being imaginary, cannot be interpreted as a mass. It, or more conveniently v, will be called the spontaneous symmetry breaking parameter. We pick one point on the hypersphere, $\phi_1^+ = \phi_2^+ = \phi_2^0 = 0$, $\phi_1^0 = v$, as the field configuration representing the vacuum. This choice breaks the symmetry: to be precise, three of the four generators of $SU(2)_L \times U(1)$ are broken, the surviving generator being precisely the combination (3.10), i.e. the electric charge.

The vacuum point can be written

$$\phi = 2^{-\frac{1}{2}} \begin{bmatrix} 0 \\ v \end{bmatrix} \tag{3.27}$$

and fluctuations about this point could be parametrized by the fields $\tilde{\phi}$:

$$\phi = 2^{-\frac{1}{2}} \begin{bmatrix} \tilde{\phi}_1^+ + i\tilde{\phi}_2^+ \\ \tilde{\phi}_1^0 + i\tilde{\phi}_2^0 + v \end{bmatrix} \tag{3.28}$$

However, a much cleverer parametrization of the four components of ϕ is

$$\phi = 2^{-\frac{1}{2}} \exp\left[\frac{i}{v}\frac{\vec{\tau}}{2}\cdot\vec{\Phi}\right] \begin{bmatrix} 0 \\ \Phi_0 + v \end{bmatrix} . \tag{3.29}$$

The vacuum corresponds to $\vec{\Phi} = 0 = \Phi_0$. The Φ fields are just as good as the $\tilde{\phi}$, in fact better, since we can use the fact that the lagrangian is gauge invariant to make an $SU(2)_L$ gauge transformation defined by

$$U = \exp\left[-\frac{i}{v}\vec{T}\cdot\vec{\Phi}\right]. \tag{3.30}$$

Acting on the scalar R, $\vec{T}$ has eigenvalue 0, so nothing happens, but acting on the isospinors L and ϕ, $\vec{T}$ becomes $\vec{\tau}/2$, and so ϕ is transformed to

$$\phi = 2^{-\frac{1}{2}}\begin{bmatrix} 0 \\ \Phi_0 + v \end{bmatrix}. \tag{3.31}$$

Only Φ_0, called the Higgs field, remains; the three fields $\vec{\Phi}$, in the new gauge, have completely disappeared. Who has eaten them? We shall see.

Now it is simply a matter of tedious algebra to insert (3.31), together with (3.12) - (3.14), into (3.24). The result can be written, together with the gauge term (3.20), in the form

$$\begin{aligned} \mathcal{L}(s) + \mathcal{L}(g) = -\tfrac{1}{4}A_{\rho\sigma}A^{\rho\sigma} - \tfrac{1}{4}Z_{\rho\sigma}Z^{\rho\sigma} + \tfrac{1}{2}m_Z^2 Z_\rho Z^\rho \\ - \tfrac{1}{2}W^\dagger_{\rho\sigma}W^{\rho\sigma} + m_W^2 W_\rho W^\rho \\ + \tfrac{1}{2}\partial_\rho\Phi_0\partial^\rho\Phi_0 - \tfrac{1}{2}\mu^2\Phi_0^2 + \text{interaction terms}, \end{aligned} \tag{3.32}$$

where

$$m_W = \tfrac{1}{2}gv, \quad m_Z = m_W \sec \Theta_W, \tag{3.33}$$

miraculously generated masses for the weak bosons, and where

$$\mu = (-2m^2)^{\frac{1}{2}}, \tag{3.34}$$

the mass of the Higgs. Remember that $m^2 < 0$, so μ is real.

A massless vector field has two physical degrees of freedom, the transverse polarization states. A massive vector, on the other hand, has moreover a third, namely a longitudinal polarization state. The three fields, W, $W^\dagger$, Z, have eaten the three fields $\vec{\Phi}$, providing them with the extra degree of freedom required by their acquisition of mass. The electromagnetic field remains massless. The Higgs field, instead of having a ridiculous imaginary mass, m, has a respectable real mass, μ. To appreciate the awesomeness of the miracle, you really have to struggle through the algebra that leads to (3.32).

The electron part of the lagrangian, (3.11), still has no mass term; but we can manufacture one from the spontaneous symmetry breaking by adding a gauge invariant coupling between the leptons and the scalar fields ϕ. Now $\overline{L}\phi$ is an $SU(2)_L$ scalar with Y = 2, hence LQR is $SU(2)_L \times U(1)$ and also Lorentz invariant. Hence an acceptable hermitian interaction lagrangian is

$$\mathcal{L}(i) = -G_i[\overline{L}\phi R + \overline{R}\phi^\dagger L], \tag{3.35}$$

where G_i is a coupling constant. After the gauge transformation (3.30), ϕ has the form (3.31), and (3.35) becomes

$$\mathcal{L}(i) = -m_e\overline{\psi}(e)\psi(e) - G_i\overline{\psi}(e)\psi(e)\Phi_0/2^{\frac{1}{2}} \tag{3.36}$$

where

$$m_e = vG_i/2^{\frac{1}{2}},$$

is the electron mass. The first term is the required mass term for the electron, with none for the neutrino, and the second term is an electron-higgs-electron interaction vertex: it turns out that the coupling is small. This algebra is easy to check: it is a mere conjuring trick, no miracle.

After some more algebra, we can rewrite the electron lagrangian (3.11) in terms of the physical fields (3.12) - (3.14) in the form

$$\begin{aligned}\mathcal{L}(e) = {} & i\overline{\psi}(\nu_e)\gamma^\rho\partial_\rho P_L\psi(\nu_e) + i\overline{\psi}(e)\gamma^\rho\partial_\rho\psi(e) - g\sin\Theta_W\overline{\psi}(e)\gamma^\rho A_\rho\psi(e) \\ & + 2^{-\frac{1}{2}}g\{\overline{\psi}(e)\gamma^\rho W_\rho P_L\psi(\nu_e) + \text{h.c.}\} \\ & - \tfrac{1}{2}g\sec\Theta_W\{\overline{\psi}(e)\gamma^\rho Z_\rho(\hat{C}_V - \gamma^5\hat{C}_A)\psi(e) - \overline{\psi}(\nu_e)\gamma^\rho Z_\rho P_L\psi(\nu_e)\}\end{aligned} \tag{3.37}$$

where

$$\hat{C}_V = \tfrac{1}{2} - 2\sin^2\Theta_W\,,\quad \hat{C}_A = \tfrac{1}{2}. \tag{3.38}$$

For physical applications, this is the most interesting result. From the coupling of the photon field, A_ρ, we see that

$$g = e\,\text{cosec}\,\Theta_W, \tag{3.39}$$

e being the charge of the positron. We see that the guess (1.20) for the W-coupling constant was not quite right: this is not e but $2^{-\frac{1}{2}}g = 2^{-\frac{1}{2}}e\,\text{cosec}\,\Theta_W$. The effective charged current-current coupling is

$$2^{-\frac{1}{2}}G = \frac{g^2}{8m_W^2} = \frac{e^2}{8m_W^2\sin^2\Theta_W}\,, \tag{3.40}$$

which gives the W-mass in terms of the Fermi universal weak coupling constant, G, and e and Θ_W.

Likewise, the guess (2.23) for the neutral current-neutrino coupling constant needs to be changed from e to

$$g\sec\Theta_W = 2\,e\,\text{cosec}\,2\Theta_W. \tag{3.41}$$

In the coupling of Z to the electron line, the coefficients that must be inserted in front of P_R and P_L in (2.24) are not equal, since $\hat{C}_A$ in (3.37) is not zero.

The whole theory can be straightforwardly extended to include the muon and its neutrino. These leptons are coupled to the same gauge vectors, A, W, Z, as are the electron and its neutrino, with the same couplings. The coupling to the isospinor of scalar fields has to be adjusted so that, after spontaneous symmetry breakdown, the muon picks up the right mass. There is no explanation of why the masses of the electron and muon are what they are; and the same things could be said for the tauon and its neutrino.

With the help of the interaction terms in (3.37), it is a

straightforward exercise to calculate the scattering amplitudes for $e^-\nu_e$ elastic scattering, by means of the graphs depicted in Fig. 3. For $e^-\nu_\mu$ scattering, only the Z boson contributes. Experimental measurements of these processes have yielded the following value for the Weinberg angle:

$$\sin^2\Theta_W = 0.215 \pm 0.012 \tag{3.42}$$

This then gives the masses

$$m_W = 83.0 \pm 2.4 \text{ GeV}$$

$$m_Z = 93.8 \pm 2.0 \text{ GeV}$$

for the quanta of the weak interaction.

4. ELECTROWEAK SEMILEPTONIC INTERACTIONS

In order to make contact with nuclear physics, we need to introduce quarks, since nuclei are made from agglomerations of quarks called nucleons. In a first attempt to mimic the electron left-handed doublet (3.4), one might be tempted to introduce

$$\begin{bmatrix} P_L\psi(u) \\ P_L\psi(d) \end{bmatrix} \tag{4.1}$$

where u is the 'up' (or proton-like) and d is the 'down' (or neutron-like) quark, and to couple this to the gauge fields. However, experimentally it is known that the charged weak current couples, with the universal coupling strength, G, not to $\psi(d)$, but rather to the linear superposition

$$\psi(d_c) = \cos\Theta_c\, \psi(d) + \sin\Theta_c\, \psi(s), \tag{4.2}$$

where s is the strange quark, and where Θ_c is the Cabibbo angle, known experimentally to be about 13^0. This motivates the replacement of (4.1) by

$$L(u) = \begin{bmatrix} P_L\psi(u) \\ P_L\psi(d_c) \end{bmatrix} \tag{4.3}$$

If L(u) is an $SU(2)_L$ weak isospinor, the current $\overline{\psi}(u)\gamma^\rho P_L\psi(d_c)$ will automatically couple to the left-handed W boson, so that the relative strengths of strangeness-conserving and strangeness-changing charged currents will agree with the experimental finding.

The orthogonal combination

$$\psi(s_c) = -\sin\Theta_c\, \psi(d) + \cos\Theta_c\psi(s), \tag{4.4}$$

was combined by Glashow, Iliopolous and Maiani with the field of the charmed quark, c, in a second left-handed doublet:

$$L(c) = \begin{bmatrix} P_L\psi(c) \\ P_L\psi(s_c) \end{bmatrix}. \tag{4.5}$$

This was done in fact before there was any experimental evidence for the charm quantum number. The right-hand projections of all four quark fields are assumed to be $SU(2)_L$ singlets, R(u), R(d), R(c), and R(s).

The assignment of weak quantum numbers is as in Table 3. The weak hypercharge, Y, has been chosen so that $Q = T_3 + \frac{1}{2}Y$. The quark part of the lagrangian, with minimal coupling, can be written

$$\begin{aligned} \mathcal{L}(q) &= i\bar{L}(u)\gamma^\rho D_\rho L(u) + i\bar{L}(c)\gamma^\rho D_\rho L(c) \\ &+ i\bar{R}(u)\gamma^\rho D_\rho R(u) + i\bar{R}(d)\gamma^\rho D_\rho R(d) + i\bar{R}(c)\gamma^\rho D_\rho R(c) \\ &+ i\bar{R}(s)\gamma^\rho D_\rho R(s), \end{aligned} \tag{4.6}$$

where the covariant derivative was defined in (3.8). Thus $\vec{T}$ becomes $\frac{1}{2}\vec{\tau}$ for the left-handed parts of $\mathcal{L}$, and zero for the right-handed parts, while the Y-values can be read off from the table.

The gauge fields are expressed in terms of the mass-eigenstates W, Z, A (3.12) - (3.14); and, after some algebra, one finds

$$\begin{aligned} \mathcal{L}(q) &= i\bar{\psi}(u)\gamma^\rho \partial_\rho \psi(u) + i\bar{\psi}(d)\gamma^\rho \partial_\rho \psi(d) + i\bar{\psi}(c)\gamma^\rho \partial_\rho \psi(c) \\ &+ i\bar{\psi}(s)\gamma^\rho \partial_\rho \psi(s) + 2^{-\frac{1}{2}} g\{\bar{\psi}(u)\gamma^\rho W_\rho^\dagger P_L \psi(d_c) + \bar{\psi}(c)\gamma^\rho W_\rho^\dagger P_L \psi(s_c) \\ &+ h.c.\} + J^\rho_{em} A_\rho + J^\rho_{neut} Z_\rho \end{aligned} \tag{4.7}$$

where the electromagnetic current is

$$\begin{aligned} J^\rho_{em} &= g \sin\Theta_W \{\tfrac{2}{3}\bar{\psi}(u)\gamma^\rho\psi(u) - \tfrac{1}{3}\bar{\psi}(d)\gamma^\rho\psi(d) \\ &+ \tfrac{2}{3}\bar{\psi}(c)\gamma^\rho\psi(c) - \tfrac{1}{3}\bar{\psi}(s)\gamma^\rho\psi(s)\} \end{aligned} \tag{4.8}$$

and since $g \sin\Theta_W = e$, the unit of charge, the expression agrees with the charge assignments in the table.

Table 3. Weak quantum numbers for the quark fields.

	$P_L\psi(u)$	$P_L\psi(d_c)$	$P_L\psi(c)$	$P_L\psi(s_c)$	$P_R\psi(u)$	$P_R\psi(d)$	$P_R\psi(c)$	$P_R\psi(s)$
T	$\frac{1}{2}$	$\frac{1}{2}$	$\frac{1}{2}$	$\frac{1}{2}$	0	0	0	0
T_3	$\frac{1}{2}$	$-\frac{1}{2}$	$\frac{1}{2}$	$-\frac{1}{2}$	0	0	0	0
Q	$\frac{2}{3}$	$-\frac{1}{3}$	$\frac{2}{3}$	$-\frac{1}{3}$	$\frac{2}{3}$	$-\frac{1}{3}$	$\frac{2}{3}$	$-\frac{1}{3}$
Y	$\frac{1}{3}$	$\frac{1}{3}$	$\frac{1}{3}$	$\frac{1}{3}$	$\frac{4}{3}$	$-\frac{2}{3}$	$\frac{4}{3}$	$-\frac{2}{3}$

The weak neutral current is

$$J^{\rho}_{neut} = \tan\Theta_W J^{\rho}_{e.m.} + \tfrac{1}{2}g \sec\Theta_W\{-\overline{\psi}(u)\gamma^{\rho}P_L\psi(u) + \overline{\psi}(d)\gamma^{\rho}P_L\psi(d) - \overline{\psi}(c)\gamma^{\rho}P_L\psi(c) + \overline{\psi}(s)\gamma^{\rho}P_L\psi(s)\}. \qquad (4.9)$$

Notice that this current is a mixture of vector parts (γ^{ρ}), and axial parts ($\gamma^{\rho}\gamma^5$), as in the leptonic case (see 3.37). An important property of this neutral current is that it does not change strangeness, which agrees with experimental findings: for example the decay $K^+ \to \pi^+ + e^+ + e^-$ has a branching ratio of 2×10^{-7} only, while that for $K^+ \to \pi^0 + e^+ + \nu_e$, which is mediated by the charged current, has a branching ratio of 4.8%. A theory without charm does have an unwanted strangeness-changing neutral current.

It is possible to generalize the lepton-scalar couplings in such a way that all the quarks obtain mass terms after the spontaneous symmetry breaking; but I suppress the details.

As a simple practical example, let us consider β^+ decay in a nucleus (see Fig. 4).
An up quark in a proton turns into a down quark, and produces therefore a neutron. The amplitude consists of the $\overline{\psi}(u)\gamma^{\rho}\psi(d_c)$ vertex from (4.7) at the top, and the $\overline{\psi}(e)\gamma^{\rho}\psi(\nu_e)$ from (3.37) at the bottom, with a W propagator (2.21) in between. Since the momentum transfer is negligible compared with m_W, we approximate the propagator by $ig_{\rho\sigma}/m_W^2$. Then, using (3.40), we find the amplitude to be

$$2^{-\frac{1}{2}}G \cos\Theta_c \overline{u}_d\gamma^{\rho}(1 - \gamma^5)u_u \, \overline{u}_{\nu}\gamma_{\rho}(1 - \gamma^5)v_e \, . \qquad (4.10)$$

This is correct only for an allowed transition with no special selection rules operating. In a spin 0 - spin 0 allowed transition, we have a pure Fermi transition, which is driven purely by the vector quark current, so instead of (4.10) we have

$$2^{-\frac{1}{2}}G \cos\Theta_c \, \overline{u}_d\gamma^{\rho}u_u\overline{u}_{\nu}\gamma^{\rho}(1 - \gamma^5)v_e \, . \qquad (4.11)$$

For a spin 1 - spin 0 pure Gamow-Teller allowed transition, for example, the spin flip indicates that only the axial quark current is operative, and so we have

$$2^{-\frac{1}{2}}G \cos\Theta_c\overline{u}_d\gamma^{\rho}\gamma^5u_u\overline{u}_{\nu}\gamma_{\rho}(1 - \gamma^5)v_e \, . \qquad (4.12)$$

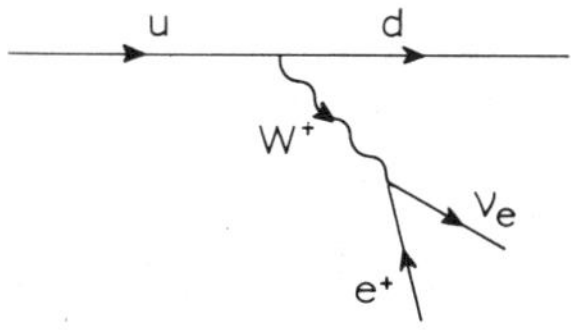

Figure 4. Feynman graph for β^+ decay.

The positrons in β^+ decay are polarized, a sure sign of the fact that the weak interaction violates parity conservation. To obtain the flux of positrons produced in a given direction with spin parallel or antiparallel to their momentum (σ_+ and σ_-), we take (4.11) for the Fermi case or (4.12) for the Gamow-Teller case, calculate the square of the modulus, sum over the quark spin states, integrate over the (unobserved) neutrino direction, and project the spin of the positron respectively parallel and antiparallel to its momentum. The polarization is defined to be

$$P = \frac{\sigma_+ - \sigma_-}{\sigma_+ + \sigma_-} \tag{4.13}$$

It works out in both Fermi and Gamow-Teller cases to be v/c, where v is the velocity of the positron. In the ultra-relativistic limit, the positrons are completely right-handed.

5. LEFT-RIGHT SYMMETRIC MODEL

The most straightforward extension of the minimal $SU(2)_L \times U(1)$ model of Weinberg and Salam to a situation in which there is exact left-right symmetry is the $SU(2)_L \times SU(2)_R \times U(1)$ theory. In this case, the left-handed fermion doublets are as before:

$$\begin{bmatrix} P_L\psi(\nu_e) \\ P_L\psi(e) \end{bmatrix}, \begin{bmatrix} P_L\psi(\nu_\mu) \\ P_L\psi(\mu) \end{bmatrix}, \begin{bmatrix} P_L\psi(u) \\ P_L\psi(d_c) \end{bmatrix}, \begin{bmatrix} P_L\psi(c) \\ P_L\psi(s_c) \end{bmatrix}. \tag{5.1}$$

To this we add, instead of right-handed singlets, a similar set of $SU(2)_R$ doublets by simply replacing P_L by P_R. Left and right weak isospin and weak hypercharge are assigned, and a minimal-coupling lagrangian is constructed, in which there are three gauge fields $\vec{A}_L$ for $SU(2)_L$, and three fields $\vec{A}_R$ for $SU(2)_R$, together with the U(1) gauge field, B. The coupling constant, g, is the same for $SU(2)_L$ and for $SU(2)_R$, so that the left-right symmetry is not spoiled. One defines W_L and W_R complex fields in terms of the first two components of $\vec{A}_L$ and $\vec{A}_R$ respectively, while the third components of these fields, together with B, are combined linearly to form the electromagnetic field, A, with pure vector coupling, and two neutral weak boson fields, Z_1 and Z_2.

The left-right symmetry is also respected in the Higgs sector of the model. As well as a left-handed doublet of complex scalar fields, we have a right-handed counterpart. In order to generate all the masses that are needed, there is a further set of left-right symmetric scalar fields, but I shall not give the details[3]. After spontaneous symmetry breakdown, the electromagnetic field remains massless, as before, and the Z- and W-fields acquire masses. The important point is that the masses of the left- and right-handed vectors do not need to be equal. If the right-handed W boson has a much larger mass than the left-handed one, $m_R >> m_L$, then the effective low-energy coupling for V + A charged currents will be much smaller than it is for V - A charged currents. The interesting question is whether the former actually occur. Experiments can produce bounds on the ratio of the masses, as we shall see.

There is however a complication: after the spontaneous symmetry breakdown, W_L and W_R are not in general mass eigenstates, i.e. there are cross-terms, $W_L^\dagger W_R + W_R^\dagger W_L$, in the quadratic part of the lagrangian. To remove these, a rotation has to be made,

$$\begin{aligned} W_1 &= W_L \cos\zeta - W_R \sin\zeta, \\ W_2 &= W_L \sin\zeta + W_R \cos\zeta, \end{aligned} \tag{5.2}$$

where W_1 and W_2 are mass eigenstates, with masses m_1 and m_2, and ζ is a mixing angle. Experimentally, ζ is small, and so W_1 is predominantly left-handed and W_2 predominantly right-handed. Likewise, the neutral vectors, A, Z_1, and Z_2 are such that there are no cross-terms. In the following, we shall concentrate exclusively on the charged currents.

The coupling of the fermions to the charged vectors can be written

$$gJ_L^\rho W_{L\rho} + gJ_R^\rho W_{R\rho}, \tag{5.3}$$

where J_L and J_R are the V - A and V + A fermionic currents. Explicitly, and in terms of the physical states (5.2), this may be written

$$2^{-\frac{1}{2}}g(\cos\zeta - \sin\zeta)\{[W_1 + \varepsilon W_2]_\rho V^\rho + [-\varepsilon W_1 + W_2]_\rho A^\rho\}, \tag{5.4}$$

where

$$\varepsilon = \frac{1 + \tan\zeta}{1 - \tan\zeta}, \tag{5.5}$$

and where the vector and axial charged currents are

$$V^\rho = \overline{\psi}(e)\gamma^\rho\psi(\nu_e) + \overline{\psi}(\mu)\gamma^\rho\psi(\nu_\mu) + \overline{\psi}(d_c)\gamma^\rho\psi(u) + \overline{\psi}(s_c)\gamma^\rho\psi(c), \tag{5.6}$$

$$A^\rho = \overline{\psi}(e)\gamma^\rho\gamma^5\psi(\nu_e) + \overline{\psi}(\mu)\gamma^\rho\gamma^5\psi(\nu_\mu) + \overline{\psi}(d_c)\gamma^\rho\gamma^5\psi(u) + \overline{\psi}(s_c)\gamma^\rho\gamma^5\psi(c). \tag{5.7}$$

The transition amplitude for β^+-decay now contains two diagrams like Fig. 4, one with an intermediate W_1^+ and one with an intermediate W_2^+. The effective coupling is therefore

$$\tfrac{1}{2}g^2(\cos\zeta - \sin\zeta)^2\{[D_1 + \varepsilon^2 D_2]^{\rho\sigma}V_\rho^\dagger V_\sigma + [\varepsilon^2 D_1 + D_2]^{\rho\sigma}A_\rho^\dagger A_\sigma + \varepsilon[D_1 - D_2]^{\rho\sigma}[V_\rho^\dagger A_\sigma + V_\rho A_\sigma^\dagger]\}, \tag{5.8}$$

where the vector propagators are as in (2.21), but with masses m_1 and m_2. For small momenta, $D_1^{\rho\sigma} \sim g^{\rho\sigma}/m_1^2$, $D_2^{\rho\sigma} \sim g^{\rho\sigma}/m_2^2$, so in this limit (5.8) becomes

$$2^{-\frac{1}{2}}G\{V^{\rho\dagger}V_\rho + \eta_{AA}A^{\rho\dagger}A_\rho + \eta_{AV}[V^{\rho\dagger}A_\rho + V^\rho A_\rho^\dagger]\}, \tag{5.9}$$

where

$$G = 2^{-\frac{1}{2}}g^2(\cos\zeta - \sin\zeta)^2\left[\frac{1}{m_1^2} + \frac{\varepsilon^2}{m_2^2}\right], \tag{5.10}$$

$$\eta_{AA} = \frac{\varepsilon^2 + \rho}{1 + \varepsilon^2\rho} \tag{5.11}$$

$$\eta_{AV} = -\frac{\varepsilon(1 - \rho)}{1 + \varepsilon^2\rho} \tag{5.12}$$

with

$$\rho = m_1^2/m_2^2\ . \tag{5.13}$$

The generalization of (4.10), the momentum-space transition amplitude for β^+ decay, is accordingly

$$2^{-\frac{1}{2}}G\cos\Theta_c\{\bar{u}_d\gamma^\rho u_u\bar{u}_\nu\gamma_\rho v_e + \eta_{AA}u_d\gamma^\rho\gamma^5 u_u\bar{u}_\nu\gamma_\rho\gamma^5 v_e$$
$$+ \eta_{AV}[\bar{u}_d\gamma^\rho\gamma^5 u_u\bar{u}_\nu\gamma_\rho v_e + \bar{u}_d\gamma^\rho u_u\bar{u}_\nu\gamma_\rho\gamma^5 v_e]. \tag{5.14}$$

The pure V - A limit corresponds to $m_2 = \infty$, $\zeta = 0$, i.e. $\rho = 0$, $\varepsilon = 1$, which yields $\eta_{AA} = 1$, $\eta_{AV} = -1$.

For pure Fermi (F) decay, the axial part of (5.14) is suppressed, and so we then have

$$2^{-\frac{1}{2}}G\cos\Theta_c\bar{u}_d\gamma^\rho u_u\bar{u}_\nu\gamma_\rho(1 + \eta_{AV}\gamma^5)v_e; \tag{5.15}$$

whereas, for a pure Gamow-Teller (GT) decay, the amplitude is

$$2^{-\frac{1}{2}}G\eta_{AV}\cos\Theta_c\bar{u}_d\gamma^\rho\gamma^5 u_u\bar{u}_\nu\gamma_\rho(1 + \frac{\eta_{AA}}{\eta_{AV}}\gamma^5)v_e. \tag{5.16}$$

To calculate the cross-sections, $\sigma_\pm$, one computes the square of the absolute value of (5.15) for F, and of (5.16) for GT decays. Here one uses standard formulae for Dirac spinors[5]:

$$\sum u_u\bar{u}_u = \frac{\gamma p_u + m_u}{2m_u} \qquad \sum u_d\bar{u}_d = \frac{\gamma p_d + m_d}{2m_d}\ , \tag{5.17}$$

where the sum is over the two spin states of the quarks. One neglects the motion of the quarks, as compared with that of the neutrino and positron, which is a good approximation, and makes projections of the positron spin respectively parallel and antiparallel to the momentum, in order to calculate the cross-sections, σ_+ and σ_-. After averaging over the unobserved neutrino direction, one finds

$$P = \frac{\sigma_+ - \sigma_-}{\sigma_+ + \sigma_-} = -\frac{2\xi}{1+\xi^2} \times \frac{v}{c}, \tag{5.18}$$

where $\xi = \eta_{AV}$ for the Fermi case, and $\xi = \eta_{AA}/\eta_{AV}$ for the Gamow-Teller case. In the V - A limit, $\xi = -1$ in both cases, and P reduces to v/c, as it should.

Using the formula (5.11) and (5.12), we find that, for pure F β^+ decay,

$$F \equiv \frac{c}{v} P = \frac{2\varepsilon}{1+\varepsilon^2} \frac{(1-\rho)(1+\varepsilon^2\rho)}{1+\varepsilon^2\rho^2}, \tag{5.19}$$

while for pure GT β^+ decay,

$$G \equiv \frac{c}{v} P = \frac{2\varepsilon}{1+\varepsilon^2} \frac{(1-\rho)(\varepsilon^2+\rho)}{\varepsilon^2+\rho^2} \tag{5.20}$$

Evidently, a measurement of both F and G would suffice in principle to determine ε and ρ, and thus the mixing angle, ζ, and the mass-ratio, m_2/m_1. If it turns out that ρ is significantly different from zero then the minimal $SU(2)_L \times U(1)$ will have been shown to be inadequate. An accurate value of ρ that is consistent with zero, on the other hand, would provide a lower bound for m_2/m_1.

In a recent experiment by a Groningen group[4], performed with the help of facilities at the Vrije Universiteit in Amsterdam, relative measurements of F (from the $0^+ \to 0^+$ β^+ decay of ^{26m}Al), and of G (from the $1^+ \to 0^+$ β^+ decay of ^{30}P), give

$$R = \frac{F}{G} = 0.986 \pm 0.038. \tag{5.21}$$

Since G is already known to 1% accuracy, this yields a value

$$F = 0.992 \pm 0.040, \tag{5.22}$$

which is a great improvement on earlier results. It is planned in future experiments to measure R to an accuracy of order 10^{-3}. Any significant deviation of R from unity (after appropriate corrections for nuclear and radiative effects) would signal a departure from the minimal $SU(2)_L \times U(1)$ theory.

REFERENCES

1. S. Weinberg, Phys. Rev. Lett. 19(1967)1264.
2. S.L. Glashow, J. Iliopolous and L. Maiani, Phys. Rev. D2(1970)1285.
3. J.C. Pati and A. Salam, Phys. Rev. D10(1974)275.
   G. Senjanović, Nucl. Phys. B153(1979)334.
4. D. Atkinson, J. van Klinken, K. Stam, W.Z. Venema, V.A. Wichers,

Groningen preprint(1982).
5. J.D. Bjorken and S.D. Drell, Relativistic Quantum Mechanics, McGraw Hill(1964).

EXERCISES

1. Obtain the Dirac equation with minimal coupling, (2.4), by applying the Euler-Lagrange equation

$$\partial_\mu \frac{\partial \mathcal{L}}{\partial(\partial_\mu \bar{\psi}_\alpha)} - \frac{\partial \mathcal{L}}{\partial(\bar{\psi}_\alpha)} = 0,$$

where $\mathcal{L}$ is the lagrangian given in (2.9) - (2.11) and where α = 1, 2, 3, 4, is a spinor index.

2. Show that

$$\mathcal{H}_{min} = m^2\phi^\dagger\phi + \lambda(\phi^\dagger\phi)^2$$

is minimized by $\phi = 0$ if $m^2 > 0$, $\lambda > 0$, and by $2\phi^\dagger\phi = -m^2/\lambda$ if $m^2 < 0$, $\lambda > 0$.

3. Calculate $\mathcal{L}(s) + \mathcal{L}(g)$, after the gauge transformation, including the interaction terms that have been omitted from Eq. (3.32).

4. Check the electron lagrangian in the new gauge, Eq. (3.37).

5. Check the form of the neutral quark current, Eq. (4.9).

6. Calculate the neutral quark current in the absence of the charmed quark. Here one assumes that $P_L\psi(s_c)$ is an $SU(2)_L$ scalar. Show that unwanted first-order neutral strangeness-changing terms are present.

7. Calculate the polarization, Eq. (4.13), of the positrons in pure Fermi and in pure Gamow - Teller β^+ decay in both the $SU(2)_L \times U(1)$ minimal theory, and in the left-right symmetric $SU(2)_L \times SU(2)_R \times U(1)$ theory.

BREAKING OF FUNDAMENTAL SYMMETRIES IN NUCLEI

E.G. Adelberger

Max-Planck-Institut für Kernphysik
D-6900 Heidelberg 1, Fed. Rep. of Germany
and
Department of Physics FM-15
University of Washington
Seattle WA 98195 USA

1. INTRODUCTION

Nuclei are interesting systems in which to study symmetry breakdowns because of the rich variety of phenomena one can observe. The nucleus "feels" the strong, electromagnetic (EM) and weak interaction which respect differing sets of fundamental symmetries so that one expects a hierarchy of symmetry breaking effects. We shall discuss the breaking of 2 symmetries-isospin (I) which is respected by the strong interaction but broken by electromagnetism, and parity (P) which is violated only by the weak interaction. These two examples in some sense span the space of symmetry-breaking effects. The ratio of the EM to the strong force felt by a pair of nucleons is roughly $Z\alpha$ so that isospin is quite obviously broken in nuclei. Nevertheless in many cases the symmetry is better than one would expect from elementary considerations. Because I-breaking effects are so obvious and easily measurable it provides a nice understandable example of a broken symmetry and we shall discuss it in some detail.

Breaking of the P symmetry in nuclei is a very small effect ($\sim 10^{-6}$) so that experiments to detect it are quite difficult. These small effects are interesting because they allow us to observe a "new" kind of nuclear force due to the weak interaction. In contrast to the case with the I symmetry - where the predominant symmetry breaking interaction is very well understood so that the effects primarily teach us about nuclear structure - the P-breaking effects in nuclei are due to an interaction which is not well understood and one hopes to learn about the interaction from the pattern of symmetry breaking. Because of time limitations we can treat only a few salient and straightforward aspects of each of the 2 symmetries.

II. ISOSPIN NONCONVERSATION

The reader interested in comprehensive reviews of this field is referred to refs. 1 and 2.

II.1 Review of Definitions

Isospin conservation is a symmetry which implies that strongly interacting particles occur in degenerate multiplets, all members of which have the same strong interactions (except for geometrical factors we discuss below). We begin by neglecting EM and other interactions which break I and review what the symmetry implies. The nucleon, N, is a multiplet with 2 members (n and p) which have identical masses. We say N is I=1/2 with two projections $I_3=+1/2$ (p) and $I_3=-1/2$ (n). Similarly the π meson is an I=1 multiplet with projections $I_3=+1$ (π^+), $I_3=0$ (π^o) and $I_3=-1$ (π^-). When applied to a system of 2 nucleons, I conservation is identical to the principle of charge independence - i.e. the statement that, in a given spin state, the strong force between two protons is identical to that between 2 neutrons or a neutron-proton pair. Consider for example the NN system at very low energies so that the orbital angular momentum L=0. We can generalize the Pauli principle to take I into account by considering n and p as two substates of a single particle N. This principle says that for two nucleons L+S+I = odd. Therefore the possible NN states are:

nn	np	pp
I=1 ~~//////~~ $^1S_0$	I=1 ~~//////~~ $^1S_0$	I=1 ~~//////~~ $^1S_0$
	I=0 ——— $^3S_1$	

where the states are labelled by $^{2s+1}L_J$. Since H_{strong} is an isospin scalar it must have the same expectation value in all 3 $^1S_0$ I=1 states and we expect all the $^1S_0$ states to have the same energy and width - i.e. the strong force to be charge independent.

II.2 Isospin Multiplets

A familiar consequence of I conservation is the occurrence of nuclear multiplets - an example of which is shown below. If I were an exact symmetry the following conditions would apply

1) all members of a given multiplet would have the same mass

$$H|\alpha,I,I_3> = E(\alpha,I)|\alpha,I,I_3>$$

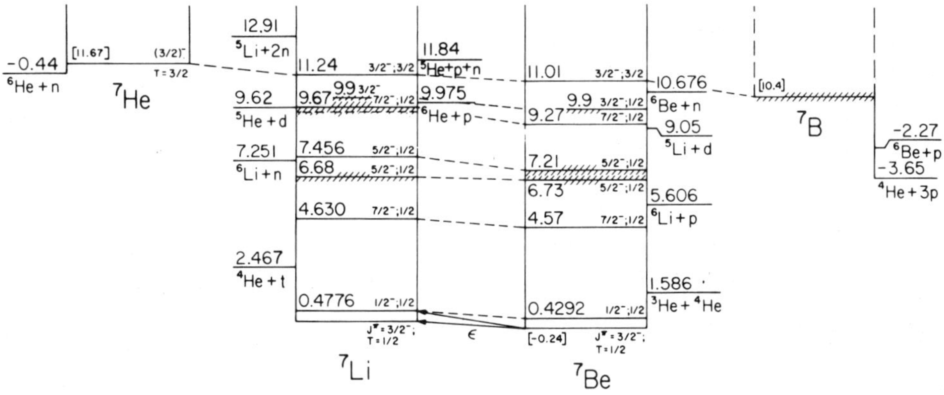

Fig. 1. The A=7 Isospin Multiplet from F.A. Ajzenberg-Selove Nucl. Phys. A320, 1 (1979)

2) all states would have a definite value of I

$$I^2|\psi\rangle = I(I+1)|\psi\rangle$$

3) all members of a given multiplet would have the same wave functions (except for neutron-proton interchanges)

$$I_{\pm}|\alpha,I,I_3\rangle = \sqrt{(I\mp I_3)(I\pm I_3+1)}\,|\alpha,I,I_3\pm 1\rangle$$

Conditions 2 and 3 imply certain selection rules on operators connecting the nuclear states.

A) Strong interactions. H_{strong} is an isospin scalar. Therefore condition 2 implies that in strong interaction processes $\Delta\vec{I}=0$. Condition 3 implies relations among matrix elements for related strong reactions. Consider for example the reactions $^{13}C(d,{}^3He)^{12}B$ and $^{13}C(d,t)^{12}C(I=1)$ where ^{12}B and $^{12}C(I=1)$ belong to the same isospin triplet and t and ^3He belong to the same isospin doublet. Since d has I=0 and ^{13}C I=1/2 the incoming channel of both reactions has I=1/2. The outgoing channels consisting of an I=1/2 and an I=1 object could have I=1/2 or I=3/2. But H is an isospin scalar so it can only connect to the I=1/2 component of the final states. Since ^{12}B and $^{12}C(I=1)$, and t and ^3He, are isospin analog pairs the ratio I=1/2 intensities in the $^{12}B + {}^3He$ and $^{12}C(I=1) + t$ final states is just given by the squares of Clebsch-Gordan coefficients, - $\sigma(d,{}^3He)/\sigma(d,t) = |C(1/2\ 1/2\ 1{-}1|1/2{-}1/2)/C(1/2{-}1/2\,1\,0|1/2{-}1/2)|^2 = 2$. It is important to note that the forces between the mass 3 and mass 12 nuclei are not charge independent since $\sigma(d,{}^3He)\neq\sigma(d,t)$. Isospin conservation is equivalent to charge independence only in the case of two "identical" I=1/2 objects.

B) EM interactions. The electric charge of any particle obeys the relation $Q=e(I_3+\frac{1}{2}Y)$ where Y is an isospin scalar called the hypercharge (which is one for nucleons). Since the EM current transforms like the electric charge it contains a piece which is an isospin scalar plus one which is an isospin vector. Hence any first order EM transitions must satisfy the selection rules $\Delta\vec{I}=0$ or 1. For each multipole the EM moments of all the various members of a given multiplet are determined by only 2 independent quantities - the isoscalar and the isovector moment. For specific multipoles more restrictive rules apply. For instance the E1 operator in the long wavelength limit is $O(E1) = \sum_i q_i\vec{r}_i = e\sum_i(\frac{1}{2} + I_{3_i})\vec{r}_i = \frac{e}{2}\sum_i\vec{r}_i + e\sum_i\vec{r}_i I_{3_i}$. The isoscalar part of the operator is proportional to the center of mass coordinate and hence cannot contribute to transitions between (orthogonal) states so that the E1 operator is $\Delta\vec{I}=1$. For magnetic (ML) transitions, the $\Delta I=1$ matrix elements are inherently roughly an order of magnitude larger than the $\Delta I=0$ matrix elements because $\mu_p-\mu_n \gg \mu_p+\mu_n$.

C) <u>β decay</u>. In the allowed approximation the Fermi and Gamow-Teller operators are $O(F)\propto\sum_i I_{\pm_i}=I_\pm$ and $O(GT)\propto\sum_i I_{\pm_i}\sigma_i$. Both satisfy the condition $\Delta\vec{I}=1$. The Fermi operator is especially interesting since it is just the total isospin raising or lowering operator and therefore only connects members of the same isospin multiplet. It is therefore very useful as a model-independent probe of I-breaking effects.

II.3 Isospin Breaking Phenomena

The existence of forces (such as EM) which are not isospin scalars causes the I symmetry to break down. We can identify 3 consequences of this breakdown:

1) the different members of an isospin multiplet acquire different masses (mass-splitting),
2) the states no longer have a definite isospin, but contain mixtures of isospins (isospin-mixing),
3) the various states of a given multiplet no longer have the same wave function (this may or may not involve isospin mixing - if not we call it "dynamic distortion"). For example consider the ^{17}O and ^{17}F ground states which have a $1d_{5/2}$ nucleon outside a I=0 ^{16}O core. The Coulomb force will cause the valence proton in ^{17}F to "stick out" further than the va-

lence neutron in ^{17}O. But this doesn't mix isospins since both levels still look like single particle (i.e. I=1/2) states outside a I=0 core. The ^{17}F state has instead been mixed with another I=1/2 state (a $2d_{5/2}$ proton state) so that it no longer has the same wavefunction as its analog in ^{17}O.

II.4 I-Breaking Interactions of Nucleons

The fact that the nucleon has I=1/2 restricts the possible forms of I-breaking interactions of nucleons. The most general possible I-breaking potential between two nucleons with isospin vectors $I=\frac{1}{2}\tau$ is $V_{ij} = V_I + V_{II} + V_{III} + V_{IV}$

where $V_I = V_0 + V_1\, \tau_i \cdot \tau_j$ is an isospin scalar ($\Delta\vec{I}=0$)

$V_{II} = V_2(\tau_{3_i}\tau_{3_j} - \frac{1}{3}\tau_i \cdot \tau_j)$ is an isospin tensor ($\Delta\vec{I}=2$)

$V_{III} = V_3(\tau_{3_i} + \tau_{3_j})$ is a symmetric isospin vector ($\Delta\vec{I}=1$)

$V_{IV} = V_4(\tau_{3_i} - \tau_{3_j}) + V_5(\tau_i \times \tau_j)_3$ is an antisymmetric isospin vector ($\Delta\vec{I}=1$).

In these expressions $V_0 \ldots V_5$ are particular scalar functions of the space and spin coordinates (see ref. 2 for details).

In nuclear physics we often employ one-body potentials to describe the effect on a single nucleon of the mean field of other nucleons. The most general one-body I-breaking potential is just

$$U_i = U_I + U_{II}$$

where $U_I = U_o + U_1\, \tau_i \cdot \tau_{core}$ is an isospin scalar ($\Delta\vec{I}=0$)

$U_{II} = U_2\, \tau_{3_i}$ is an isospin vector ($\Delta\vec{I}=1$),

and U_0, U_1 and U_2 are scalar functions of the space and spin coordinates.

From this we can see that, to lowest order, ΔI=1 phenomena in a nucleus of mass A can occur via the mean field of A-1 nucleons while ΔI=2 phenomena cannot.

Now consider the sources of I nonconservation in nuclei. The most important of these is the Coulomb force

$$V_C = \frac{e^2}{4r_{ij}} (1 + \tau_3^i)(1 + \tau_3^j)$$

which contributes to V_I, V_{II}, V_{III}, U_I and U_{II} but not to V_{IV} and has the symmetry that $V_{II} = V_{III}$. Next in importance are magnetic forces which can contribute to V_{IV}.

In addition to these direct EM effects one also expects a number of subtle indirect EM effects due to EM interactions of the mesons which mediate the nuclear force. In contrast to the direct EM effects which are long-range forces the indirect EM effects have a short range characteristic of the nuclear force so we call them "charge dependent nuclear forces". One example is the effect on the NN force of the "Coulomb energy" of the pion ($m_{\pi^\pm}$= 139.6 MeV but m_{π^0} = 135.0 MeV). Since $\pi^\pm$ exchange contributes only to pn interactions, while π^0 exchange occurs in nn, np and pp interactions, the $m_{\pi^\pm} - m_{\pi^0}$ difference will cause the "strong" pn force to differ a little bit from the "strong" nn and pp forces. This effect therefore contributes to V_{II}. Other effects, such as isospin mixing of the ρ^0 and ω^0 mesons contribute, in principle, to V_{III} and V_{IV}.

For completeness we must also include one other mechanism for I-breaking - which enters via the $\frac{p^2}{2m}$ term in the Hamiltonian rather than in the potential. Although the I symmetry requires that m_n=m_p, m_n is greater than m_p by 1.29 MeV or ∿0.14%. This disagrees with the naive expectation that the Coulomb energy of the proton would cause m_p-m_n>0 (recall that $m_{\pi^\pm}$-m_{π^0}>0). If we model the p as uud and the n as udd where u and d are quarks with charges +2/3 e and -1/3 e respectively Coulomb effects still produce a mass difference of the wrong sign. The np mass difference can be explained as a fundamental ΔI=1 isospin violation - the mass of the d quark is greater than that of the u. (This ud mass difference does not produce any mass splitting of the pions but will mix a little bit of I=0 into the π^o).

II.5 I-Breaking in the Two Nucleon System

The low energy NN system is a delicate probe of I breaking because the ^1S_o state is barely unbound so that the scattering lengths, a, are very sensitive to small changes in the potential. In the NN system the potentials V_I, V_{II} and V_{III} do not contribute to I-mixing matrix elements - i.e. the matrix elements $<I=0|V_i|I=1>$ vanish if i=I, II or III. The potential V_{IV} does produce I mixing in the NN system, but its matrix element vanishes in L=0 states since the $^3S_1$ and ^1S_o states can't mix by angular momentum conservation. The observed scattering lengths are $a_{nn} = -16.4 \pm 1.2$ fm, $a_{np} = -23.715 \pm 0.015$ fm and the Coulomb corrected $a_{pp} = -17.2 \pm 3.0$ fm. There is clear evidence for a short range V_{II} potential ($a_{nn} \neq a_{pp}$) which has the same sign as the long range V_{II} from the Coulomb force and corresponds to V_{np} being ∿2% more attractive than V_{nn}. On the other hand there is no evidence for V_{III} potentials since there is no significant a_{nn} - a_{pp} difference. One cannot learn about V_{IV} from the scattering lengths since V_4 and V_5 vanish in S states. One can probe V_{IV} by studying np scattering at medium energies and trying to detect the mixing of the $^3P_1$ and $^1P_1$ channels in polarization experiments. The predicted effects are very small and have not yet been observed. In a complex nucleus I mixing can be generated from (and is dominated by) the diagonal NN matrix elements of V_{II} and V_{III} with the off-diagonal NN matrix elements of V_{IV} playing a minor role.

II.6 Mass Splitting of Nuclear Multiplets

One of the most obvious manifestations of I-breaking is the removal of the degeneracy of the nuclear I multiplets. In lowest order perturbation theory $\Delta E(I_3)=\langle I,I_3|\sum_{ij} V_{ij}|I,I_3\rangle$. Since V_{ij} contains terms which carry $\Delta\vec{I}=0,1$ and 2 $\Delta E(I_3)$ is predicted to have the form $\Delta E(I_3)=a+bI_3+cI_3^2$. This simple first-order expression provides a remarkably accurate account of the observed mass splittings. For example in multiplets with $I\geq 3/2$ one can test with great precision (see ref, 3 for a nice review) whether extra terms ($aI_3^3 + eI_3^4$ etc) are required to fit the results. Non-zero values of these extra terms are required in only 2 (mass 8 and 9) out of 29 cases and even there the extra coefficients are very small. Why should the first-order expression work so well especially in heavier nuclei where $Z\alpha$ is certainly not a small number? For example in the A=37 nuclei $d=-2\pm5$ keV compared to $b=-6202$ keV so that $d/b=(0.3\pm1.0)\times10^{-3}$ which is small compared to $(\overline{Z}\alpha)^2=0.018$. The simplest answer is probably that the main charge dependent forces are Coulomb and the "exact" result for the classical Coulomb energy of Z protons forming a uniformly charged sphere of radius r is $\propto Z(Z-1)/r$. If r is constant this has the form of our first order perturbation theory result. Because of the repulsion of the protons r will expand somewhat with increasing I_3 but the calculated effect produces very small departures from the quadratic form. In fact the observed b and c coefficients follow closely the behaviour expected for a uniformly charged sphere

$$b = -0.6\ \left|(A-1)\frac{e^2}{r}\right| + m_n - m_p$$
$$c = +0.6\ \frac{e^2}{r}$$

with b increasing strongly with A while c decreases somewhat since $r\tilde{\propto}A^{1/3}$ (see fig. 2).

Clearly to learn more about the forces responsible for I violation (or alternatively about the nuclear wavefunctions) we have to look at the <u>magnitudes</u> of the b and c coefficients rather than at the <u>form</u> of the mass splitting equation. The situation is simplest for the c-coefficient. If the charge dependent components of V_{ij} are

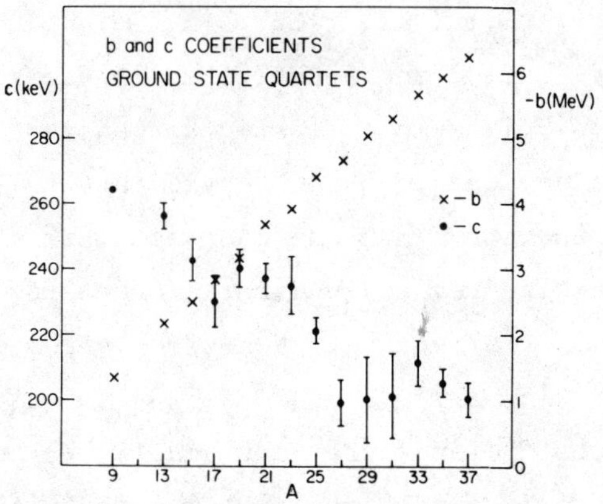

Fig. 2. The b and c coefficients of I=3/2 ground states (from ref. 3).

treated as perturbations then, in lowest order, the c-coefficient arises only from the V_{II} term.

Since c is quite small it seems likely that 1st order perturbation theory should be a reasonable approximation - detailed calculations support this conjecture. An analysis[4] of the c-coefficients has been made for the 1p shell nuclei. With V_{II} given by direct EM effects (primarily the Coulomb interaction) the RMS discrepancy of theory and experiment was Δc=35.1 keV. After including a short range contribution to V_{II} (with $V_{np} \approx 2\%$ more attractive than V_{nn} in the S=0 state) Δc dropped to 18.1 keV. We see that the effects of the short range charge dependent forces are quite small and only become apparent after careful analysis.

Calculations of the b-coefficient are much trickier since b is large and depends on U_{II} as well as on V_{III} and V_{IV}. As we shall see below U_{II} depends in a complicated way upon the binding energy and nlj values of the orbit and so that perturbation calculations are not likely to be reliable and analyses of b are notoriously tricky. However, Nolen and Schiffer[5] have made careful single-particle analysis of displacement energies $\Delta E = M(\alpha,I,I_3+1) - M(\alpha,I,I_3)$ and find that direct EM effects systematically underpredict ΔE by $\sim 7\%$. The discrepancy is already present in mass 3! This effect could possibly be due to V_{III} (or less likely V_{IV}) terms in the short range charge dependent interaction. If so why didn't it show up in the NN scattering lengths? Understanding the Nolen-Schiffer anomaly is an important item on the agenda of current research - but it goes well past the level of these lectures.

II.7 I-Mixing in Complex Nuclei

In complex nuclei I-mixing is heavily dominated by the Coulomb force. In the next two sections, therefore, we make the simplification that $V_{IV}=0$ and $V_{II}=V_{III}$ as would occur if the only I-breaking forces were Coulomb. We also assume these can be treated as perturbations.

A "super naive" argument might us to expect INC effects of order $Z\alpha$ but this would be a gross overestimate. In even the heaviest nuclei we see states ψ and ψ' which are good isospin analogs in the sense that $I_+|\psi\rangle = c|\psi'\rangle$.

On the other hand there are measurements which might lead us to think that I was an extraordinarily good quantum number in nuclei, namely the strengths of I-forbidden Fermi $\beta^{\pm}$ decays which are shown schematically in fig. 3. These directly measure the intensity

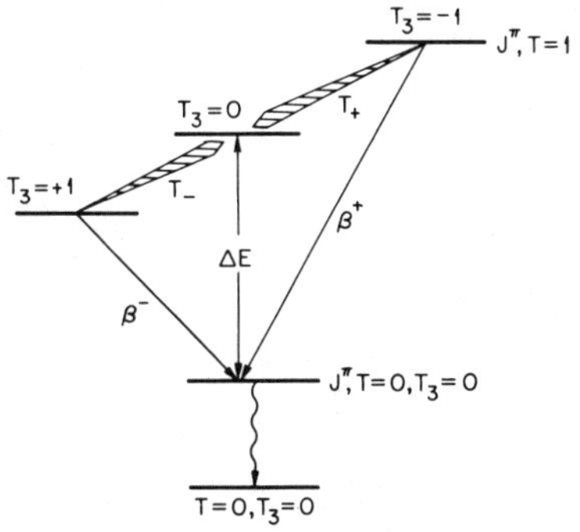

Fig. 3. Schematic diagram of I-forbidden Fermi $\beta^{\pm}$ transitions.

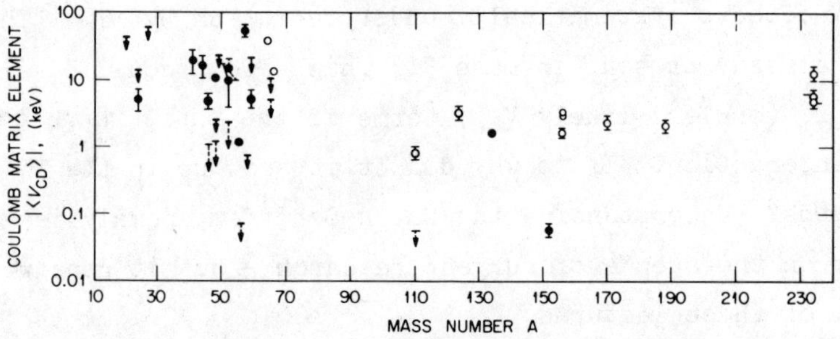

Fig. 4. I-mixing matrix elements inferred from I-forbidden Fermi $\beta^{\pm}$ decays (from ref. 6).

of isospin I in a state with nominal isospin I$\pm$1, and can be used to infer values for I-mixing matrix elements between two specific states. Results are reviewed in ref. 6 from which Fig. 4, showing the extracted values of the I-mixing matrix elements is taken. The average matrix element is $\sim$10 keV. The average intensity of the I impurity is thus $\sim(\frac{10\ \mathrm{keV}}{E})^2$. An average ΔE is $\sim$5 MeV which leads to an average impurity of only 4×10^{-6}. This, of course, does not refer to the <u>total</u> impurity but only to the small portion consisting of the admixture of single state so it gives an unreasonably small estimate of the average size of isospin impurities.

We can gain quite a bit of insight into the I-mixing problem from a very simple schematic model (see ref. 7). Consider, for example, the particle-hole excitations of an I=0 nucleus whose ground state we assume has filled 4-fold degenerate orbitals (2 spin x 2 isospin states for each space orbit). We label the active orbits i and j. An example could be the $1p_{3/2}^{-1}\ 1p_{1/2}$ excitation of ^{12}C where the particle and hole are coupled to $J^{\pi}=1^{+}$. There are 4 such states which, if I is exact, form 3 analog I=1 levels and 1 antianalog I=0 state which we show schematically as

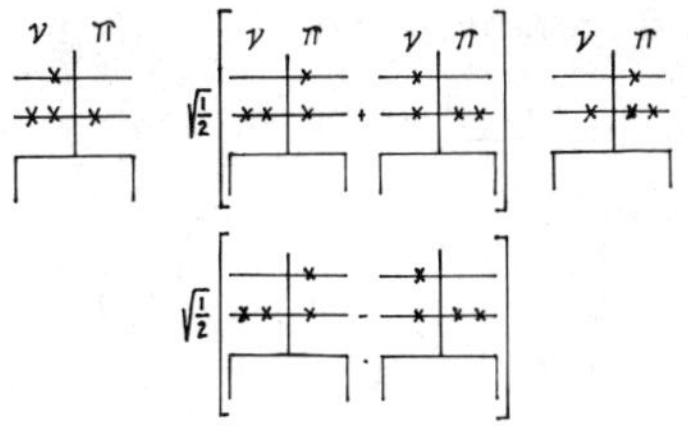

Note that our core is ^{8}Be so that we always deal with particles, rather than holes and particles. The Coulomb force mixes the I=1 and I=0 states in the I_3=0 nucleus

$$|U\rangle = |1\rangle + \varepsilon|0\rangle$$
$$|L\rangle = |0\rangle - \varepsilon|1\rangle$$

where the I impurity amplitude $\varepsilon = \dfrac{\langle 10|H^c|00\rangle}{E_x(I=1)-E_x(I=0)}$ with $\langle 10|H^c|00\rangle = \frac{1}{2}(E^c_{ii}-E^c_{ij})$.

In this notation E^c_{ij} is the Coulomb energy of a pair of protons - one in orbit i and the other in orbit j. E^c_{ij} has single-particle contributions U^c_i and U^c_j due to the Coulomb interaction of the proton with the core and two-body contributions V^c_{ij} due to the interaction of the valence protons. It is easy to show that

$$\langle 10|H^c|00\rangle = \frac{1}{2}|2U^c_i + V^c_{ii} - (U^c_i + U^c_j + V_{ij})|$$
$$= \frac{1}{2}|(U^c_i - U^c_j) + (V^c_{ii} - V^c_{ij})|$$

while the mass splitting coefficients are

$$b = \frac{1}{2}|m(1,1) - m(1,-1)| = \frac{1}{2}|U^c_i + U^c_j + 2V^c_{ij} + V^c_{ii}| - \Delta$$

and

$$c = \frac{1}{2}|m(1,-1) - 2m(1,0) + m(1,1)| = \frac{1}{2} V_{ij}$$

where $\Delta = m_n - m_H$.

Note that the isospin mixing involves <u>differences</u> of charge dependent single-particle energies and <u>differences</u> of charge-dependent two-body matrix elements while the mass splitting parameters do not.

This is an important and general result. Hence if $U_i^c = U_j^c$ and $V_{ij}^c = V_{ii}^c$ the isospin mixing would vanish even though the mass splitting could be very large.

In fact the differences $|U_i^c - U_j^c|$ are much smaller than the U^c's themselves. This explains why the isospin impurities in nuclear states, especially in heavy nuclei, are typically much smaller than one would guess from simple $Z\alpha$ arguments. To a rough approximation the Coulomb force simply raises the energy of all proton states by the same amount. Another factor tending to preserve the isospin purity of heavy nuclei is the large neutron excess which is automatically in a state of good isospin, $I_{excess} = \frac{1}{2}(N-Z)$.

Although the various U_i^c's are roughly the same it is interesting to focus, for a moment, on the "fine structure". In Table 1 are listed the excitation energies of some low-lying I=1/2 levels in A=13 (see also Fig. 7).

Table 1. Coulomb Energy Shifts for A=13

J^π	$E_x(^{13}C)$	$E_x(^{13}N)$	$E_x(^{13}N)-E_x(^{13}C)$
$1/2^-$	0 keV	0 keV	0 keV
$1/2^+$	3 089	2 365	- 724
$3/2^-$	3 684	3 511	- 173
$5/2^+$	3 854	3 547	- 307
$5/2^+$	6 864	6 364	- 500
$5/2^-$	7 547	7 376	- 171
$1/2^-$	8 860	8 918	+ 58

It is quite obvious that the Coulomb energies vary from state to state. There is a general trend for levels in ^{13}N to have lower E_x than in ^{13}C. This is a finite-well effect. Because of the Coulomb energy states in ^{13}N are less tightly bound than in ^{13}C. As E_x rises the states "spread out" more and the Coulomb energy drops. The effect is most pronounced for s-states since they have no angular momentum barrier to inhibit the spreading.

Now return to the schematic model of the mixing of an I=1 analog and an I=0 anti-analog level. There are quite a few cases where one can identify such pairs of states and measure their mutual mixing. Interestingly enough in these cases one finds much larger I-mixing matrix elements than were observed in Fig. 4. The most famous example occurs in ^{8}Be. In the region where the lowest 2^+ T=1 level is expected there are two 2^+ levels (at 16.6 and 16.9 MeV) with very curious properties. The two levels have roughly equal widths to $\alpha+\alpha$. The bottom level has a single-particle parentage which is almost purely ^{7}Li+p while the upper level is almost purely ^{7}Be+n. These results (and many others) are consistent with almost complete isospin mixing of the levels. Apparently the strong interaction produces nearly degenerate I=0 and I=1 states of virtually the same configuration - i.e. an analog-antianalog pair. Because of this near degeneracy the Coulomb force has exceptionally large effects and produces the almost completely mixed states $|16.9\rangle$ and $|16.6\rangle$ where

$$|16.9\rangle = \alpha|1\rangle + \beta|0\rangle \qquad \alpha^2 + \beta^2 = 1$$

$$|16.6\rangle = -\beta|1\rangle + \alpha|0\rangle$$

$$\alpha\beta = \frac{\langle 1|H^c|0\rangle}{E_{16.9}-E_{16.6}}$$

We can evaluate α and β from the alpha and nucleon widths of the levels and deduce that $\langle 0|H|1\rangle = -145$ keV. The minus sign indicates that the lower level looks predominantly like a proton excitation while the upper level is predominantly a neutron excitation. Qualitatively similar matrix elements are observed for the 1^+ states of ^{12}C(8), the 2^- states of ^{16}O(9) and the 4^+ states of ^{24}Mg(10).

In favorable cases I-mixing matrix elements between pairs of states can be obtained using the selection rule that $\Delta\vec{I}=1$ for E1 transitions in $I_3=0$ nuclei. If the "I=1" state has a very strong E1 transition then it can be a good approximation to account for the E1 speed of the "I=0" level by an admixture of the"I=1" state. Such arguments yield I-mixing matrix elements for the lowest 0^- state in ${}^{14}N$ and the lowest 1^- state in ${}^{16}O$. In ${}^{16}O(1^-)$ the sign of H^c is determined by a subtle interference effect in inelastic electron scattering[11]. The H^c for the 2^- state in ${}^{14}N$ is evaluated from the M2 γ-ray transitions. All these "experimental" values are summarized in Table 2.

We can use our simple schematic model to relate these isospin-mixing matrix elements in mass A to mass differences in masses A-1, A and A+1. We assume that the A-1 and A+1 ground states are described by the configurations $\begin{array}{c|c} & \\ \hline xx & x \end{array}$ and $\begin{array}{c|c} & \\ \hline x & xx \end{array}$ for A-1, and $\begin{array}{c|c} x & \\ \hline xx & xx \end{array}$ and $\begin{array}{c|c} & x \\ \hline xx & xx \end{array}$ for A+1 and that the U^c's and V^c's do not depend on A. Then we obtain relations such as the following for ${}^{12}C$

$$\langle 0|H^c|1\rangle = \frac{1}{2}|({}^{13}N-{}^{13}C)-({}^{11}C-{}^{11}B)-({}^{12}B-2.{}^{12}C^*+{}^{12}N)|$$

where, for example, ${}^{13}N$ refers to the mass of the ${}^{13}N$ ground state.

An even simpler case occurs for the negative parity states in the A=4n+2 nuclei. Consider, for example, the 0^- I=0 and I=1 states in A=14 consisting of one particle in the $1p_{1/2}$ orbit and one in the $2s_{1/2}$ orbit. These ${}^{14}N$ levels look like:

$$|I=1\rangle = \frac{1}{\sqrt{2}}\left(\begin{array}{c|c} x & \\ \hline & x \end{array} + \frac{1}{\sqrt{2}}\begin{array}{c|c} & x \\ \hline x & \end{array}\right)$$
$$|I=0\rangle = \frac{1}{\sqrt{2}}\left(\begin{array}{c|c} x & \\ \hline & x \end{array} + \frac{1}{\sqrt{2}}\begin{array}{c|c} & x \\ \hline x & \end{array}\right)$$

and we get a very simple relation for the I-mixing in ${}^{14}N$: for example $\langle 0|H^c|1\rangle = \frac{1}{2}|U_2^c-U_1^c| = \frac{1}{2}|E_x({}^{13}N)-E_x({}^{13}C)|$ where E_x refers to the excitation energy of the low-lying $1/2^+$ state in A=13. Here the whole mixing is due to the single particle Coulomb energy differences.

The predicted matrix elements estimated in this fashion are

listed in Table 2. The agreement of our simple schematic model with experiment is very good except for the case of mass 8. Since finite-well and A-dependent effects are very large in the lightest nuclei we shouldn't be too disappointed that our "idiot-model" breaks down here. (Recall that the $a+bI_3+cI_3^2$ law began to fail in the lightest nuclei as well).

Table 2. Isospin Mixed Doublets

Nucleus	E_x(MeV)	$-H^c_{exp}$(keV)	$-H^c_{model}$(keV)
^8Be 2^+	{16.6, 16.9}	145	88
^{12}C 1^+	{12.7, 15.1}	130 ± 26	125
^{14}N 0^-	{4.9, 8.8}	395 ± 74[a]	362
^{14}N 3^-	{5.8, 8.9}	≤103	154
^{16}O 1^-	{7.1, 13.1}	591 ± 54	494
^{16}O 2^-	{12.5, 13.0}	≥155 ± 30	163
^{24}Mg 4^+	{8.4, 9.5}	106 ± 40	71

a Experiment doesn't fix the sign of H^c but it is expected to be negative.

Note that in all 6 cases where the sign has been measured it is negative - which in our convention means that the lower-lying state has an excess of proton excitations. This is easy to understand. In a self-conjugate nucleus it costs less energy to make a proton particle-hole excitation (which breaks a repulsive Coulomb pairing interaction) than it does to excite a neutron. Hence the low-lying state acquires an excess of proton excitations and the upper level an excess of neutron excitations (remember that two interacting levels repel each other).

Let's recap what we have learned about I-mixing of low-lying states. The mixtures are a result of the low-lying level trying to reduce its energy even further by rearranging its charge distribution. It can do this by changing a proton from one orbit to another and it is only the differences of Coulomb energies between the two orbits which causes the mixing. There are two kinds of differences - in the U^c and the V^c. Typically U^c drops with increasing orbit energy. This drives a self-conjugate nucleus to have an excess of proton excitations - especially if the excited orbit is an s state. In the A=4n nuclei another effect favors proton excitations. V_{ii} is greater than V_{ij} because of the Coulomb pairing energy. The interplay of all these effects can be seen in Table 2.

II.8 Dynamic Distortion in Complex Nuclei

"Dynamic distortion" refers to a form of I-breakdown in which states do not have isospin admixtures, but nevertheless are no longer related by $I_\pm|\alpha,I,I_3\rangle = C|\alpha,I,I_3\pm1\rangle$. This has both simple and subtle aspects. The simplest form of "dynamic distortion" is the distortion of single particle radial wavefunctions by Coulomb binding energy differences. This causes a proton wavefunction to have a longer tail than the corresponding neutron wavefunction and leads to differences between corresponding neutron and proton matrix elements. A famous example is the β decay of ^{12}B and ^{12}N to ^{12}C (see Fig. 5). If I were a perfect symmetry the ^{12}B and ^{12}N decays would have identical matrix elements. However, in the ^{12}B decay a neutron bound by 3.37 MeV decays to a proton bound by 15.96 MeV. In the ^{12}N decay a proton bound by only 0.60 MeV decays to a neutron bound by 18.72 MeV. The larger difference in binding energies in ^{12}N decays reduces the radial overlap in the ^{12}N decay matrix element compared to that in the ^{12}B decay - such effects presumably cause the observed ^{12}N decay matrix element to be ∿5% less than that for ^{12}B decays. (This extreme single-particle model of the nuclear states is oversimplified - note that the effect has the opposite sign for decays to ^{12}C (4.4)- but the essential physics is correct).

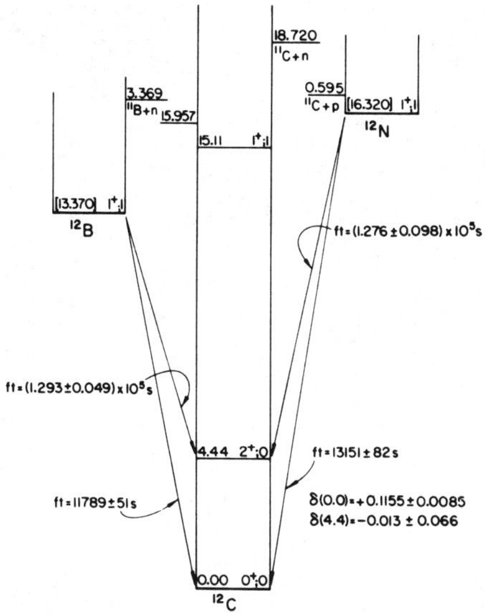

Fig. 5. Charge Asymmetries in the A=12 Beta Decays.

More subtle effects also occur. For example, consider the lowest two 0^+ states in the mass 42 I=1 nuclei (see Fig. 6). If the I-symmetry were exact the states would be $|A\rangle$ and $|B\rangle$. Now "turn on" the Coulomb force. The new states become

$$|0_1^+\rangle = |A\rangle - \varepsilon(I_3)|B\rangle \qquad \text{where } \varepsilon(I_3) = \frac{\langle A|H^c|B\rangle}{E_B - E_A}$$
$$|0_2^+\rangle = |B\rangle + \varepsilon(I_3)|A\rangle$$

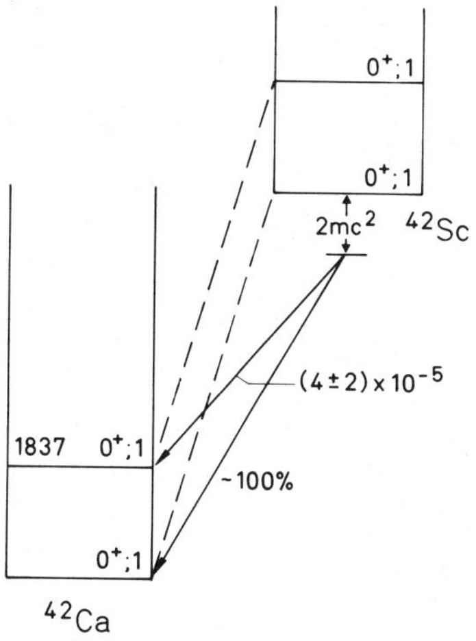

Fig. 6. I forbidden β^+ Decay of ^{42}Sc.

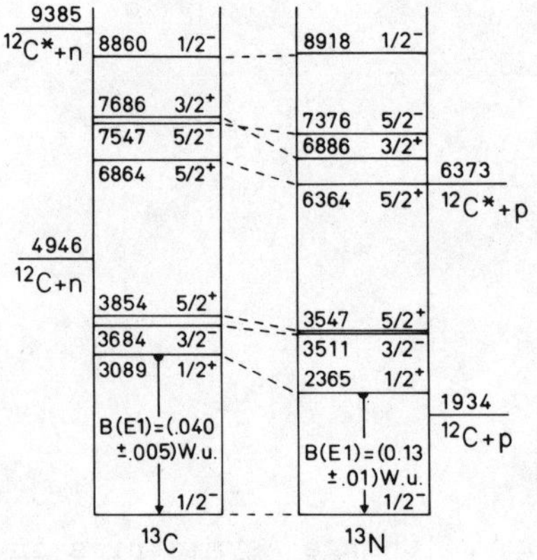

Fig. 7. Charge asymmetry of mirror E1 transitions in A=13.

Note that ε is a function of I_3 because in general H_c depends on the charge. We are thus considering I_3 dependent configuration mixing. The Fermi β decay operator provides a sensitive and model independent way to detect this mixing since the β^+ decay of $^{42}Sc(0_1^+)$ to $^{42}Sc(0_2^+)$ is energetically allowed. The matrix element is

$$M_F = \langle 0_2^+(I_3=-1)|I_-|0_1^+(I_3=0)\rangle$$
$$= \sqrt{2}|\varepsilon(-1)-\varepsilon(0)|$$

This decay has been observed[12] and the measured value corresponds to $|\varepsilon(-1)-\varepsilon(0)| \approx (1.6\pm0.4)\times10^{-2}$. It would be wrong to leave you with the impression that dynamic distortion effects are always small - in the A=13 nuclei (see Fig. 7) the distortion causes mirror E1 transitions to differ in strength by a factor of $\sim$3. This big effect is due to charge dependent configuration mixing and not to the distortion of the single-particle radial wavefunctions themselves. For details of this exceptional case see ref. 13.

III. PARITY NONCONSERVATION

III.1 Introduction and Definitions

The parity symmetry, P, is an inversion of the 3 space axes

and conservation of parity is equivalent to the statement "if a process is allowed so is its mirror image". Under P $\vec{r} \to -\vec{r}$, $\vec{p} \to -\vec{p}$ but $\vec{L} = \vec{r}x\vec{p} \to + \vec{L}$. Quantities such as $\vec{r}$ and $\vec{p}$ are vectors, $\vec{L}$ is a pseudovector, p^2 a scalar and L•p a pseudoscalar. Since P^2 is the identity operator the eigenvalues of P which we denote by π are +1 or -1. If a Hamiltonian contains only scalars (such as p^2 and no pseudoscalars (such as σ•p) then a non-degenerate state which starts out with a definite parity π will have π as a constant of the motion. Conversely if a non-degenerate state starting out with a definite parity = π acquires a component with parity = -π one can infer that the Hamiltonian contains some pseudoscalar terms. In what follows we shall always restrict ourselves to non-degenerate systems since degenerate states with opposite π apparently do not occur in atoms, nuclei or particles.

We should not expect to observe P-violation at very low energies ($\frac{v}{c} \to 0$) since one cannot make sensible Hamiltonians with non-vanishing pseudoscalar terms at $\frac{v}{c} = 0$. However it is well known that at high energies where $\frac{v}{c} \to 1$ the elementary charged-current weak interaction violates P maximally - the pseudoscalar terms in the weak Hamiltonian become equal to the scalar terms and observables contain equal mixtures of positive and negative parities. On the other hand even at very high energies the strong and EM interactions do not show detectable P violation so we assume these interactions conserve parity exactly. We expect the states of systems with EM or strong interactions (such as atoms, nuclei and particles) to be almost pure eigenstates of parity with only very small amplitudes for wrong parity admixtures caused by the weak interaction. On the other hand systems prepared in pure weak processes (such as the β and $\bar{\nu}_e$ in β^- decay) will (as v/c → 1) violate parity maximally.

The weak interactions of the pointlike quarks and leptons are apparently well understood. These particles interact via weak currents which are mixtures of vector (V) and axial vector (A) and the parity violation is a result of the VA interferences. The weak interaction theory of Weinberg, Salam and Glashow accounts for all experimental results in the leptonic sector. On the other hand the weak interactions of strongly interacting particles are not so well understood. The strong interaction modifies the effective weak Hamiltonian in a way which is not easy to understand - even generating approximate symmetries (the $\Delta\vec{I}=1/2$ rule) for which there is no accepted explanation. Hence it is interesting to study weak interactions between hadrons.

Hadronic weak processes are expected to fall into two categories - those in which the strangeness S (another quantum number respected by strong and EM forces but not conserved in weak processes) changes by 1 unit, and those for which $\Delta S=0$. $|\Delta S|=1$ processes are easy to observe since there are particles with $|S|=1$ (such as the Λ baryon are K meson) whose only open hadronic decay channels are to S=0 particles ($\Lambda \to N\pi$, $K \to \pi\pi$ or $K \to \pi\pi\pi$). Since there are no allowed strong interactions, the hadronic weak decays complete only with other weak decays (such as $\Lambda \to pe^- \bar{\nu}_e$ and $K^+ \to \mu^+\nu_\mu$). On the other hand $\Delta S=0$ hadronic weak processes are much harder to observe since the $\Delta S=0$ weak interaction (for instance in the $np \to np$ process) has to complete with the enormously greater strong interaction. One can isolate the weak interaction contribution to $\Delta S=0$ processes by studying P-violating amplitudes to which the strong and EM interactions cannot contribute. At present the only case where this can be done is in the P-nonconserving NN force).

III.2 P-Breaking NN Interactions

Because weak interactions do not conserve I the weak NN force is considerably more complicated than the familiar strong NN interaction. We shall therefore adopt a simplification which is appropriate for describing P-breaking in nuclei - namely restriction to the low-energy limit of the P-nonconserving (PNC) NN interaction. The weak interaction has a short range. Therefore as $\frac{v}{c} \to 0$ the only important PNC amplitudes are those connecting S and P states of relative motion, the contributions of D and higher states being suppressed because of the centrifugal barrier and the "hard-core" of the strong NN interaction. There are 5 $S \leftrightarrow P$ transitions consistent with the Pauli principle:

$^3S_1 \leftrightarrow {}^1P_1$ $\Delta\vec{I}=0$ which contributes only to np
$^3S_1 \leftrightarrow {}^3P_1$ $\Delta\vec{I}=1$ which contributes only to np
$^1S_0 \leftrightarrow {}^3P_0$ $\Delta\vec{I}=0$ which contributes to nn, np and pp
$^1S_0 \leftrightarrow {}^3P_0$ $\Delta\vec{I}=1$ which contributes to nn and pp
$^1S_0 \leftrightarrow {}^3P_0$ $\Delta\vec{I}=2$ which contributes to nn, np and pp

Current research is directed toward

1) measuring effects sensitive to the amplitudes of the 5 $S \leftrightarrow P$ transitions. Clearly this requires enough experiments to sample 5 different linear combinations of the amplitudes,
2) analysing the measurements in terms of the 5 $S \leftrightarrow P$ amplitudes, and
3) predicting the strengths of the 5 $S \leftrightarrow P$ transitions using the fundamental weak interaction theory and models of hadron structure.

Item 3 above is a challenging problem in particle physics and we do not have time to pursue it here. We confine ourselves to some observations about the meson exchange theory the PNC NN force. The P-violation is achieved by having one meson (M) NN vertex strong (P-conserving) and the other weak (PNC). At low energy the exchange of the lightest meson (the pion) plays an especially important role since it has the longest range and is not much suppressed by the hard core. Weak interactions respect the combined symmetry C times P, where C (charge conjugation) turns a particle into its antiparticle without affecting its spin or momentum. (For example the helicity of electrons emitted in β decay is $-\frac{v}{c}$ while that of positrons is $+\frac{v}{c}$. Operating with CP on an electron having helicity $-\frac{v}{c}$ turns it to a positron with helicity $+\frac{v}{c}$). CP conservation forbids the PNC emission of π^o mesons, although PNC $\pi^{\pm}$ emission is allowed. To see this transform the $N \rightarrow N\ \pi^o$ process into the $N\bar{N} \rightarrow \pi^o$ process. The π has $J^{\pi}=0^-$. Therefore in a PNC process the $N\bar{N}$ must be in a $J^{\pi}=0^+$ state. The $N\bar{N}$ = $p\bar{p}$ or $n\bar{n}$ system can be described by the quantum numbers L, S and π^c where π^c is the C eigenvalue which must be ±1 since $C^2=1$. The Pauli principle says that the $N\bar{N}$ state must be odd under the exchange of the two fermions - i.e. under exchange of space, spin and particle-antiparticle labels. Therefore $\pi^c = (-)^{L+S}$ which has the value $\pi^c = +1$ in the 0^+ state. Therefore the $N\bar{N}$ system is CP=+1. The π^o has C =+1 (it decays into two γ rays) and therefore CP=-1. Hence $N\bar{N} \rightarrow \pi^o$ is forbidden. Since only $\pi^{\pm}$ exchange is allowed, weak pion exchange occurs solely in the np system. As a charge exchange process it must have ΔI=1 and therefore contributes only to the $^3S_1 \rightarrow {}^3P_1$ amplitude.

The exchange of heavier mesons such as ρ $(J^{\pi};I=1^-;1)$ and $\omega(J^{\pi};I=1^-;0)$ contributes to most of the S↔P transitions. The only simplification is that the $\Delta\vec{I}=2$ amplitude does not receive contributions from exchange of the I=0 ω meson since the strong (I-conserving) vertex requires that two of the nucleons be coupled to I=0.

At this point you may wonder why we only consider PNC one-boson exchange. Shouldn't we also include the PNC exchange of two bosons, especially two pions? After all the exchange of two pions coupled to $J^{\pi};I=0^+;0$ (the "sigma meson") plays an important role in the strong NN force. Fortunately PNC σ exchange is prohibited by CP conservation just as was π^o exchange. Thus we do not have to worry about 2π in a relative S-state. The contribution of 2π in a P-state is presumably largely included in the single ρ exchange. Hence it is hoped that neglect of 2 boson exchange is a reasonable approximation.

The isospin properties of the resulting PNC NN potential have the form

$$V^{PNC} = V_I^* + V_{II}^* + V_{III}^* + V_{IV}^* + V_\pi^*$$

where
$$\begin{aligned} V_I^* &= V_0^* + V_1^* \tau_i \cdot \tau_j \\ V_{II}^* &= V_2^* (1 - \tfrac{1}{3} \tau_i \cdot \tau_j) \\ V_{III}^* &= V_3^* (\tau_i + \tau_j)_3 \\ V_{IV}^* &= V_4^* (\tau_i - \tau_j)_3 \\ V_\pi^* &= V^* (\tau_i \times \tau_j)_3 \end{aligned}$$

where $V_0^* \ldots V_4^*$ and V_π^* are pseudoscalar functions of the space and spin coordinates. For more details see ref. 14.

Pion exchange contributes to V_π, and vector meson exchange to V_0 through V_4. The strengths of the PNC potentials $V_0^* \ldots V_\pi^*$ have recently been estimated[14] by Desplanques, Donoghue and Holstein (DDH) using the Weinberg-Salam weak-interaction theory and the quark model of hadron structure. There are many uncertainties in such calculations so they give quite broad acceptable ranges for the parameters along with "best estimates" for each.

III.3 P-Violating Observables

Consider a transition between states A and B induced by operator O. Let A have predominant parity π_A with a small mixture of parity π'_A $|A\rangle = |\pi_A\rangle + \varepsilon|\pi'_A\rangle$ and for simplicity assume B has pure parity $|B\rangle = |\pi_B\rangle$. In general, there will be two transition amplitudes connecting A and B

$$A_{reg} = <\pi_B|0|\pi_A>$$
$$A_{irreg} = \varepsilon<\pi_B|0|\pi'_A>$$

which we label regular and irregular. There are two categories of observables which are sensitive to the P-violation in state A -

i) if $A_{reg} \neq 0$ one can detect the interference of A_{reg} and A_{irreg} by studying a pseudoscalar observable. The effects will be $\propto \varepsilon$ and if non zero are proof that P-violation has occurred since a mirror image experiment would give the opposite result.

ii) if $A_{reg} = 0$ one can detect the rate of $A \rightarrow B$ transitions. This rate is $\propto \varepsilon^2$ and if non-zero we can infer that P-violation has occurred. However a non-vanishing rate does not necessary imply P-violation because it does not prove that $|\pi_A>$ and $|\pi'_A>$ were coherent. However this is a fine point and can usually be neglected.

Three kinds of pseudoscalar observable have been used to detect nuclear P-nonconservation

i) helicity dependence of transition rates. Consider a process such as A+B→ something, where B is unpolarized and A is polarized longitudinally ($\sigma_A \cdot P_A \neq 0$). One tries to detect a difference (which is P-violating) for the rates when $\sigma_A \cdot P_A = +1$ and when $\sigma_A \cdot P_A = -1$,

ii) asymmetry measurements. Consider a process such as A→b+B where A is spin polarized. One tries to detect an asymmetry $\sigma_A \cdot p_b$ in the emission of b with respect the spin of A,

iii) neutron spin rotation. Consider the transmission through a slab of matter of a beam of transversely polarized slow neutrons. The beam is in a coherent superposition of $\vec{\sigma}_n \cdot \hat{p}_n = +1$ and $\vec{\sigma}_n \cdot \hat{p}_n = -1$ states. Its coherent interaction with the slab nuclei can be described by an index of refraction n where $n-1 \propto \lambda^2 f_o$ with $\lambda = h/p_n$ and f_o is the forward neutron scattering amplitude. After traversing a slab of length l the neutron has acquired a phase $\varphi = \frac{2\pi l n}{\lambda}$. If there is a P-violating component of the interaction of the neutron with the slab nuclei then n will have a term $\propto \lambda^2\ \sigma_n \cdot p_n$, which produces a difference Δn in the index for the two neutron helicity states. This produces a P-violating rotation of the plane of polarization by an angle $\Delta\theta = \frac{\Delta}{2} = \frac{\pi l}{\lambda} \Delta n \propto \lambda f_o$. It is interesting that in n spin rotation one can detect P-violation at zero energy because the factor of λ in the expression for $\Delta\theta$ cancels the p_n factor in f_o.

III.4 P-Violation in the NN System

Experiments have been reported on 3 different pseudoscalar observables in the NN system.

i) the helicity dependence, A_L, of proton-proton scattering has been studied[15,16] at $E_P = 15$ and $E_P = 45$ MeV. An unpolarized proton target is bombarded by longitudinally polarized protons and one detects $A_L = \frac{\sigma_+ - \sigma_-}{\sigma_+ + \sigma_-}$ where $\sigma_\pm$ are the cross sections for $S\cdot\hat{p} = \pm 1$ protons. Both results (A_L a few parts in 10^7) are in accord with the DDH "best estimates".[14]

ii) the circular polarization, P_γ, of the 2.2 MeV γ-ray emitted when unpolarized thermal n's are captured by unpolarized protons has been measured. The result[17] ($P_\gamma = (1.3\pm0.45)\times10^{-6}$) is $\sim$20 times larger than the "best estimates",[14] but the authors of ref. 17 have found, on the basis of new results, reason to doubt this value.

iii) the asymmetry, A_γ, of the 2.2 MeV γ-ray emitted when polarized cold n's are captured by unpolarized protons has been investigated[18] but the experimental sensitivity was not small enough to test the DDH "prediction".[14]

These experiments are extraordinarily difficult (the quoted error bar on the P_γ measurement corresponds to an experimental effect of only 2.2×10^{-8}!) and we shouldn't worry about the disagreement between the P_γ experiment and theory until the measurement has been repeated. Nevertheless it is interesting to note that these three observables sample quite different combinations of the 5 $S\leftrightarrow P$ transitions, so the results cannot be mutually contradictory. For example the p+p experiment samples all three of the $^1S_0 \leftrightarrow {}^3P_0$ amplitudes since the initial and final states in p+p scattering are I=1. The two np→dγ experiments which detect the interferences of a small E1 PNC amplitude with the dominant M1 amplitude are more subtle. These M1 and E1 amplitudes are shown in Fig. 8. Consider first the P_γ measurement. The P-allowed transition is predominantly $^1S_0 \leftrightarrow {}^3S_1 +$ M1. The PNC force can mix $^3P_0$ into the initial state, or $^3P_1$ or $^1P_1$ into the final state. However the $^3P_1$ admixture in the final state is not connected to the $^1S_0$ initial state by the E1 operator (which, in the long-wavelength limit, cannot flip spins and is an isovector. Hence the P_γ experiment samples the $^1S_0 \leftrightarrow {}^3P_0$ amplitudes with ΔI=0 and ΔI=2 and the $^3S_1 \leftrightarrow {}^1P_1$ amplitude with ΔI=0. The A_γ experiment requires a J=1 amplitude in the incoming channel since an isolated J=0 state must decay isotropically. There is some $^3S_1$ in the incoming channel (the deuteron has ∿4% $^3D_1$ component so some $^3S_1$ strength must be unbound). In this case the E1 matrix elements only couple to the $^3S_1 \leftrightarrow {}^3P_1$ matrix elements so the experiment is sensitive only to ΔI=1 PNC amplitudes.

III.5 P-Violation in Light Nuclei

The existing NN experiments do not completely specify the weak NN interaction. For example the constraint on the strength of V_π^* (set by the A_γ measurement) is almost useless. To learn more about the PNC NN interaction we must turn to P-violation in complex nuclei.

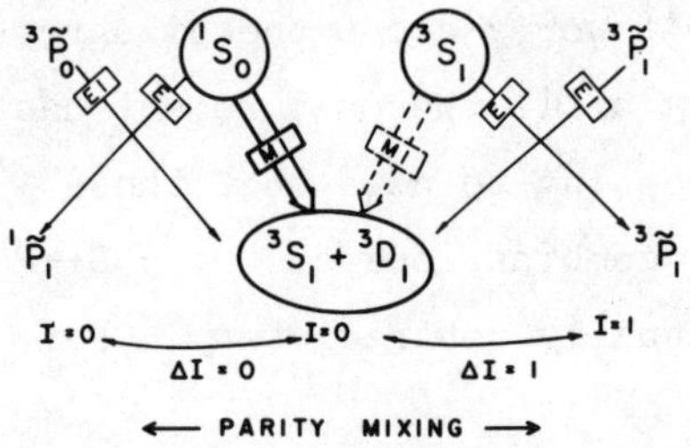

Fig. 8. Amplitudes involved in the P_γ and A_γ measurements in the np→dγ reaction at low energies.

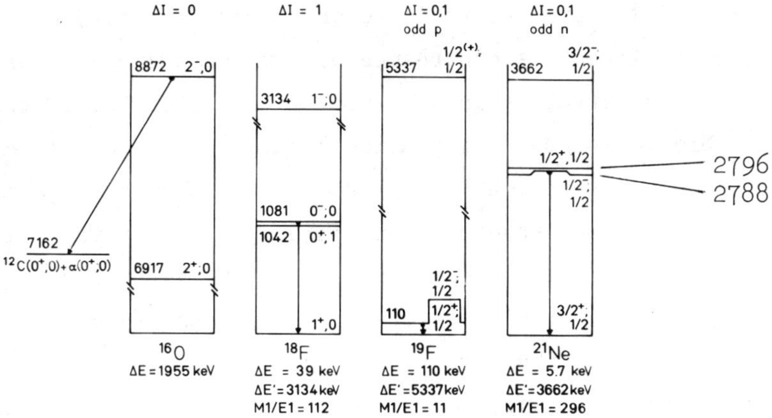

Fig. 9. Parity mixed doublets in light nuclei. ΔE is the doublet splitting, while ΔE' is the splitting to the next nearest state which could participate in the parity impurities in the doublet. M1/E1 is the EM enhancement factor.

Here, in favorable cases, nuclear structure can greatly amplify the P-violating effects. This makes experiments more practical, but at the same time introduces uncertainties into the extraction of the PNC NN parameters from the experimental results. We shall concentrate on understanding the origin of the amplification and on the methods which have been used to minimize the uncertainties in the interpretation of the results.

The most interesting cases of PNC in complex nuclei are the so-called "two-level systems" - light nuclei having nearly degenerate pairs of levels with the same spin but opposite parities.[19-22] Some examples are shown in Fig. 9. The energy splitting of these doublets can be so small (down to 5.7 keV in ^{21}Ne) that it is a very good approximation to describe the parity admixtures in the two states of the doublet as simple two-level mixing so that the physical states $|A\rangle$ and $|B\rangle$ become

$$|A\rangle = |+\rangle + \varepsilon|-\rangle \qquad |B\rangle = |-\rangle - \varepsilon|+\rangle \qquad \text{with } \varepsilon = \frac{\langle +|H_{PNC}|-\rangle}{E_+ - E_-}$$

where $|+\rangle$ and $|-\rangle$ are the positive and negative parity eigenstates. The mixing amplitudes ε can be measured unambiguously by detecting pseudoscalar observables in the γ-decay of one member of the parity-mixed doublet. Two pseudoscalar observables are commonly detected - the circular polarization P_γ of γ-rays emitted by an unpolarized state, or the forward-backward anisotropy, A_γ, of γ-rays emitted by a polarized state. It is easy to see how the γ-ray decays can greatly amplify the observable effects of a small parity mixture ε. Consider, for example, the case of low-lying J=0 doublet in ^{18}F (see Fig. 9). The isospin-favored M1 decays of the "0^+" level are very fast ($\tau^+ \propto 2.5\text{x}10^{-15}$ sec, where $\tau^+ \propto |\langle g.s.|M1|+\rangle|^{-2}$) while the isospin-forbidden E1 decays of the "0^-" level are quite slow ($\tau^- = 2.8\text{x}10^{-11}$ sec, where $\tau^- \propto |\langle g.s.|E1|-\rangle|^{-2}$). The parity mixing of the 0^+ and 0^- levels induces a circular polarization P_γ of the γ-ray deexiting 0^- state where $P_\gamma = 2\varepsilon\frac{\langle g.s.|M1|+\rangle}{\langle g.s.|E|-\rangle}$. The term $\frac{\langle g.s.|M1|+\rangle}{\langle g.s.|E1|-\rangle}$ is an amplification factor due to the electromagnetic transition rates. In this case the amplification factor is very large ($\approx$110) because the observable is sensitive to the PNC admixtures of a rapidly decaying state into a slowly decaying one. The corresponding amplification factors in ^{19}F and ^{21}Ne are 11 and 296. These electromagnetic amplification factors together with the small energy denominators are responsible for the relatively large effects in nuclei with A>2 compared to those in the two nucleon system.

Parity mixtures in light nuclei also provide favorable circumstances for extracting the isospin structures of the PNC NN interaction as can be seen from Fig. 9. The PNC α-decay[19] of ^{16}O probes only the $\Delta I=0$ amplitudes since the initial ($^{16}O\ 2^-$) and final ($^{12}C + \alpha$) states both have I=0. The parity mixture[20] in ^{18}F is pure $\Delta\vec{I}=1$ since one state of the doublet is I=0 and the other I=1. In ^{19}F[21] and ^{21}Ne[22] the parity mixtures are sensitive to both $\Delta\vec{I}=0$ and $\Delta\vec{I}=1$ amplitudes but the interference in ^{19}F (an odd-proton nucleus) is predicted to have the opposite sign of that in ^{21}Ne (an odd-neutron nucleus).

Table 3. Parity Violating Observables in Light Nuclei[a]

system	quantity	expt	theory	$\|\langle +\|H_{PNC}\|-\rangle_{exp}\|$
p+p	A_L(15 MeV)x10^7	-1.7 ± 0.8	-1.6	
	A_L(45 MeV)x10^7	-2.3 ± 0.8	-2.9	
p+d	A_L(15 MeV)x10^7	-0.35 ± 0.85	-1.6	
^{16}O	Γ_αx10^{10}(eV)	1.93 ± 0.28	∿0.6	
^{18}F	P_γx10^3	-0.8 ± 1.2	±2.0[b]	0.14 ± 0.21 eV
^{19}F	A_γx10^5	-7.9 ± 2.3	-8.9	0.40 ± 0.12 eV
^{21}Ne	P_γx10^3	0.8 ± 1.4	±0.5[b]	< .025 eV

a experimental and theoretical results for p+p and p+d are summarized in ref. 16
experimental results for A=16,18,19 and 21 are found in refs. 19-22 - for theory see text.

b the predicted sign of the PNC observable depends on the sign of a highly retarded E1 matrix element whose magnitude (but not sign) is fixed by a measured life-time.

The measured PNC effects in light nuclei between 16 ≤ A ≤ 21 are shown in Table I together with the corresponding PNC matrix elements. Three different techniques were employed in the measurements. In ^{16}O the very small ($\Gamma \sim 10^{-10}$eV) parity nonconserving α-decay width, sensitive to the intensity of the parity admixture, was detected.[19] The ^{19}F experiment[21] observed the asymmetry of γ-decays of polarized $^{19}F^*$ nuclei formed in a nuclear reaction induced by polarized protons, while in the ^{18}F and ^{21}Ne measurements[20,22] the circular polarization of the deexcitation γ-rays from unpolarized nuclei was detected. The ^{18}F system is particularly important since it is sensitive only to the ΔI=1 amplitudes and can be analysed (see below) with very little uncertainty due to nuclear structure. The combined results of 3 concordant experiments yield only an upper limit on P_γ but a new experiment which should measure P_γ with considerably improved sensitivity is in progress[23].

We now turn to the analyses of these PNC nuclear matrix elements in terms of PNC NN interaction. Since the relevant levels lie at low excitation energies in light nuclei one might expect that shell model calculations would give a reliable account of the nuclear wavefunctions. Such model computations using a limited number of empericcally fitted parameters (single-particle energies and residual two-body interactions) have, with only few exceptions, been very successful in accounting for a vast body of data (level schemes, magnetic moments, and M1 and Gamow-Teller transition rates) in the 1p [24] ($5 \leq A \leq 15$) and 2s1d [25] ($17 \leq A \leq 39$) shell nuclei. These calculations, in which the active particles are allowed to occupy all possible configurations lying anywhere within one major shell are called the $0\hbar\omega$ shell models. The observables, such as M1 moments, which are so well described by the $0\hbar\omega$ shell model do not connect configurations within the $0\hbar\omega$ model space to those having particles in the next major shell. Therefore corrections due to the next major shell enter only in second order. Calculations of PNC observables, which involve transition between states whose parity changes, clearly require model spaces in which the particles can lie in two opposite-parity major shells. The minimum acceptable basis for the "unnatural parity" state is the $1\hbar\omega$ model space - i.e. one particle is allowed to be promoted into the next major shell in all possible ways while the remaining particles have all possible $0\hbar\omega$ configurations. The $1\hbar\omega$ predictions, for example of low-lying E1 and forbidden β-decay rates, are notably less accurate than those of M1 transitions by the $0\hbar\omega$ model. The reason for this is known. The wavefunctions of the natural and unnatural parity states which are predominantly $0\hbar\omega$ and $1\hbar\omega$ respectively contain small components which are $2\hbar\omega$ and $3\hbar\omega$ respectively. These $2\hbar\omega$ (or $3\hbar\omega$) components do not greatly alter the predicted M1 and allowed beta decay observables since they are not connected to the dominant $0\hbar\omega$ (or $1\hbar\omega$) configuration by the M1 and beta decay operators. But the situation is quite different for parity changing observables such as PNC matrix elements and E1 transi-

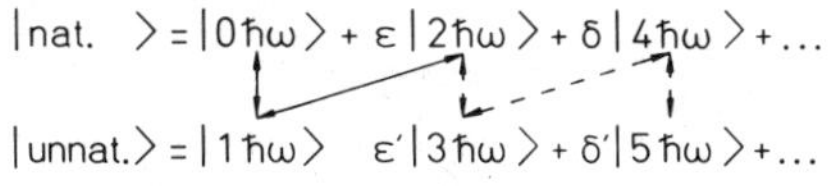

Fig. 10. Transitions between opposite parity states.

tion rates. These operators directly connect the small $2\hbar\omega$ component of the natural parity state to the dominant $1\hbar\omega$ component of the unnatural parity state (see Fig. 10). For the E1, forbidden β decay, and PNC operators the $2\hbar\omega \rightarrow 1\hbar\omega$ amplitude generally interferes destructively with the $0\hbar\omega \rightarrow 1\hbar\omega$ amplitude which explains the well-known suppression of low-lying E1 transitions and forbidden β decays. However, because of the extremely large model spaces, complete $2\hbar\omega$ calculations are very time consuming and have been done only in a limited number of cases.

Fortunately there is a way to reduce greatly the uncertainties in the analysis of the PNC matrix elements. It relies on similarities between the PNC operator and that for first-forbidden β decay. Both operators transform as $\Delta J^{\pi}=0^{-}$: this transformation is exact for the PNC operator and a very good approximation for the forbidden β decay operator since it is dominated by the time component of the axial current. This similarity between H_{β} and H_{PNC} is especially close for the $\Delta I=1$ parity mixing operator which is dominated by $\pi^{\pm}$ exchange. The corresponding forbidden β decay operator receives a large (and calculable using low-energy theorems - see refs. 26 and 27) contribution from pion exchange currents. These two pion exchange operators have identical dependence on the nuclear coordinates - for H_{β} the operator is known "exactly" while H_{PNC} contains the unknown amplitude, f_{π}, for weak $\pi^{\pm}$ emission by nucleons.

We remove most shell model uncertainties in this extraction of the PNC NN parameters by comparing experiment and theory for the ratio

$$R = \frac{\langle f|H_{PNC}|i\rangle}{\langle f|H_\beta|i\rangle}$$

For example consider the forbidden β decays of ^{18}Ne and ^{19}Ne shown in Fig. 11. These connect the same (to within an isospin rotation) pairs of states involved in the parity mixing in ^{18}F and ^{19}F. The prediction for R is much less sensitive to details of the shell model than are the absolute predictions of $\langle f|H_{PNC}|i\rangle$ or $\langle f|H_\beta|i\rangle$ themselves. This follows because the matrix element for any one-body operator O which transforms as ΔJ^π, ΔI can be expanded as

$$\langle f|O|i\rangle = \sum_{l,m} OBTD_{l,m} \cdot SPME_{l,m}$$

where l,m denote single-particle states. The $SPME_{l,m}$ are single-particle matrix elements which depend on the operator O but not upon the shell model wavefunctions, while the $OBTD_{l,m}$ (which are functions of ΔJ^π and ΔI) are one-body transition densities which carry all of the shell model dynamics. Both H_β and H_{PNC} behave similarly under the parity operation and under space and isospace rotations. Therefor the β decay and PNC operators sample virtually the same set of transition densities, so that the ratio of matrix elements becomes quite insensitive to the shell model wavefunctions. Haxton[27] has tested these ideas in mass 18. As he makes successively more elaborate shell model calculations (from an incomplete $0\hbar\omega + 1\hbar\omega$ model

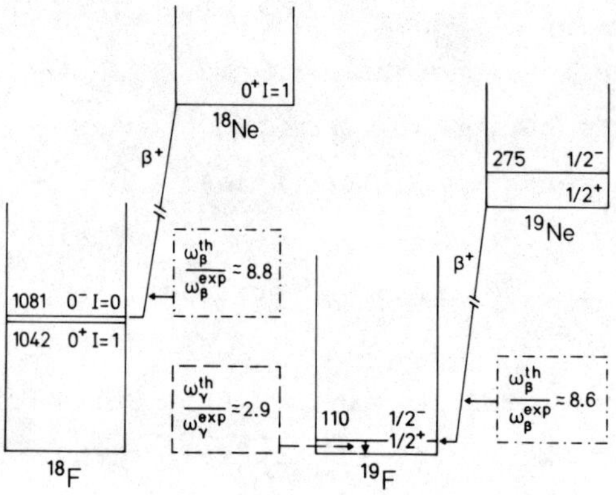

Fig. 11.
β and γ transitions corresponding to the parity mixing in ^{18}F and ^{19}F. The observed suppressions compared to the predictions of the $0\hbar\omega + 1\hbar\omega$ shell model are shown.

to a complete $2\hbar\omega$ + $1\hbar\omega$ model) $\langle 0^-|H_\beta|0^+\rangle$ drops by a factor of 4. However the ratio $\frac{\langle 0^-|H_{PNC}|0^+\rangle}{\langle 0^-|H_\beta|0^+\rangle}$ changes by only 10%.

The rates, ω_β, for the forbidden β decays of ^{18}Ne and ^{19}Ne were recently measured[28] so that one can compare experiment and theory[29] for the ratios of matrix elements. For A=19 we have

$$\left.\frac{A_\gamma}{\sqrt{\omega_\beta}}\right|_{exp} = -(3.6\pm1.1)\times10^{-2}\,\text{sec}^{-1} \text{ and } \left.\frac{A_\gamma}{\sqrt{\omega_\beta}}\right|_{th} = -4.1\times10^{-2}\,\text{sec}^{-1}$$

while for A=18 we have $\left.\frac{P_\gamma}{\sqrt{\omega_\beta}}\right|_{exp} = (-0.3\pm0.4)\,\text{sec}^{-1}$ and $\left.\frac{|P_\gamma|}{\sqrt{\omega_\beta}}\right|_{th} = 0.68\,\text{sec}^{-1}$

Again we see agreement between experiment and the DDH predictions. The theory is particularly "clean" for the case of the pure $\Delta\vec{I}=1$ PNC transition in ^{18}F. The matrix element is strongly dominated by π exchange. Pion exchange currents also play a major role in the corresponding forbidden β^+ decay of ^{18}Ne - the fractional contribution of π exchange to the decay rate being almost independent of nuclear models[14,15]. The experimental ratio $\frac{|P_\gamma|}{\sqrt{\omega_\beta}}$ directly gives the amplitude for PNC π emission in terms of well-known quantities[27].

It is interesting that the values of $|\langle f|H_\beta|i\rangle|$ measured in ^{18}Ne and ^{19}Ne decays are a factor of 2.96±0.21 and 2.93±0.24 smaller than $1\hbar\omega$ calculations. This suggests that the effects of $2\hbar\omega$ components in the $0\hbar\omega$ state can be accounted for by a "pseudoscalar effective charge" of 1/2.95. The idea is similar to the familiar case of E2 γ-ray transitions where effective E2 charges are very successful in accounting for the omission of $2\hbar\omega$ components in the wavefunctions. Theoretical predictions[29,30] for ^{16}O, ^{18}F, ^{19}F and ^{21}Ne, incorporating this "pseudoscalar effective charge", are given in Table 3. They give quite a reasonable account of the data. Theory yields the correct sign and magnitude in ^{19}F, where $\Delta\vec{I}=0$ and $\Delta\vec{I}=1$ PNC matrix elements are predicted to interfere constructively, and correctly predicts a very small matrix element in ^{21}Ne where the $\Delta\vec{I}=0$, $\Delta\vec{I}=1$

interference is predicted to be destructive.

The preponderance of evidence from a variety of experimental results is internally consistent and consistent with the theoretical estimates of Desplanques, Donoghue and Holstein[14]. Data ranges from p+p scattering where there are no nuclear physics uncertainties, to light nuclei where one can do detailed shell model calculations, and on to heavy nuclei (see a summary in ref. 31) where so far we have only relatively simple single-particle analyses of the nuclear wavefunctions.

What are the loose ends? A major "fly in the ointment" is a recent measurement of A_L in the p+α reaction which gave a result[32] smaller than expected and consistent with zero. This seems to indicate a $\pi^{\pm}$ exchange potential of the opposite sign compared to that observed in ^{19}F and heavy nuclei. Clearly we need more experimental results on parity admixtures to make sure the picture is understandable and consistent. It will be particularly important to have new results in the np and ^{18}F systems. On the theoretical front we can look forward to real improvements in the accuracy of shell model predictions stimulated in large part because of the parity-mixing problem. And finally, now that "best estimates" of ref. 14 appear to give a reasonable account of the data, it is essential to "tighten up" the predictions of the PNC NN force in order to learn more about the interplay between strong and electroweak interactions.

REFERENCES

1. Isospin in Nuclear Physics ed. by D.H. Wilkinson, North-Holland Publ. Co. Amsterdam, 1969.
2. E.M. Henley and G.A. Miller in Mesons in Nuclei I ed. by M. Rho and D.H. Wilkinson, North-Holland Publ. Co., Amsterdam 1979, p. 407
3. W. Benenson and E. Kashy, Rev. Mod. Phys. 51, 527 (1979).
4. R.D. Lawson, Phys. Rev. C19 , 2359 (1979).
5. J.A. Nolen and J.P. Schiffer, Ann. Rev. Nucl. Sci. 19, 471 (1969).

6. S. Raman, T.A. Walkiewicz and H. Behrens, At. Nucl. Data Tables 16, 451 (1975).
7. A.B. McDonald and E.G. Adelberger, Phys. Rev. Lett. 40, 1692 (1978).
8. J.B. Flanz et al, Phys. Rev. Lett. 43, 1922 (1979).
9. G.W. Wagner et al, Phys. Rev. C16, 1271 (1977).
10. C.D. Hoyle et al to be published and C.D. Hoyle, PhD thesis (unpublished), Univ. of Washington, 1981.
11. H. Miska et al, Phys. Lett. 59B, 441 (1975).
12. P.M. Endt and C. Van der Leun, Nucl. Phys. A310, 1 (1978).
13. D. Kurath, Phys. Rev. Lett. 35, 1546 (1975).
14. B. Desplanques, J.F. Donoghue, and B.R. Holstein, Ann. of Physics 124, 449 (1980).
15. D.E. Nagle et al, in High Energy Physics with Polarized Beams and Targets, ed. by G.H. Thomas, AIP Conf. Proc. No. 52 (AIP, New York 1978), p. 224.
16. W. Haeberli in Polarization Phenomena in Nucl. Phys., ed. G.G. Ohlsen et al, AIP conf. Proc. No. 69 (AIP, New York, 1981) 1340.
17. V.M. Lobashov et al, Nucl. Phys. A197, 241 (1972) and V.M. Lobashov, comment at Neutrinos 82 Conf., Lake Bolotin, Hungary.
18. J.F. Cavaignac, B. Vignon and R. Wilson, Phys. Lett. 67B, 148 (1977).
19. K. Neubeck, H. Schober, and H. Wäffler, Phys. Rev. C10, 320 (1974).
20. C.A. Barnes et al, Phys. Rev. Lett. 40 840 (1978); P.G. Bizetti et al, Lett. Nuovo Cimento, 29, 167 (1980); G. Ahrens et al, Nucl. Phys. A (to be published).
21. E.G. Adelberger et al, Phys. Rev. Lett. 34, 402 (1975).
22. K.A. Snover et al, Phys. Rev. Lett. 41, 145 (1978); E.D. Earle private communication.
23. H.-B. Mak, private communication.
24. S.D. Cohen and D. Kurath, Nucl. Phys. 73, 1 (1965).
25. B.A. Brown, W. Chung, and B.H. Wildenthal C22, 774 (1980).
26. M. Rho and G.E. Brown, Comments Nucl. Part. Phys. 10, 201 (1981).
27. W.C. Haxton, Phys. Rev. Lett. 46, 698 (1981).
28. A.M. Hernandez and W.W. Daehnick, Phys. Rev. C25, 2957 (1982).
29. W.C. Haxton, B.F. Gibson, and E.M. Henley, Phys. Rev. Lett. 45, 1677 (1980).
30. B.A. Brown, W.A. Richter, and N.S. Godwin, Phys. Rev. Lett. 45, 1681 (1980).
31. B. Desplanques, Nucl. Phys. A335, 147 (1980).
32. R. Henneck et al, Phys. Rev. Lett. 48, 725 (1982).

MICROSCOPIC BASIS OF COLLECTIVE SYMMETRIES

Akito Arima

Department of Physics, Faculty of Science
University of Tokyo
Hongo, Bunkyo-ku, Tokyo, Japan

Abstract

I would like to discuss mainly the seniority scheme, i.e. SU(2) symmetry, in single closed shell nuclei. Conditions will be shown for an effective interaction to conserve seniority. Then an approach to a microscopic basis of the Interacting Boson Model will be very briefly sketched.

1. Two-Body Interaction

There has been a great deal of work which attempts to derive an effective interaction from a nucleon-nucleon interaction such as the Hamada-Johnston potential.[1)] However one does not yet have an ideal interaction which reproduces the correct binding energies and level structures of nuclei. In this lecture I will not go into any detailed discussions of the effective interaction. Instead, I will point out a simple feature of it.

If single particle wave functions are approximated by harmonic oscillator functions, they are proportional to

$$\exp\{-\frac{r^2}{2b^2}\}$$

where r is the radial coordinate of a particle and b is the size parameter of the harmonic oscillator potential, $b \approx A^{1/6}$ fm. Now a two-particle wave function is proportional to

$$\exp\{-\frac{1}{2b^2}(r_1^2 + r_2^2)\} \quad . \tag{1}$$

Using the relative coordinate, r, and the center-of mass coordinate, R,

$$\vec{r} = \vec{r}_1 - \vec{r}_2$$

$$\vec{R} = \frac{1}{2}(\vec{r}_1 + \vec{r}_2),$$

one can easily rewrite the function (1) as

$$\exp(-\frac{R^2}{b^2})\cdot\exp(-\frac{r^2}{4b^2}) \quad .$$

Thus the probability of finding the two particles at a relative distance r is proportional to

$$r^2\exp(- \frac{r^2}{2b^2}) \tag{2}$$

Here, r^2 from the volume element is taken into account. This function (2) has its maximum at

$$r_m = \sqrt{2}\ b = \sqrt{2}\ A^{1/6}\ \mathrm{fm}.$$

The nucleus ^{210}Pb has two valence nucleons in the $1g_{9/2}$ shell. For this nucleus, r_m takes the value

$$r_m = 3.4\ \mathrm{fm},$$

which is much larger than the range of the one-pion-exchange potential. This means that the one-pion-exchange potential is almost a $\delta(r)$ function, if this potential is observed from the two-nucleon system.

One thus introduces the delta-function interaction as an effective interaction:

$$V(r) = - g\ \delta(r). \tag{3}$$

The matrix element of this interaction between two antisymmetrized states $|j_1 j_2\ JM\rangle$ and $|j_1' j_2'\ JM\rangle$ can be easily calculated. The result is as follows

$$\langle j_1 j_2\ JM|-g\delta(r)|j_1' j_2'\ JM\rangle$$

$$= -G\ \frac{\sqrt{(2j_1+1)(2j_2+1)(2j_1'+1)(2j_2'+1)}}{(2J+1)\sqrt{(1+\delta_{j_1 j_2}\delta_{\ell_1 \ell_2})(1+\delta_{j_1' j_2'}\delta_{\ell_1' \ell_2'})}}$$

$$\times\ (j_1\ \tfrac{1}{2}\ j_2\ -\tfrac{1}{2}|J0)(j_1'\ \tfrac{1}{2}\ j_2'\ -\tfrac{1}{2}|J0)\ \times\ \frac{1+(-1)^{\ell_1+\ell_2-J}}{2} \tag{4}$$

where

$$G = \frac{g}{4\pi}\int R_{j_1}(r)\ R_{j_2}(r)\ R_{j_1'}(r)\ R_{j_2'}(r)\ r^2 dr.$$

Here R_j is the radial wave function of a nucleon in the j orbit.

Applying this expression to ^{210}Pb, one finds the following interaction energy of the two valence $1g_{9/2}$ nucleons:

$$E(J) = \langle (1g_{9/2})^2\ JM|-g\delta(r)|(1g_{9/2})^2\ JM\rangle$$

$$= -G\cdot\ \frac{10\cdot 10}{2(2J+1)}\ (9/2\ 1/2\ 9/2\ -1/2|J0)^2. \tag{5}$$

The coupling constant G can be estimated by using the binding energies of ^{208}Pb, ^{209}Pb and ^{210}Pb as follows.[2)] The observed interaction energy in the ground state of ^{210}Pb is given as

$$\begin{aligned} E^{obs}(0) &= -\ [BE(^{210}Pb) - BE(^{208}Pb) \\ &\quad - 2\{BE(^{209}Pb) - BE(^{208}Pb)\}] \\ &= -\ 1.24\ \text{MeV}. \end{aligned} \tag{6}$$

Equating this with E(0) given by equation (5), one obtains the following value of G:

$$G = -0.25\ \text{MeV}. \tag{7}$$

Using (5), one can immediately predict the excitation energies of ^{210}Pb;

$$\Delta E(J) = E(J) - E(0).$$

The calculated excitation energies are shown in Fig. 1, together with their observed values. One sees that this simple interaction explains reasonably well the observed energy levels of ^{210}Pb.

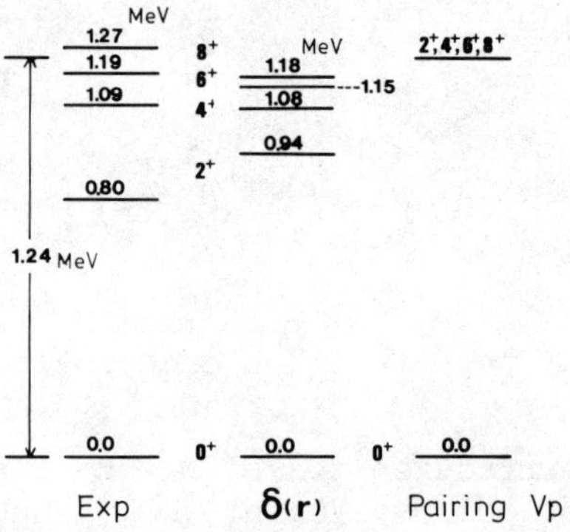

Fig. 1 Levels of ^{210}Pb

2. Seniority - SU(2) Symmetry

Figure 1 suggests that an interaction simpler than the δ interaction can be used to study the level structure of the Pb isotopes. This interaction is called the pairing interaction V_p which is defined by

$$\langle j_1 j_1 \; J | V_p | j_2 j_2 J \rangle = - \frac{\sqrt{(2j_1+1)(2j_2+1)}}{2} G \, \delta_{J0}. \tag{8}$$

The purpose of this section is to study the level structure of single closed shell nuclei such as the Pb isotopes under the assumption of the pairing interaction.

The model Hamiltonian of n identical nucleons in a single

j-shell (or many degenerate j-shells $j_1, j_2, \ldots j_k$) is

$$H = \sum V_{p,i,j}$$

where constant single particle energies are omitted. In order to handle this Hamiltonian, we introduce the concept of "quasi-spin".[3)]

A creation operator $a_m^{(j)+}$ satisfies the following commutation relations with the components of the angular momentum operator J;

$$[J_\pm, a_m^{(j)^+}] = \sqrt{(j \mp m)(j \pm m + 1)}\, a_{m\pm 1}^{(j)^+}$$

$$[J_z, a_m^{(j)^+}] = m\, a_m^{(j)^+}. \tag{9}$$

These relations confirm that $a_m^{(j)^+}$ is an m component of a spherical tensor of half integer rank j. The modified annihilation operators

$$\tilde{a}_m^{(j)} = (-1)^{j-m}\, a_{-m}^{(j)} \tag{10}$$

are also components of a spherical tensor of half-integer rank j. These creation and annihilation operators are assumed to obey the following anticommutation relations appropriate for fermion operators:

$$\{a_m^{(j)^+}, a_{m'}^{(j')^+}\} = \{\tilde{a}_m^{(j)}, \tilde{a}_{m'}^{(j')}\} = 0$$

$$\{\tilde{a}_m^{(j)}, a_{m'}^{(j')^+}\} = (-1)^{j-m}\ \delta(j,j')\delta(-m,m') \tag{11}$$

We can now define the tensor product of these creation and annihilation operators as usual; for example

$$[a^{(j_1)^+} \times a^{(j_2)^+}]_M^{(J)} = \sum_{m_1} (j_1 m_1 j_2 m_2 | JM) a_{m_1}^{(j_1)^+}\, a_{m_2}^{(j_2)^+}.$$

For the sake of convenience, we define the following three tensor products:

$$A^+(j_1 j_2 JM) = \frac{1}{\sqrt{1+\delta_{j_1,j_2}}}\, [a^{(j_1)^+} \times a^{(j_2)^+}]_M^{(J)} \tag{12}$$

$$\tilde{A}(j_1j_2JM) = -\frac{1}{\sqrt{1+\delta_{j_1,j_2}}}[\tilde{a}^{(j_1)} \times \tilde{a}^{(j_2)}]^{(J)}_M \tag{13}$$

and

$$U(j_1j_2JM) = [a^{(j_1)^+} \times \tilde{a}^{(j_2)}]^{(J)}_M . \tag{14}$$

With these operators, a two-body interaction V is given by

$$V = \sum_{j_1j_2j_3j_4} \sqrt{2J+1}\, G_J(j_1j_2j_3j_4)[A^+(j_1j_2J)\times\tilde{A}(j_3j_4J)]^{(0)}_0 \tag{15}$$

where

$$G_J(j_1j_2j_3j_4) = \frac{(1+\delta(j_1j_2))(1+\delta(j_3j_4))}{4} \langle j_1j_2\ J|V|j_3j_4\ J\rangle . \tag{16}$$

The wave functions $|j_1j_2\ J\rangle$ are properly antisymmetrized and normalized.

We first confine ourselves to a single shell configuration j^n of identical nucleons (single closed shell nuclei). The following three operators are defined:

$$\begin{aligned} S_+ &= \sqrt{\Omega}\, A^+(jjoo) \\ &= \frac{1}{2}\sum_m (-1)^{j-m} a^{(j)^+}_m a^{(j)^+}_{-m} , \end{aligned} \tag{17-a}$$

$$S_o = \frac{1}{2}(N-\Omega), \tag{17-b}$$

$$\begin{aligned} S_- &= \sqrt{\Omega}\, \tilde{A}(jjoo) \\ &= \frac{1}{2}\sum_m (-1)^{j-m} a^{(j)}_{-m} a^{(j)}_m , \end{aligned} \tag{17-c}$$

$$N = \sum_m a^{(j)^+}_m a^{(j)}_m ,$$

and

$$\Omega = (2j+1)/2.$$

Notice that these three operators are spherical tensors. It is very trivial to prove that these operators satisfy the same commutation relations as angular momentum operators satisfy;

$$[S_+, S_-] = 2S_0$$

$$[S_0, S_\pm] = \pm S_\pm \; . \tag{18}$$

Thus, these operators will henceforth be called <u>quasi-spin</u> operators.

Using the quasi-spin operators, one defines the interaction

$$V = -G\, S_+ S_-$$

which will be shown to be equal to the pairing interaction. The eigenvalues of the quadratic invariant $S^2 = S_+S_- + S_z^2 - S_z$ are $s(s+1)$, while those of S_z are $s_z=(n-\Omega)/2$. The eigenvalues of V which are written as $-G\,Q(s,s_z)$ are found to be

$$-G\,Q(s, s_z) = -G\,(s+s_z)(s-s_z+1). \tag{19}$$

The vacuum (n=0) has of course

$$s_z = -\frac{\Omega}{2} \; .$$

Because no smaller value of s_z exists, s must be $\frac{\Omega}{2}$. Therefore the vacuum $|0\rangle$ can be labelled by $s_z = -\frac{\Omega}{2}$ and $s = \frac{\Omega}{2}$. The vacuum has angular momentum zero, being expressed as

$$|s = \frac{\Omega}{2},\; s_z = -\frac{\Omega}{2};\; J=0\; M=0\rangle \equiv |0\rangle$$

An operation with S_+ on the vacuum creates two particles which are coupled to zero angular momentum. This operation cannot change the total quasi-spin s but increases s_z by one unit. The two particle state thus created is

$$|s = \frac{\Omega}{2},\; s_z = -\frac{\Omega}{2} + 1,\; J=0\rangle = |j^2\; J=0\rangle$$

$$= \frac{1}{\sqrt{\Omega}}\, S_+|0\rangle. \tag{20}$$

The eigenvalue of V in this state is

$$\langle j^2\; J=0|V|j^2\; J=0\rangle = -G\Omega$$

which is identical to the matrix element of the pairing interaction according to (8). It is thus proved that

$$V_p = -G\, S_+ S_- .$$

The two particle system j^2 has, of course, many other states which have non-vanishing even angular momenta J. The eigenvalues of s_z for these states are $(2-\Omega)/2$ and the wave functions can be written as

$$|s,\, s_z = (2-\Omega)/2,\, J\neq 0> \tag{21}$$

where s may be found as follows. Operating with S_- on these states must produce zero. The operator S_- annihilates two particles but cannot change the value of J which is not zero. The resultant states have zero particles but the vacuum has only J=0 and hence no such states exist; namely

$$S_-|s,\, s_z = \frac{2-\Omega}{2},\, J=0>=0 \tag{22}$$

Thus $\frac{2-\Omega}{2}$ is the minimum value of s_z for this s which must therefore be $\frac{\Omega-2}{2}$;

$$|s = \frac{\Omega-2}{2},\, s_z = \frac{2-\Omega}{2},\, J=0>. \tag{23}$$

We now introduce the concept of "seniority number".[4) 5)] If a v particle state

$$|j^v\, J> \tag{24}$$

vanishes when one operates with S_-, i.e.

$$S_-|j^v\, J>=0, \tag{25}$$

the value of s of this state is written as

$$s = \frac{\Omega-v}{2}\,. \tag{26}$$

Here v is called the "seniority number" of this state. Then the state (24) can be rewritten as

$$|j^v vJ>\equiv|j^v\, J> = |s= \frac{\Omega-v}{2},\, s_z = \frac{v-\Omega}{2},\, J>. \tag{27}$$

For example the vacuum has v=0 and the states (23) have v=2.

Starting with the state (27) and operating $(n-v)/2$ times with S_+, one can construct an n particle wave function which has the same s,

$$|j^n\ v\ J\rangle \equiv \sqrt{\frac{(\Omega-\frac{n+v}{2})!}{(\frac{n-v}{2})!(\Omega-v)!}}\ (S_+)^{\frac{n-v}{2}}\ |\frac{\Omega-v}{2},\ \frac{v-\Omega}{2},\ J\rangle. \tag{28}$$

For example states with v=0 J=0 are constructed as

$$|j^n\ v=0,\ J=0\rangle = \sqrt{\frac{(\Omega-\frac{n}{2})!}{(\frac{n}{2})!\Omega!}}\ (S_+)^{n/2}|0\rangle. \tag{29}$$

These states have the pairing energies given by

$$-G\ Q(n,\ 0) = -\frac{G}{4}\ n\ (2\Omega-n+2).$$

If $n \ll \Omega$, the energies can be approximately written as

$$-G\ Q(n,\ 0) \simeq -G\Omega\cdot \frac{n}{2}\ .$$

This equation means that the pairing energies are just $\frac{n}{2}$ times $G\Omega$ which is the pairing energy of a 0^+ pair. Similarly, states with seniority two are written as

$$|j^n\ v=2\ J\rangle = \sqrt{\frac{(\Omega-\frac{n+2}{2})!}{(\frac{n-2}{2})!(\Omega-2)!}}\ (S_+)^{(n-2)/2}|j^2\ v=2\ J\rangle, \tag{30}$$

which have the pairing energies

$$-G\ Q(n,\ 2) = -\frac{G}{4}\ (n-2)(2\Omega-n).$$

The excitation energies of the seniority 2 states $\Delta E(n,\ v=2)$ are given by the difference between $-G\ Q(n,2)$ and $-G\ Q(n,0)$;

$$\begin{aligned}\Delta E(n,\ 2) &= -G\ Q(n,2) + G\ Q(n,0)\\ &= G\Omega\end{aligned}$$

which is equal to the pairing energy of a pair. It is interesting to observe that $\Delta E(n,2)$ is independent of n. This is generally true for all seniorities. Observed excitation energies of single closed shell nuclei are almost independent of n as often pointed out by Talmi[6]. The pairing interaction explains this feature.

We have to ask ourselves the following question; how good is the seniority quantum number? In section 4, we will see that seniority is very good in single closed shell nuclei. If seniority is good or the Hamiltonian conserves seniority, the Hamiltonian must commute with the quasi-spin operators $S_{\pm}$, S_0. This means that the system has SU(2) (quasi-spin) symmetry.

3. Reduction Formulae for One-Body and Two-Body Operators in the Seniority S

We have seen from eq.(28) in the last section that a state of n particles and seniority v may be constructed from a state of v particles by repeated application of S_+. It might not be surprising, then, that matrix elements of n particle states with seniority v may be related to matrix elements of v particles, a result which would be of considerable importance since v is usually small for the states of physical interest. Such results[4),5)] are expressed by reduction formulae, so-called because they effectively reduce the labour involved in the calculation of matrix elements. The formulae are further useful in that the n dependence of matrix elements of physical quantities is then immediately apparent.

In this section several reduction formulae will be derived for one-body operators. To carry out this program, the following observation is crucial:[3)] the creation and annihilation operators $a_m^{(j)\dagger}$ and $\tilde{a}_m^{(j)}$ - in addition to being irreducible spherical tensors in angular momentum space - are components of an irreducible spherical tensor of rank 1/2 with respect to the quasi-spin angular momentum eq.(17) as a consequence of the commutation relations

$$\begin{aligned}
&[S_+, a_m^{(j)\dagger}] = 0 \;; \quad [S_+, \tilde{a}_m^{(j)}] = a_m^{(j)\dagger} \\
&[S_0, a_m^{(j)\dagger}] = \frac{1}{2} a_m^{(j)\dagger}; [S_0, \tilde{a}_m^{(j)}] = -\frac{1}{2} \tilde{a}_m^{(j)} \\
&[S_-, a_m^{(j)\dagger}] = \tilde{a}_m^{(j)} \quad ; [S_-, \tilde{a}_m^{(j)}] = 0.
\end{aligned} \tag{31}$$

It is convenient to define double tensors,

$$S_{1/2\ m}^{(1/2,j)} = a_m^{(j)\dagger} \qquad S_{-1/2\ m}^{(1/2,j)} = \tilde{a}_m^{(j)}; \tag{32}$$

i.e., under quasi-spin transformations the operator $S^{(1/2,j)}$ transforms as a spherical tensor of rank 1/2 and under the usual angular momentum transformations, $S^{(1/2,j)}$ is a spherical tensor of rank j. Any operator consisting of products of creation and annihilation operators may be expressed as a double tensor. For example, consider an operator defined by eq.(14) with k odd

$$U(jjkq) \equiv [a^{(j)\dagger} \times \tilde{a}^{(j)}]^{(k)}_q$$
$$= \sum_m (jmjm'|kq)\, a^{(j)\dagger}_m \tilde{a}^{(j)}_{m'} . \qquad (33)$$

Using the symmetry of the Clebsch-Gordan coefficients, one can rewrite the right hand side as follows,

$$\frac{1}{2} \sum (jmjm'|kq)(a^{(j)\dagger}_m \tilde{a}^{(j)}_{m'} + a^{(j)\dagger}_m \tilde{a}^{(j)}_{m'})$$

$$= \frac{1}{2} \sum (jmjm'|kq)(a^{(j)\dagger}_m \tilde{a}^{(j)}_{m'} - \tilde{a}^{(j)}_{m'} a^{(j)\dagger}_m)$$

$$+ \frac{1}{2} \sum (jmjm'|kq)(-1)^{j-m}\delta(m, -m')$$

$$= \frac{\sqrt{2}}{2} \sum (jmjm'|kq) \frac{1}{\sqrt{2}} (S^{(\frac{1}{2}j)}_{\frac{1}{2}m} \cdot S^{(\frac{1}{2}j)}_{-\frac{1}{2}m'} - S^{(\frac{1}{2}j)}_{-\frac{1}{2}m} \cdot S^{(\frac{1}{2}j)}_{\frac{1}{2}m'})$$

$$= \frac{\sqrt{2}}{2} [S^{(\frac{1}{2}j)} \times S^{(\frac{1}{2}j)}]^{(0,k)}_{0,q} .$$

Here the $[\quad]^{(\lambda,k)}_{\rho\cdot q}$ are double tensors; λ and ρ refer to the rank and z-component of the operator in quasi-spin space, and k and q are the rank of the ordinary irreducible tensor and its z-component. Hence, U(jjkq) for k odd is a double tensor, scalar in quasi-spin space and in a new notation,

$$S^{(0,k)}_{0,q}(jj) = \sqrt{2}\, U(jjkq) = [S^{(1/2\ j)} \times S^{(1/2\ j)}]^{(0,k)}_{0,q} . \qquad (34)$$

The same type of manipulation will give other examples of double tensors:

$$S^{(1,k)}_{1,q}(jj) \equiv -A^{\dagger}(jjkq)$$

$$S^{(1,k)}_{0,q}(jj) \equiv -\ \{U(jjkq) + \sqrt{\tfrac{\Omega}{2}}\delta(k,0)\} \qquad (35)$$

$$S^{(1,k)}_{-1,q}(jj) \equiv \tilde{A}(jjkq)$$

where k is even.

Generally, a one-body operator $\sum f_i^{(k)}$ is proportional to $S^{(1,k)}(jj)$ if its sign is not changed by the time-reversal operation, and is proportional to $S_{0,q}^{(0,k)}(jj)$ if it changes sign. The best example of the latter is the magnetic dipole operator and of the former the electric quadrupole operator.

An operator which is a quasi-spin scalar cannot connect two states of different s or s_z. Its matrix elements are therefore diagonal in seniority and, further, cannot depend on either s_z or n because the operator is invariant under rotations in quasi-spin space. More formally, a scalar quasi-spin operator commutes with the quasi-spin raising and lowering operators. We then find

$$<j^n v\alpha J \| S_0^{(0,k)}(jj) \| j^n v'\alpha' J'> = \delta_{vv'} <j^v v\alpha J \| S_0^{(0,k)}(jj) \| j^v v\alpha' J'> \quad (36)$$

or, more explicitly,

$$<j^n v\alpha J \| \sum_{i=1}^{n} f_i^{(k)} \| j^n v'\alpha' J'> = \delta_{vv'} <j^v v\alpha J \| \sum_{i=1}^{v} f_i^{(k)} \| j^v v\alpha' J'> , \quad (37)$$

assuming $\sum_i f_i^{(k)}$ is a quasi-spin scalar.

On the other hand operators which are quasi-spin vectors can change s by one unit or v by two units. The Wigner-Eckart Theorem may then be usefully applied to the matrix elements in quasi-spin space (now the Clebsch-Gordan coefficients which appear are just the familiar ones since the quasi-spin operators obey the same commutation relations as the ordinary angular momentum operators or, to be erudite, form an SU(2) algebra);

$$<j^n v J \| S_0^{(1,k)}(jj) \| j^n v'\alpha' J'> =$$

$$\frac{(\frac{1}{2}(\Omega-v')1\ \frac{1}{2}(n-\Omega)\ 0|\frac{1}{2}(\Omega-v)\ \frac{1}{2}(n-\Omega))}{\sqrt{2(\frac{\Omega-v}{2})+1}}$$

$$\times <s = \frac{\Omega-v}{2}\ \alpha J \||| S_{(jj)}^{(1,k)} \||| s = \frac{\Omega-v'}{2}\ \alpha' J'>$$

where $<|||\quad|||>$ means that this matrix element is reduced in both quasi-spin and angular momentum space. The n dependence is confined to the Clebsch-Gordan coefficient. Taking the ratio of two Clebsch-Gordan coefficients, one for $s_z = \frac{n-\Omega}{2}$ and the other for $s_z = \frac{\bar{v}-\Omega}{2}$, one may rewrite the reduced matrix element as

$$<j^n v\alpha J|| S_0^{(1,k)}(jj)|| j^n v'\alpha' J'> = \frac{f_{10}(n)}{f_{10}(\bar{v})} <j^{\bar{v}} v\alpha J|| S_0^{(1,k)}(jj)|| j^{\bar{v}} v'\alpha' J'> \quad (38)$$

where $f_{10}(n) = (\frac{1}{2}(\Omega-v')1\ \frac{1}{2}(n-\Omega)0|\ \frac{1}{2}(\Omega-v)\ \frac{1}{2}(n-\Omega))$ and

$v'=v$, $v\pm2$ and $\bar{v} = \max(v,v')$.

More explicitly, if $f^{(k)}$ is a quasi-spin vector, one has the following reduction formulae:

$$<j^n v\alpha J|| \sum_{i=1}^{n} f_i^{(k)} || j^n v\alpha' J'> = \frac{\Omega-n}{\Omega-v} <j^v v\alpha J|| \sum_{i=1}^{v} f^{(k)} || j^v v\alpha' J'> \quad (39)$$

$$<j^n v\alpha J|| \sum_{i=1}^{n} f_i^{(k)} || j^n v-2\alpha' J'> = \sqrt{\frac{(n-v+2)(2\Omega-n-v+2)}{4(\Omega-v+1)}}$$

$$<j^v v\alpha J|| \sum_{i=1}^{v} f_i^{(k)} ||j^v v-2\alpha' J'> \quad (40)$$

Just as the matrix elements of one-body operators may be simplified using reduction formulae, so may the matrix elements of two-body operators. Such formulae can be obtained for scalar two-body operators, the prime example of which is the two-body interaction. Only the results are shown here.

A two-body interaction may be expressed in terms of double tensors,

$$V = -\sum_{\lambda,J} \sqrt{2J+1}\, G_J [S^{(1,J)}(jj) \times S^{(1,J)}(jj)]_{0,0}^{(\lambda,0)} (111-1|\lambda 0). \quad (41)$$

We introduce a new interaction defined by[7]

$$\bar{V} = 2\sum_{k} (-1)^{1-k} \sqrt{2k+1}\, F_k\, [A^+(jjk) \times \tilde{A}(jjk)]_0^{(0)}$$

$$= -2\sum_{\lambda,k} (-1)^{1-k} \sqrt{2k+1}\, F_k [S^{(1,k)}(jj) \times S^{(1,k)}(jj)]_{0,0}^{(\lambda,0)} (111-1|\lambda 0) \quad (42)$$

where

$$F_k = \sum_J (2J+1)\, w(jjjj;\, Jk) G_J . \tag{43}$$

Using this $\bar{V}$ interaction, one can write down the following reduction formulae;[7)]

$$\langle j^n v\alpha J|V|j^n v-4\alpha' J\rangle$$

$$= \sqrt{\frac{(2\Omega+2-n-v)(2\Omega+4-n-v)(n-v+2)(n-v+4)}{8(2\Omega+2-2v)(2\Omega+4-2v)}}\langle j^v v\alpha J|\tfrac{1}{3}(V-\bar{V})|j^v v-4\alpha' J\rangle, \tag{44}$$

$$\langle j^v v\alpha J|\tfrac{1}{3}(V-\bar{V})|j^v v-4\alpha' J\rangle = \langle j^v v\alpha J|V|j^v v-4\alpha' J\rangle,$$

$$\langle j^n v\alpha J|V|j^n v-2\alpha' J\rangle$$

$$= \frac{\Omega-n}{2(\Omega-v)}\sqrt{\frac{(n-v+2)(2\Omega+2-n-v)}{(\Omega+1-v)}}\langle j^v v\alpha J|\tfrac{1}{3}(V-\bar{V})|j^v v-2\alpha' J\rangle, \tag{45}$$

$$\langle j^v v\alpha J|\tfrac{1}{3}(V-\bar{V})|J^v v-2\alpha' J\rangle = \langle j^v v\alpha J|V|j^v v-2\alpha' J\rangle,$$

and

$$\langle j^n v\alpha J|V|j^n v\alpha' J\rangle$$

$$= \{\frac{(\Omega-2v)(2\Omega-n-v)}{4(\Omega-v)(\Omega-v-1)}(G_0+2F_0) - F_0\}(n-v)\delta(\alpha,\alpha')$$

$$+ \frac{(\Omega-v)(\Omega-v-2)+(n-\Omega)^2}{2(\Omega-v)(\Omega-v-1)}\langle j^v v\alpha J|V|j^v v\alpha' J\rangle$$

$$+ \frac{(\Omega-v)^2-(n-\Omega)^2}{2(\Omega-v)(\Omega-v-1)}\langle j^v v\alpha J|\bar{V}|j^v v\alpha' J\rangle. \tag{46}$$

The derivations of these formulae can be found in reference (7).

4. Conservation of Seniority in Single Closed Shell Nuclei

Equations (44) and (45) guarantee that the seniority- or SU(2) -symmetry is conserved when

$$\bar{V} = V + aS_+S_- + C \tag{47}$$

where a and C are constants. Another way to write the condition is as follows

$$V = V_1 + bS_+S_- + d \qquad (47')$$
$$\bar{V}_1 = V_1$$

where b and d are constants. As discussed in section 2, the pairing interaction produces excitation energies independent of n. A constant interaction of course gives the same energy to all states which belong to a nucleus. For any interaction V_1 satisfying the condition (47'),

$$G_J + 2F_J = 0$$

which is easily confirmed by comparing (15) and (42). Therefore the first term in (46) becomes very simple as

$$-F_0(n-v)\delta(\alpha,\alpha').$$

The $(n-\Omega)^2$ terms in the second and third terms are cancelled out for this interaction V_1. Thus we conclude that any interaction V satisfying the condition (47') or (47) not only conserves seniority but also produces excitation energies independent of n. Experimentally as mentioned already, excitation energies of single closed shell nuclei are almost independent of n. This fact suggests that an effective interaction should satisfy approximately the condition (47').

A multipole-multipole interaction is often used as a simple example of a phenomenological interaction. A λ-pole interaction is defined by

$$V_\lambda = (2\lambda+1)k^{(\lambda)} \sum(u_i^{(\lambda)}\cdot u_j^{(\lambda)}) \qquad (48)$$
$$k^{(\lambda)} = \text{const.}$$

where $u_i^{(\lambda)}$ is a unit (spherical) tensor defined by

$$\langle j\| u^{(\lambda)} \| j'\rangle = \delta_{jj'} \qquad (49)$$

The λ-pole interaction may be rewritten as

$$V_\lambda = \frac{1}{2}(2\lambda+1)k^{(\lambda)}(U^{(\lambda)}\cdot U^{(\lambda)}) - \frac{1}{2}(2\lambda+1)k^{(\lambda)}\sum_i(u_i^{(\lambda)}\cdot u_i^{(\lambda)})$$

with

$$U_q^{(\lambda)} = \sum_i u_{q,i}^{(\lambda)} ,$$

the sum being over all particles. The expectation value of the single particle term is found by applying the Wigner-Eckart theorem and eq.(49).

$$<j\ m \mid (u_i^{(\lambda)} \cdot u_i^{(\lambda)}) \mid j\ m>$$

$$= \sum_q <j\ m \mid u_{q,i}^{(\lambda)} \mid j\ m'>(-1)^q <j\ m' \mid u_{-q,i}^{(\lambda)} \mid j\ m> = \frac{1}{2\Omega}\ .$$

The second quantized form of $U_q^{(\lambda)}$ is proportional to the operator already defined by eq.(33);

$$U_q^{(\lambda)} = -\frac{1}{\sqrt{2\lambda+1}}[a^{(j)+} \times \tilde{a}^{(j)}]_q^{(\lambda)}$$

$$= -\frac{1}{\sqrt{2(2\lambda+1)}} S_{0,q}^{(k,\lambda)} = -\frac{1}{\sqrt{2\lambda+1}} A_{0,q}^{(k,\lambda)}$$

where k=0, λ=odd or k=1, λ=even.

If λ=odd, V can be expressed in terms of quasi-spin scalar operators $S_{0,q}^{(0,\lambda)}$ or $A_{0,q}^{(0,\lambda)}$

$$V_\lambda = \frac{1}{2}\sqrt{2\lambda+1}\ k^{(\lambda)}\ (-1)^\lambda [A^{(0,\lambda)}A^{(0,\lambda)}]_{0,0}^{(0,0)} - \frac{1}{4\Omega}k^{(\lambda)}(2\lambda+1)\hat{N}.$$

We then defined the following operator

$$V^{odd} \doteq \sum_{i,j}\sum_\lambda (2\lambda+1)k^{(\lambda)}(u_i^{(\lambda)} \cdot u_j^{(\lambda)})\ \frac{1-(-1)^\lambda}{2} \qquad (50)$$

As only odd values of λ appear in (50), it will be called an odd-rank tensor interaction[5]. By construction V^{odd} is scalar in the quasi-spin space and therefore cannot change the seniority of a wave function. In other words V^{odd} conserves the SU(2) symmetry.

It is then anticipated that for V^{odd} the relation (47) holds exactly (a = C = 0). Indeed it is not difficult to prove this fact. For V^{odd}, the matrix elements which are diagonal in seniority become very simple;

$$<j^n v\alpha J|V^{odd}|j^n v'\alpha' J>$$

$$=\delta_{vv'}[\frac{1}{2}(n-v)G_0\delta_{\alpha\alpha'} + <j^v v\alpha J|V^{odd}|j^v v\alpha' J>].$$

We now define the following interaction

$$V^{even} \equiv \sum_{i>j}\sum_{\lambda}(2\lambda+1)k^{(\lambda)}(u_i^{(\lambda)}\cdot u_j^{(\lambda)})\frac{1+(-1)^{\lambda}}{2}$$

which is called an even-rank tensor interaction.[5] This interaction, in general, does not conserve the seniority quantum number. However there are some exceptions. One exception is a $\delta(r)$ interaction. This interaction, however, can be shown to be equivalent to an odd-rank tensor interaction (see appendix). A typical example of the even-even tensor interaction is a $Q\cdot Q$ interaction which is defined as

$$V_Q = \kappa Q\cdot Q$$

where

$$Q_m = \sum_i r_i^2 Y_m^{(2)}(\theta_i\phi_i).$$

It is easily shown that this interaction does not satisfy the condition (47');

$$\bar{V}_Q \neq V_Q.$$

Thus the n-independence of observed excitation energies implies that the strength of V_Q between two identical nucleons should be weak.

How good is the seniority quantum number in real nuclei? Five matrix elements of the effective interaction between two $0g_{9/2}$ protons have been determined by Gloeckner and Serduke.[8] They used single closed nuclei in which N=50, Z≥39; i.e. ${}^{89}_{39}Y_{50}$, ${}^{90}_{40}Zr_{50}$, ${}^{91}_{41}Nb_{50}$, ${}^{92}_{42}Mo_{50}$, ${}^{93}_{43}Tc_{50}$. The matrix elements $G_J(9/2\ 9/2, 9/2\ 9/2)$ are as follows;

$$G_0 = -1.702 \text{ MeV}, G_2 = -0.644 \text{ MeV}, G_4 = 0.243 \text{ MeV}, G_6 = 0.351 \text{ MeV}, G_8 = 0.588 \text{ MeV}.$$

Using these matrix elements and eq.(43), one can calculate the F_J, which turn out to be given as

$$F_0 = -1.18,\ F_2 = 0.78,\ F_4 = 0.52,\ F_6 = 0.21,\ F_8 = 0.25.$$

Then $G_J + 2\,F_J$ are

$$G_0 + 2\,F_0 = -4.06,\ G_2 + 2\,F_2 = 0.92,\ G_4 + 2\,F_4 = 1.29,$$

$$G_6 + 2\,F_6 = 0.78,\ G_8 + 2\,F_8 = 1.09.$$

After adding $aS_+S_- + C$ where $5a = -5.08$ and $C = -1.02$, one finds

$$G_0 + 2\,F_0 + 4.06 = 0,\ G_2 + 2\,F_2 - 1.02 = 0.10,\ G_4 + 2\,F_4 - 1.02$$

$$= -0.27,\ G_6 + 2\,F_6 - 1.02 = 0.24,\ G_8 + 2\,F_8 - 1.02 = -0.07,$$

which are very small. This fact guarantees that the seniority quantum number is approximately conserved.

Equation (39) shows that the matrix elements of a quadrupole operator (λ=2) which are diagonal in seniority vanish in the middle of a shell when the seniority quantum number is conserved.

Recently Morinaga and his collaborators[9)] measured the life times of 10^+ states in ${}^{148}_{66}Dy_{82}$, ${}^{150}_{68}Er_{82}$ and ${}^{152}_{70}Yb_{82}$. These 10^+ states are supposed to consist of $h_{11/2}$ protons outside of ${}^{146}_{64}Gd_{82}$ which seems to satisfy certain conditions as a doubly magic nucleus.[10)] The nucleus ${}^{152}Yb$ has 6 protons in the $h_{11/2}$ orbit. We then expect that the matrix elements of the quadrupole operator should vanish between two states with v=2; i.e. 10^+ and 8^+. The B(E2; $10^+ \rightarrow 8^+$) values observed by the Munich group are shown in table 1, together with the calculated values which are normalized to the B(E2; $10^+ \rightarrow 8^+$) of ${}^{148}_{66}Dy_{82}$. In the calculation, the reduction formula (39) is used. The observed values are very well explained by this simple formula. Thus, the experiment confirms that the seniority (SU_2) scheme is very good.

The observed B(E2; $8^+ \rightarrow 6^+$) in nuclei with N=50 also show similar behavior. However, the mixture of two configurations, $p_{1/2}{}^2(0)\ g_{9/2}{}^n$ and $g_{9/2}{}^{n+2}$, obscures the n dependence of B(E2; $8^+ \rightarrow 6^+$).

Table 1. $B(E2;\ 10^+ \to 8^+)$ of nuclei with N=82 (in W. U.)

The calculated value for ^{148}Dy is normalized to the experimental value.

	$^{148}_{66}Dy_{82}$	$^{150}_{68}Er_{82}$	$^{152}_{70}Yb_{82}$
obs	0.94	0.23	0.02
cal	0.94*	0.23	0
n	2	4	6
$\left(\frac{6-2}{6-2}\right)^2$	1	1/4	0

Table 2. Life times of 8^+ states in nuclei with N=50 (in ns)

	$^{90}_{40}Zr_{50}$	$^{92}_{42}Mo$	$^{94}_{44}Ru$
obs	180±9	275±10	$(102\pm7)\times10^3$
cal	180*	2008	1831
n	2	4	6
$\left(\frac{5-n}{5-2}\right)^2$	1	1/9	1/9

The seniority scheme is badly broken for nuclei which have both protons and neutrons in their valence orbits. The delta interaction, for example, no longer conserves seniority, because the following matrix element is generally large;

$$\langle j_\pi^2(2)j_\nu^2(2)\ 0\ |-\sum_{i,j} g\ \delta(r_{ij})|\ j_\pi^2(0)j_\nu^2(0)\ 0\rangle,$$

which turns out to be well approximated by the matrix element of the $(Q_\pi \cdot Q_\nu)$ interaction.

5. SU(6) Symmetry of Nuclear Collective Motion

It is well known that in almost all even-even nuclei, the first excited 2^+ levels strongly decay into their ground states, emitting E2 γ rays.

In the single j space such 2^+ states can be easily constructed by operating with a quadrupole operator

$$Q_m = \sum_i r_i^2 Y_m^{(2)}(\theta_i \phi_i)$$

$$= \langle j \| r^2 Y^{(2)} \| j \rangle U_m^{(2)}$$

on the ground states. Since the ground states have seniority 0, the 2^+ states thus constructed have seniority 2;

$$|j^n\ v=2\ 2^+M\rangle = N_2 Q_M |j^n\ v=0\ 0^+\rangle$$

$$= N_2 Q_M \sqrt{\frac{(\Omega - \frac{n}{2})!}{\Omega!(\frac{n}{2})!}} (S^+)^{n/2} |0\rangle \qquad (51)$$

where N_2 is the normalization constant. The right hand side of this equation can be rewritten as

$$\text{R H S} = N_2' \sqrt{\frac{(\Omega - \frac{n}{2})!}{\Omega!(\frac{n}{2})!}} \frac{n}{2}(S^+)^{(n-2)/2} D^+|0\rangle \qquad (52)$$

where

$$D_M^+ = A^+(jj;\ 2M)$$

$$= -[U_M^{(2)}, S_+]$$

and

$$N_2' = -\langle j \| r^2 Y^{(2)} \| j \rangle N_2 .$$

Equation (52) means that the wave functions of 2^+ states can be constructed by using D^+ and S^+ as building blocks;

$$|j^n\ v=2\ 2^+M\rangle = N_2''(S^+)^{(n-2)/2} D^+|0\rangle . \qquad (53)$$

Operating once more with the Q operator on the wave functions of the first excited states (53), one obtains states with seniority four,

$$|j^n\ v=4\ J\ M\rangle = N_{4,J}(S^+)^{(n-4)/2}\ (D^+)^2_{J,M}|0\rangle.$$

In this process, we have omitted a term involving $(S^+)^{(n-2)/2}$ $[Q_{M'}, D^+]|0\rangle$, because this term has seniority 2.

Using S^+ and D^+ as building blocks, one can construct a set of wave functions which span a subspace of the original shell model space. This subspace is called the SD subspace.[11] It was found that the SD subspace provides a good way of truncating the gigantic shell model space.[11] [12] We then map a state belonging to the SD subspace

$$|j^n(S^{(n-v)/2}\ D^{v/2})\alpha JM\rangle = N(S^+)^{(n-v)/2}\ (D^+)^{v/2}_{\alpha JM}|0\rangle \qquad (54)$$

to a boson state

$$|s^{(n-v)/2}\ d^{v/2}\ \alpha JM\rangle = N_V(s^+)^{(n-v)/2}\ (d^+)^{v/2}_{\alpha JM}|0\rangle, \qquad (55)$$

which belongs to the sd boson space.[11] By this mapping, one can deal with the nuclear quadrupole collectivity. Since the total number of s and d bosons corresponds to a half of nucleon number, the total boson number must be conserved. Thus we have the SU(6) invariance of the boson system. Now the nuclear quadrupole collective motion seems to be well described by the SU(6) symmetry.[13] Since s and d bosons interact among themselves, this SU(6) model is called the Interacting Boson Model.[13]

If we do not distinguish protons from neutrons, the model is called the Interacting Boson Model 1.[13] In the Interacting Boson Model 2[14] we introduce both protons bosons and neutron bosons. This model has a close relation with the nuclear shell model.

As an example of application of the Interacting Boson Model 2, the electron inelastic scattering of ^{134}Ba is discussed here. According to this model, the operator which is responsible for exciting a nucleus from its ground state to any 2^+ excited states is written as

$$M^{(2)}_{\mu} = M^{(2)}_{\mu,\pi} + M^{(2)}_{\mu,\nu}$$

$$M^{(2)}_{\mu,\rho} = \alpha_\rho(r)\,(d^+_\mu s + s^+\tilde{d}_\mu)_\rho + \beta_\rho(r)\,[d^+\tilde{d}]^{(2)}_{\mu,\rho}\ , \ \rho = \pi \text{ or } \nu.$$

In order to calculate the form factor of the $0^+_1 \to 2^+_1$ excitation, one must know α_π, β_π, α_ν and β_ν. However, Otsuka[15] found numerically that the β terms contribute very little and then can be ignored. The other two functions $\alpha_\pi(r)$ and $\alpha_\nu(r)$ can be experimentally obtained from the data on ${}^{138}_{56}Ba_{82}$ and ${}^{128}_{50}Sn_{72}$. Here we assumed that the structure of proton (neutron) bosons in ${}^{134}_{56}Ba_{78}$ is the same as that in ${}^{138}_{56}Ba_{82}$ (${}^{128}_{50}Sn_{78}$).

The Sendai[15] group has recently performed experiments on ${}^{138}_{56}Ba$ ${}^{138}_{56}Ba$, ${}^{124}_{50}Sn_{74}$ (instead of ${}^{128}_{50}Sn_{78}$) and ${}^{134}_{56}Ba_{78}$. The functions $\alpha_\pi(r)$ and $\alpha_\nu(r)$ experimentally determined are shown in Fig. 2. Using them, one can calculate the form factor for the $0^+_1 \to 2^+_1$ transition. The result is shown in Fig. 3, together with the observed form factor. One sees a good agreement between them.

Otsuka[15] then calculated microscopically the two functions $\alpha_\pi(r)$ and $\alpha_\nu(r)$, assuming $g_{7/2}$, $d_{5/2}$, $d_{3/2}$, $s_{1/2}$ and $h_{11/2}$ orbits. His results are reasonable for large values of r which are important for describing the form factor. There is, however, found as shown in Fig. 2 a discrepancy between the calculated and experimental results in small values of r. It is an interesting future problem to dissolve this discrepancy. The Broken Pair Approximation, too, seems very promising for deriving those functions.[16]

Appendix

A delta function δ(r) can be expanded in terms of unnormalized spherical harmonics $C^{(k)}_q(\theta,\phi)$ as follows;

$$\delta(r_{12}) = \frac{\delta(r_1-r_2)}{r_1^2} \sum_k (2k+1)(C^{(k)}(\theta_1,\phi_1)\cdot C^{(k)}(\theta_2,\phi_2)).$$

Here, k=even. Since

$$\langle jm'|C^{(k)}_q|jm\rangle = \langle j\|C^{(k)}\|j\rangle\ \langle jm'|U^{(k)}_q|jm\rangle,$$

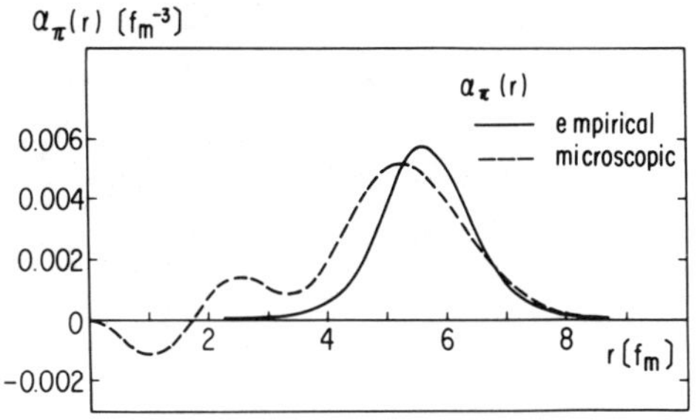

Fig. 2a

The function $\alpha_\pi(r)$ of $^{138}_{56}Ba$; observed by the Sendai group and calculated by Otsuka.

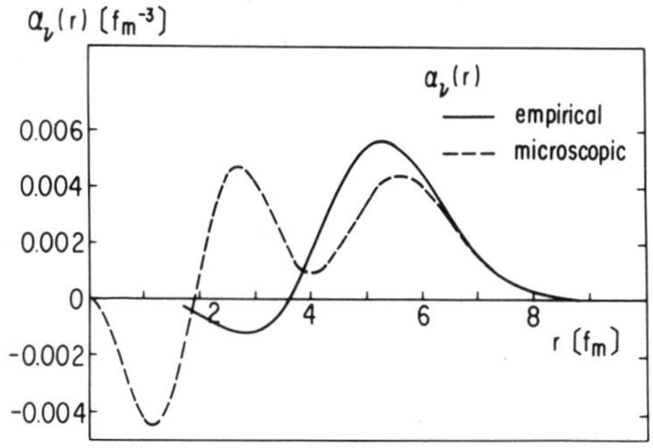

Fig. 2b

The function of $\alpha_\nu(r)$ of $^{124}_{50}Sn_{74}$ observed by the Sendai group and calculated by Otsuka.

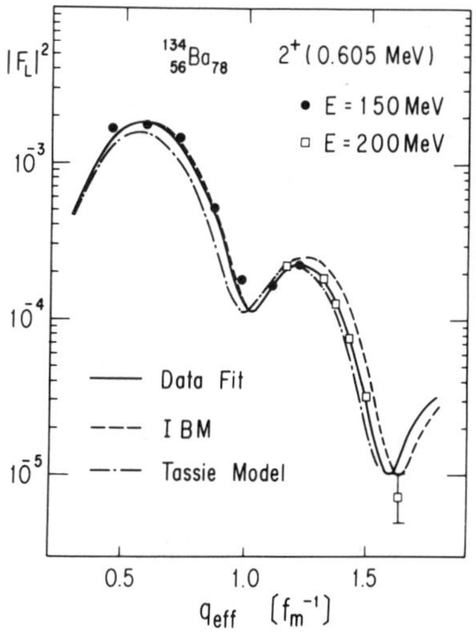

Fig. 3

The form factor of $0_1^+ \rightarrow 2_1^+$ excitation by the electron inelastic scattering.

in the single j configuration, one can rewrite $\delta(r_{12})$ as

$$\delta(r_{12}) = \frac{\delta(r_1-r_2)}{r_1^2} \sum_k (2k+1)<j\| C^{(k)} \| j>^2 (U_1^{(k)} \cdot U_2^{(k)}).$$

Therefore the $\delta(r)$ interaction is an even-rank tensor interaction.

The $\delta(r)$ interaction is non-vanishing for two identical nucleons only when they have S=0. This is because the wave function of the two nucleons with S=1 should be antisymmetric in its spatial part and therefore cannot have a relative s state. Then one has the following identity (See ref. 17),

$$\delta(r_{12}) \frac{1+(\sigma_1 \cdot \sigma_2)}{4} = 0$$

or

$$\delta(r_{12}) = - (\sigma_1 \cdot \sigma_2) \delta(r_{12}). \tag{56}$$

Recoupling angular momenta, one rewrites the right hand side of eq.(56) as follows;

$$\delta(r_{12}) \propto - \sum_{\lambda,k} <11(0)kk(0);0|1k(\lambda)\ 1k(\lambda);0>(2k+1)$$

$$(t_1^{(1k;\lambda)} \cdot t_2^{(1k;\lambda)}),$$

where $<11(0)kk(0);0|1k(\lambda)1k(\lambda);0> = \sqrt{(2\lambda+1)/3(2k+1)}$

and

$$t^{(1k;\lambda)} = \sum (1\nu kq|\lambda\mu)\sigma_\nu \cdot U_q^{(k)}.$$

In the single j space, $t^{(1k;\lambda)}$ vanishes unless λ=odd. This is easily proved as follows. One observes that

$$<j\| t^{(1k;\lambda)} \| j> \propto \begin{Bmatrix} \frac{1}{2} & \frac{1}{2} & 1 \\ \ell & \ell & k \\ j & j & \lambda \end{Bmatrix}.$$

The 9-j symbol on the right hand side is non-vanishing when $1+k-\lambda$= even. Because k=even, λ must be odd.

We thus find that the $\delta(r)$ interaction can be expressed in terms of only the odd-rank tensor $t_q^{(1k,\lambda)}$, which is proportional to

$U_q^{(\lambda)}$. Namely the $\delta(r)$ interaction is equivalent to an odd-rank tensor interaction which conserves the seniority quantum number.

References

1) A. Arima, Proc. Int. Conf. on Nucl. Phys. (1980) p.19 (Ed. R.M. Diamond and J.O. Rasmussen).
2) A.H. Wapstra and N.B. Gove, Nucl. Data Tables, 9 (1971).
3) R.D. Lawson and M.H. Macfarlane, Nucl. Phys. 66:80 (1965). R.D. Lawson, Theory of the shell Model (Oxford, 1980).
4) G. Racah, Phys. Rev. 62:438 (1942); 63:367 (1943); 76:1352 (1949).
5) A. de Shalit and I. Talmi, Nuclear Shell Theory (Academic Press, 1963).
6) I. Talmi, Nucl. Phys. A178:1 (1971); S. Shlomo and I. Talmi, Nucl. Phys. A198:81 (1972).
7) A. Arima and M. Ichimura, Prog. Theoret. Phys. 36:296 (1966). A. Arima and H. Kawarada, J. Phys. Soc. Japan 19:1768 (1964).
8) D.H. Gloeckner and F.J.D. Serduke, Nucl. Phys. A220:477 (1974).
9) E. Nolte et al., Proc. 4th Intern. Conf. on Nuclei Far from Stability (Helsingor, 1981) p.253; Private communication.
10) M. Ogawa, R. Broda, K. Zell, P.J. Daly and P. Kleinheinz, Phys. Rev. Lett. 41:289 (1978).
11) T. Otsuka, A. Arima and F. Iachello, Nucl. Phys. A309:1 (1978).
12) T. Otsuka, Nucl Phys. A368:244 (1981),Phys. Rev. Lett. 46:710 (1981).
13) A. Arima and F. Iachello, Phys. Rev. Lett. 35:1069 (1975); Ann. Phys. NY. 99:253 (1976); 111:201 (1978); 123:468 (1979).
14) T. Otsuka, A. Arima, F. Iachello and I. Talmi, Phys. Lett. 66B:205 (1977);76B:139 (1978).
15) T. Otsuka, Y. Mizuno and K. Mori, private communication.
16) Y.K. Gambhir, A. Rimini and T. Weber, Phys. Rev. 188:1573 (1969). Y.K. Gambhir, P. Ring and P. Schuck, Phys. Rev. C25:2858 (1982). K. Allaart and E. Boeker, Nucl. Phys. A168:630 (1971).
17) G. Racah and I. Talmi, Physica 18:1093 (1952).

SYMMETRY ASPECTS OF THE SHELL MODEL

P.W.M. Glaudemans

Fysisch Laboratorium
Rijksuniversiteit
Utrecht, The Netherlands

I. ROTATIONAL SYMMETRIES

1. INTRODUCTION

In principle the shell model can be applied to investigate spectroscopic data for all nuclei throughout the nuclear mass region. Obviously this includes properties of a single-particle nature but also collective properties. In practice there are several severe limitations, however. The main restriction is imposed by the maximum size of the configuration space that can be handled numerically. This means that the number of the various components in the calculated wave functions must be kept limited. The amplitudes of the components result from a diagonalization of the Hamiltonian matrix. With present day large computers one can reasonably well store and diagonalize matrices of dimension N×N, with N up to about 5000.

The size of the configuration space depends strongly on (i) the number of active particles (protons and/or neutrons) i.e those that do not belong to a closed core, (ii) on the number of orbits which can be occupied by the active particles, (iii) on the j-values of these active orbits and (iv) on the spin J and isospin T of the state to be calculated. This is illustrated in Table 1 for some cases with n active particles in the p, sd and fp shell coupled to a given J and T.

It follows from Table 1 that only the A = 4-16 and A = 16-40 nuclei can be treated in a complete major-shell configuration. For heavier nuclei the dimensions are obviously far too large to be handled this way. Shell-model calculations can be applied to

Table 1 Size of the configuration space for n particles in one major shell

mass region	major shell	l_j-values[a)]	n,J,T	size[b)]
A = 4-16	p-shell	p_1 p_3	n=6,J=2,T=1	14
			n=8,J=2,T=1	7
A = 16-40	sd-shell	s_1 d_3 d_5	n=12,J=3,T=1	6.706
			n=12,J=0,T=1	1.372
A = 40-80	fp-shell	p_1 p_3 f_5 f_7	n=20,J=6, T=1	81.804.784
			n=20,J=30,T=0	1

a) The j-values are given as 2j.
b) The first row gives the maximum size in each major shell.

heavier nuclei, but then only for a rather limited number of active particles and/or for a restricted occupation of the various active orbits of a major shell. A truncation of the model space will be necessary but it may give difficulties in separating off certain unphysical effects due to center-of-mass motions of the nucleus. This subject will be treated in the next lecture, however. For heavy nuclei with many active particles occupying several orbits, an interpretation of the experimental data in terms of collective motions of the nucleons may be much more appropriate.

The shell model and collective model depart from seemingly quite different assumptions about the behaviour of nucleons. As stated before both models may be able to describe the same properties, however. In the shell model it is assumed that in zeroth order each nucleon moves freely in a (harmonic oscillator) potential which represents the interaction with the other nucleons. The perturbation derives from a supposedly small residual interaction between the nucleons. The latter is assumed to describe the effects that could not be absorbed in the average potential. In the collective model the nucleons are strongly correlated. This gives rise to rotations or vibrations of the nucleus (Bohr and Mottelson,ref.1). Recent studies, in particular with the Interacting Boson Model (Iachello, ref.2), give a better understanding of the relation between these two different assumptions.

Presently we will follow a phenomenological approach to investigate this relation. We will discuss nuclei that are light enough for a rather detailed shell-model treatment but also heavy enough to clearly show some collective features. It will be demonstrated that fp-shell nuclei may satisfy these conditions. We will use results from the very recent shell-model study of Mooy et al.(ref.3) on the A = 52-60 nuclei. It should be remarked that these nuclei are not expected to build up strongly deformed shapes only. They are close to the doubly magic nucleus $^{56}_{28}Ni_{28}$, which favours spherical shapes. However, this core also implies that a limited number of valence particles may suffice for a reasonable description in terms of the shell model. Moreover, the j = 1/2, 3/2, 5/2 and 7/2 values of the fp-shell orbits are large enough to obtain states of rather high spin. This is advantageous to identify collective properties.

We will follow the approach of an experimentalist to identify rotational bands. Therefore various observables, e.g. excitation energies and electromagnetic properties are calculated first with the shell model. Subsequently these theoretical values are treated as if it were experimental data. The complete set of "experimental" data is used to investigate whether a specific collective model, in our case that of an axially symmetric rotor (cigar), is able to reproduce some of these shell-model values.

2. BRIEF REVIEW OF THE SHELL-MODEL APPROACH

The main principles of a shell-model calculation may be summarized as follows. Suppose one has a core of closed (sub)shells, e.g. $^{4}_{2}He_2$, $^{16}_{8}O_8$, $^{28}_{14}Si_{14}$, $^{40}_{20}Ca_{20}$, $^{56}_{28}N_{28}$, $^{90}_{40}Zr_{50}$, $^{146}_{64}Gd_{82}$ or $^{208}_{82}Pb_{126}$. Let n active particles, occupying the orbit ρ, be coupled to spin J_n and isospin T_n. Similarly let m particles in orbit λ be coupled to J_m and T_m. The configuration of the n+m particles can then be denoted as $\{\rho^n\lambda^m\}_{JT}$ with $\vec{J} = \vec{J}_n + \vec{J}_m$ and $\vec{T} = \vec{T}_n + \vec{T}_m$.

The binding energy of a nucleus is defined as the negative value of the total energy needed to decompose the nucleus into free protons and neutrons. It can be calculated from an expression which describes a step by step approach to accomplish this decomposition (deShalit and Talmi,ref.4; Brussaard and Glaudemans,ref.5; Lawson, ref.6).This expression is given by (see also the schematic illustration)

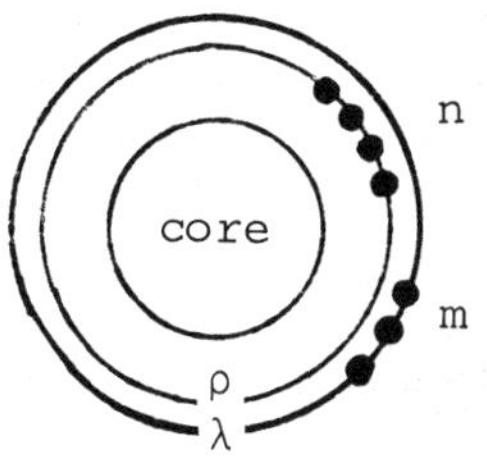

$$E^b\{\rho^n\lambda^m\}_{JT} = \sum_{i<j}^{n+m} H_{JT}^{(ij)} + ne_\rho + me_\lambda + E^b(\text{core}) + E(\text{Coulomb}). \quad (1.1)$$

(i) The first term (describing the residual interaction) represents the energy needed to separate the n+m particles from each other but not from the core. Note that its value depends on J and T.

(ii) The terms e_ρ and e_λ take care of the energy needed to remove a particle from the orbits ρ and λ, respectively. It is assumed that this "single-particle energy" does not depend on the number of particles in each active orbit.

(iii) The term E^b (core) denotes the binding energy of the core. It is assumed to be constant.

(iv) The last term represents the Coulomb energy contribution to the binding energy. This term is usually calculated from simple expressions or is taken constant for an (n+m)-particle system.

Excitation energies are obtained by subtracting the ground-state binding energy calculated with (1.1) from that of the various other (J,T) states of the n+m particles. It is clear that the constant value E^b (core) and a constant E(Coulomb) do not affect excitation energies and thus can be ignored. The single-particle energies are usually treated as parameters. Their values can be obtained from experiment, e.g. from the E^b of a nucleus with the same core plus one active particle.

The evaluation of the first term in (1.1) describing the residual interaction is by far the most complex. We will not discuss the procedures to express this interaction between n+m particles in terms of two-body matrix elements. The latter can be evaluated by making certain assumptions about the effective nucleon-nucleon force or can be obtained from a least-squares fit to experimental data.

Sofar it is assumed that a nuclear state is decribed by a simple $\{\rho^n\lambda^m\}_{JT}$ configuration. More realistic wave functions are obtained when configuration mixing is taken into account. This means that a wave function ψ_{JT} should be written as

$$\psi_{JT} = \sum_{i=1}^{N} a_i \, \phi_{JT}^{(i)} , \quad (1.2)$$

where each $\phi_{JT}^{(i)}$ denotes a given configuration $\{\rho^n\lambda^m\}_{JT}$. The amplitudes a_i follow from solving the eigenvalue problem $H\psi = E\psi$.

Each element of the Hamiltonian matrix H can be expressed in terms of the two-body matrix elements mentioned before. The square of a_i gives the probability of finding the nucleus in the state $\phi_{JT}^{(i)}$.

3. THE FP-SHELL NUCLEI

We will use a set of wave functions constructed for about 20 nuclei in the mass region A = 52-60 (Mooy et al.,ref 3). States of odd- and even-mass nuclei with spin values up to J = 16 and excitation energies up to E_x = 16 MeV are taken into account. The wave functions are generated with a mass-independent effective interaction. The latter is obtained from a least-squares fit to the experimental spectra. The experimental levels are usually known for the low-excitation low-spin region only.

3.1 The shell-model space and effective interaction

The configuration space for the wave functions is restricted to the four fp-shell orbits. The configurations taken into account can be specified by $f^{-n}\ r^m$, where f denotes the $f_{7/2}$ orbit and r stands for any of the other fp-shell orbits $p_{3/2}$, $f_{5/2}$, $p_{1/2}$. The numbers n and m, defined with respect to the doubly magic ^{56}Ni core, represent the minimum number of holes and particles, respectively, needed to describe the ground state of a specific nucleus. For instance one has (n,m) = (3,0) for $^{53}_{26}Fe_{27}$ and (n,m) = (1,3) for $^{58}_{27}Co_{31}$. It has been shown (Metsch and Glaudemans,ref.7) that this space is too limited for a good description of many fp-shell observables and should be enlarged by the inclusion of configurations with an additional $f_{7/2}$ hole. Therefore the particle-hole excited configurations given by $f^{-n-1}\ r^{m+1}$ are also included.

The requested two-body matrix elements and single-particle energies althogether involve 19 parameters which are determined from a least-squares fit to the data. No mass dependence of these parameters is assumed. The result obtained is that the about 140 experimental excitation energies used for the fit of the 19 parameters are well reproduced. Also the levels not included, because of the large dimensions of the matrices involved, were well accounted for.

As an example we show in Fig.1.1 the excitation energies calculated for the lowest four levels of each J in ^{52}Fe together with the experimentally known states. It is seen that the theory reproduces the experimental points very well. From this picture it follows immediately that the $J^\pi = 12^+$ state in ^{52}Fe at E_x = 6.8 MeV is an isomer (τ_m(exp) = 66±1s). The state is known to decay by beta emission to a $J^\pi = 11^+$ state in ^{52}Mn.

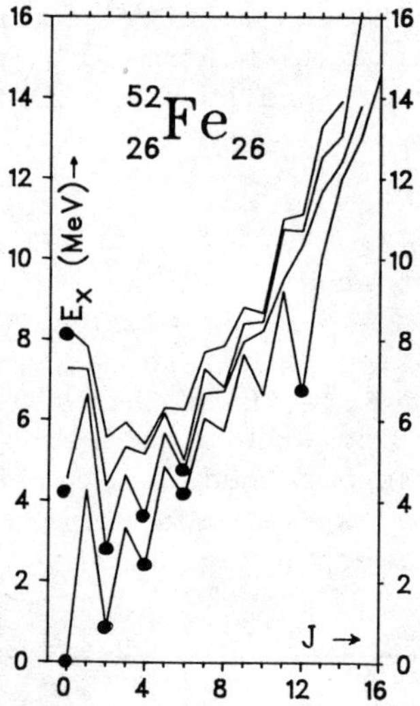

Fig.1.1 The spectrum of ^{52}Fe. The calculated excitation energies of the lowest four states of each J are connected by lines. The corresponding experimental states are indicated by dots.

3.2 The quality of the wave functions

The reliability of the calculated wave functions can be tested most critically by comparing the calculated electromagnetic properties with the experimental data. Some 25 measured g-factors are known for the A = 52-60 nuclei. In Fig.1.2 these data are compared with the present shell-model values. The latter are calculated using bare-nucleon g-factors. In this and the following two figures exact agreement between theory and experiment yields a point on the diagonal (dotted) line. It is seen that in almost all cases the agreement between theory and experiment is very good.

For the calculation of electric properties effective charges (Brussaard and Glaudemans, ref.5) are used, with the usual assumption that for protons and neutrons the same extra charge Δe can be taken. Such an effective charge is assumed to compensate to a large extent the effects due to the restricted configuration space which is used in all shell-model calculations. The effective charge employed in

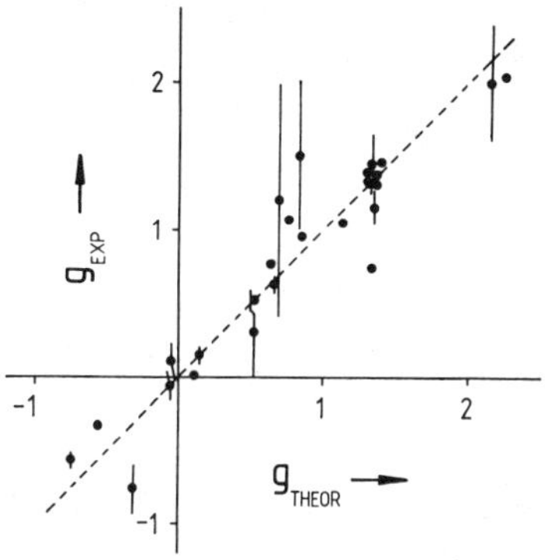

Fig.1.2 Comparison of theoretical and experimental g-factors for A = 52-60 nuclei.

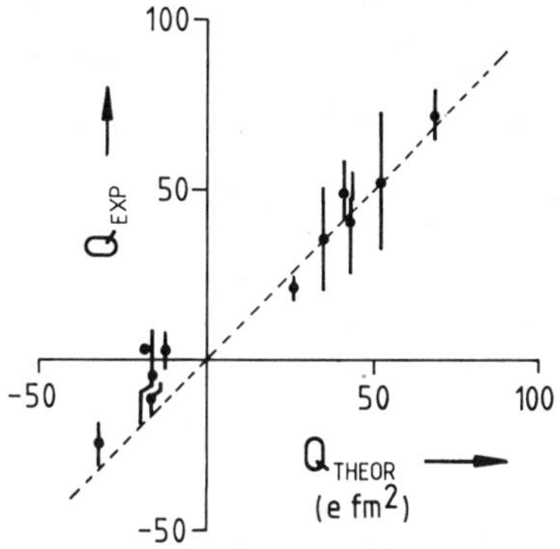

Fig.1.3 Comparison of theoretical and experimental quadrupole moments in A = 52-60 nuclei.

the present calculations is obtained from the expression $\Delta e = (0.6 + f)e$. This relation has been obtained empirically from a fit to experimentally known strong E2-transition rates and Q-pole moments. The number f with $0 \leqslant f \leqslant 1$ denotes the fraction of the intensity of the particle-hole excited configurations $f^{-n-1}\, r^{m+1}$ in the wave functions, see Sect.3.1. It may be remarked that the electric properties will play a crucial role in the investigation of collective structures as will be discussed later.

In Fig.1.3 the comparison is shown between the calculated and experimental quadrupole moments. It is seen that nearly all measured Q-pole moments are reproduced whithin the experimental error. Note that even the largest moment measured for the A = 52-60 nuclei, i.e. $Q = +72 \pm 7$ efm^2 for the $J^{\pi} = 10^+$ state in ^{54}Fe at $E_x = 6.53$ MeV, is very well accouted for by the shell model.

To further illustrate the accuracy of the wave functions we finally show the comparison between theory and experiment for E2-transition strengths in Fig.1.4. It is gratifying to see that again the very large E2 strengths are reproduced correctly.

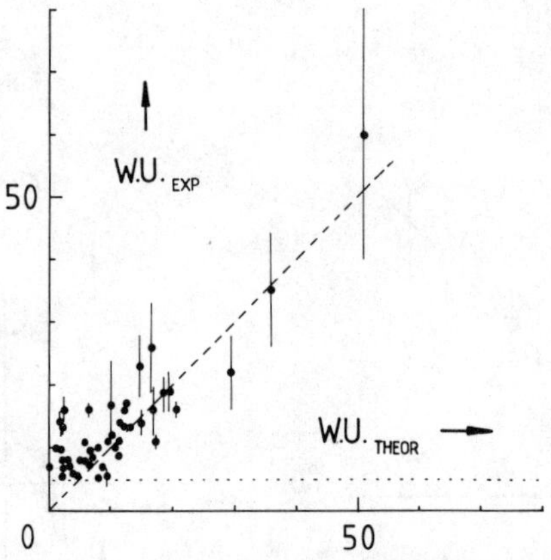

Fig.1.4 Comparison of theoretical and experimental E2-transition rates for experimental strengths > 5 W.u. (Endt,ref.8) in A = 52-60 nuclei.

4. SELECTION OF ROTATIONAL BANDS

4.1 The rotational model

The A = 52-60 shell-model results are scrutinized in an attempt to search for a possible collective interpretation of the calculated observables. The collective model investigated presently is that of the axially-symmetric (cigar-shaped) rotor. The Hamiltonian is assumed to be separable in an intrinsic and a collective part. The projection of the spin J onto the symmetry axis is given by K. For a K = 0 band one has the even values J = 0, 2, 4,..., whereas for a K ≠ 0 band the spins are given by J = K, K+1, K+2,... . The following formulas apply, see e.g. de Shalit and Feshbach,ref.9.

The dependence of the energy E on the spin J for a given value of K follows from

$$E(J,K) = E_0(K) + \frac{\hbar^2}{2\Theta}\{J(J+1) + a(-1)^{J+\frac{1}{2}}(J+\tfrac{1}{2})\delta(K,\tfrac{1}{2})\}. \qquad (1.3)$$

Here $E_0(K)$ represents the energy of the band head, Θ denotes the moment of inertia for an axis perpendicular to the symmetry axis and a the decoupling constant. Note that the a-dependent term does not contribute for $K \neq 1/2$. The quadrupole moments are given by

$$Q(J) = Q_0 \frac{3K^2 - J(J+1)}{(J+1)(2J+3)} \quad \text{for } K \neq 1 \quad , \tag{1.4}$$

where Q_0 is the intrinsic quadrupole moment. The electric quadrupole transition strengths expressed in W.u. can be calculated from

$$M_W^2(E2, J_i \rightarrow J_f) = \frac{125}{36} (1.2\, A^{1/3})^{-4}\, Q_0^2 (2J_f+1) \begin{pmatrix} J_f & 2 & J_i \\ -K & 0 & K \end{pmatrix}^2 . \tag{1.5}$$

Dipole moments are obtained from

$$\mu(J) = g_R J + (g_\Omega - g_R) \frac{K^2}{J+1} \{1 + b(-1)^{J+\frac{1}{2}} (2J+1)\,\delta(K,\tfrac{1}{2})\} \quad , \tag{1.6}$$

where g_R is the gyromagnetic ratio for the collective rotation, g_Ω is the g-factor for the intrinsic motion of the nucleus and <u>b</u> is a decoupling constant. The <u>b</u>-dependent term vanishes for $K \neq 1/2$.

4.2 <u>The selection procedure</u>

The excitation energies, electric-quadrupole and magnetic-dipole moments as well as the M1- and E2-transition strengths are calculated with the shell model for the lowest four levels of each (A,J,T) combination. The levels up to the highest spin possible (J = 16 in the present model space) are taken into account. The method used for the selection of rotational bands from the shell-model results can be described as follows.

Rotational motion is connected with large quadrupole deformation of the nucleus and thus with large values for the matrix elements of the electric quadrupole operator, see (1.4) and (1.5). Therefore a search is made for sequences of levels connected by enhanced E2-transition rates. For these sequences to be candidates for a rotational band, the observables calculated from the shell-model are required to fulfil at least four criteria:

(i) The E2 strengths must be larger than 5 W.u.

(ii) The excitation energies must increase gradually with J except for a possible K = 1/2 band, see (1.3).

(iii) The quadrupole moments must change monotonically as a function of J, see (1.4).

(iv) The number of levels must exceed the number of parameters needed in the rotational model to describe the band. Thus K = 1/2 bands consist of four or more spin-consecutive levels and $K \neq 1/2$ bands consist of three or more spin-consecutive levels, as follows from (1.3).

More than 40 band-like structures are identified in the A = 52-60 mass region according to this procedure. For a further evaluation of the purity of the bands with respect to the rotational model the numerical values of the observables are taken into account. The inclusion of electromagnetic properties in the selection procedure of rotational bands appeared to be necessary. Often the quality of the J(J+1) energy dependence and strong E2 transition rates are taken as a measure for the quality of the rotational character of a band. It turns out that these observables are not sufficient in our case, however. Thus, for the present shell-model results, a perfect J(J+1) energy spectrum and enhanced E2-transition strengths do not imply that moments and M1 transitions follow nicely the rotational model predictions. The procedure is as follows. Firstly with (1.3), (1.4) and (1.6) least-squares fits are performed to the excitation energies, electric quadrupole moments and g-factors of the shell-model levels of each band. From these fits the rotor parameters $E_0(K)$, Θ, $\underline{a}$, Q_0, g_R, g_Ω and $\underline{b}$ were determined. Secondly restrictions are imposed on the discrepancies between the values of the observables obtained in the rotational model and those obtained in the shell model.

The values of the electromagnetic transition strengths are used for the selection procedure as well. The rotational E2 strengths are calculated from (1.5) with the parameter Q_0, obtained from the fit to the microscopic quadrupole moments. In general the microscopic E2 strengths agree very well with the rotational values. For most bands the microscopic value lies between 0.7 and 1.5 times the rotational value. Shell-model bands with E2 strengths outside this range are not considered to be good rotors and are thus rejected. Also the shell-model bands with M1 strengths that differ strongly from the rotor values are excluded. Finally about 20 good rotor bands remain for a more detailed analysis.

5. RESULTS

First we will discuss some general aspects and later treat a specific case (^{52}Fe) in more detail. For a derivation of the formula used below the reader is referred to, e.g., deShalit and Feshbach, ref.9.

5.1 General aspects

Two important rotor parameters are the moment of inertia Θ and the deformation parameter β. With the assumption of a uniformly charged speroid the deformation parameter can be obtained from Q_0 with the relation

$$Q_0 = \frac{3}{\sqrt{5\pi}} Z R_0^2 \beta (1 + 0.16\beta) \qquad (1.7)$$

where Z denotes the number of protons and R_0 the radius of the un-

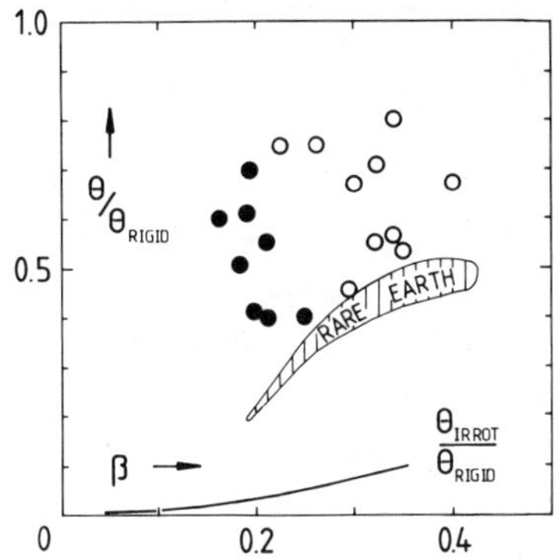

Fig.1.5 Values of Θ/Θ_{rigid} versus β for rotor bands in A= 52-60 nuclei.

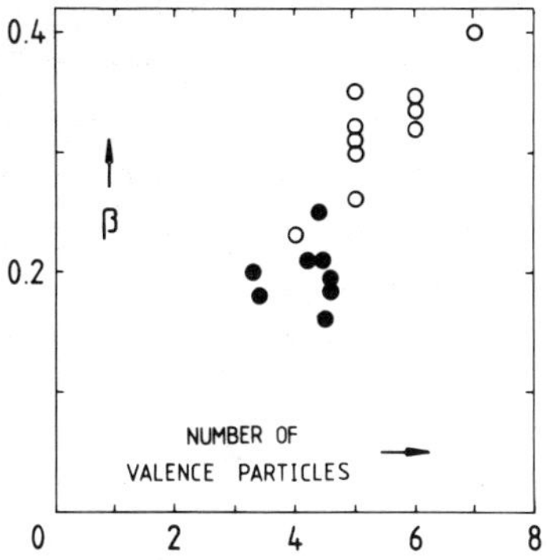

Fig.1.6 Value of β versus number of valence particles.

deformed nucleus given by $R_0 = 1.2\ A^{1/3}$fm. The upper limit for the moment of inertia is given by the rigid-body value

$$\Theta_{rigid} = \frac{2}{5}\,A\,M\,R_0^2\,(1 + 0.31\beta), \tag{1.8}$$

where A denotes the number of nucleons and M the nucleon mass. A lower limit for Θ is given by the assumption of irrotational flow of the nuclear matter

$$\Theta_{irrot} = \frac{9}{8\pi}\,A\,M\,R_0^2\,\beta^2. \tag{1.9}$$

Thus the moments of inertia should fulfil the relation

$$0 < \frac{\Theta_{irrot}}{\Theta_{rigid}} < \frac{\Theta}{\Theta_{rigid}} < 1. \tag{1.10}$$

All bands selected according to the criteria discussed in Sect. 4 appear to satisfy (1.10). A plot of Θ/Θ_{rigid} versus β is given in Fig.1.5. The points for the fp-shell nuclei belong to the region with $0.4<\Theta/\Theta_{rigid}<0.8$ and $0.13 < \beta < 0.4$. The experimental data for the rare-earth region given for comparison in Fig.1.5 show that quite surprisingly the deformation expected in the fp-shell nuclei is as large as that of the well-known strongly deformed rare-earth nuclei. In Figs.1.5 and 1.6 states dominated by $f^{-n}r^{m}$ and $f^{-n-1}r^{m+1}$ configurations, see Sect.3.1, are denoted by filled and open circles, respectively.

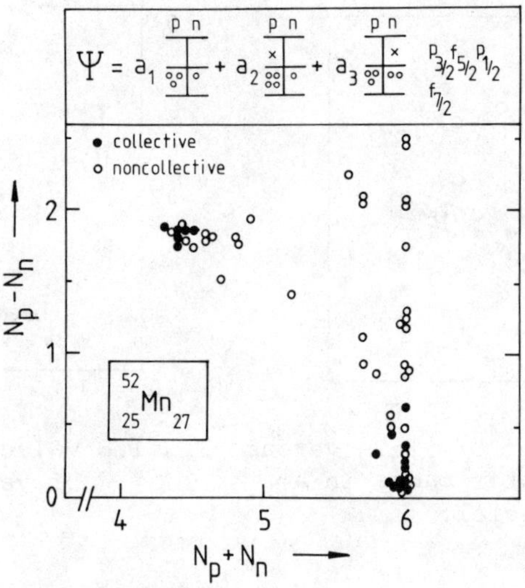

Fig. 1.7 Relation between numbers of active protons N_p and neutrons N_n in ^{52}Mn. Black dots represent collective states.

One expects the deformation parameter β to increase with the number of active particles. This correlation is found indeed as is shown in Fig. 1.6, where the number of valence particles or active particles is defined as the sum of the holes in the $f_{7/2}$ orbit and the particles in the upper three fp-shell orbits. Watt et al., ref.10, performed calculations in the full sd shell and showed that the shell-model states belonging to a given band can be characterized by their subshell occupancies. It was found that the subshell occupation numbers vary only gradually with excitation energy for the levels of a rotational band. We also find that this number remains nearly constant. It should be remarked that states which differ by more than one particle in any given subshell can not be connected by strong E2 transitions, since one deals with one-body transition operators.

The proton-neutron interaction is mainly responsible for collective properties, see e.g. deShalit and Goldhaber, ref.11. This can be demonstrated nicely with ^{52}Mn. In the simplest configuration $^{52}_{25}Mn_{27}$ has three $f_{7/2}$ -proton holes (N_p=3) and one $f_{7/2}$ -neutron hole (N_n=1). One thus has N_p+N_n=4 active nucleons. We find that the wave functions dominated by this configuration produce only two weakly deformed ($\beta \simeq 0.2$) band-like structures with K=1 and K=6. These states with $N_p + N_n \simeq 4$ and $N_p - N_n \simeq 2$ are indicated by black dots in Fig. 1.7. The model space for ^{52}Mn also includes components with five $f_{7/2}$ holes, as outlined in Sect. 3.1. For these components we have $N_p + N_n = 6$. The wave functions dominated by the a_2 term, i.e. with mainly active protons, see Fig.1.7, do not show collectivity. For these states we have $N_p - N_n \simeq 5-1=4$. The five-hole states showing collective properties are all dominated by the a_3 term shown in Fig.1.7. Only the latter term yields $N_p - N_n \simeq 3-3=0$ and thus an almost equal number of protons and neutrons. Thus the p-n interaction is most effective in the last term of the wave function. One sees in Fig.1.7 that indeed for $N_n + N_p \simeq 6$ and $N_p - N_n \simeq 0$ the collective states are found.

5.2 Applications to ^{52}Fe

A very interesting nucleus for investigations on collective behaviour is provided by $^{52}_{26}Fe_{26}$, which we will treat in some detail. For this nucleus we can restrict the requested observables to excitation energies, E2-transition rates, magnetic-dipole and electric-quadrupole moments. The M1 strengths are all very small, since for

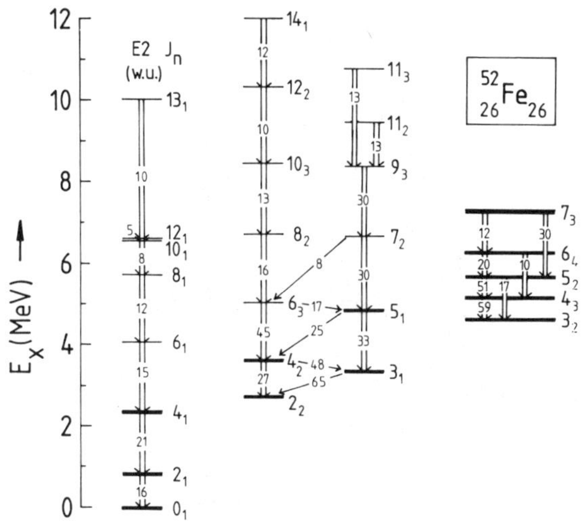

Fig.1.8 Calculated levels in ^{52}Fe connected by E2 strengths >5 W.u. Bold lines indicate collective states, see Figs. 1.9-1.11.

a T_z=0 nucleus only the (weak) isoscalar part contributes (Brussaard and Glaudemans, ref.5). The calculated energy levels of ^{52}Fe have already been presented in Fig.1.1. They were seen to agree well with the scarce experimental data. One should realize that experimentally it is very difficult to determine a rather complete set of observables, such as spins and parities, electromagnetic transition rates and moments for many levels. Hence from now on we consider the calculated observables as "experimental" data.

From an investigation of the E2 strengths it follows that almost half of the calculated levels in ^{52}Fe are connected by rather strong (>5 W.u.) E2 transitions. These levels as well as their transition rates are presented in Fig.1.8. The first impression may be that these levels belong to four "bands" as is shown in Fig.1.8. A further investigation reveals that this is not true, however.

The K=0 band (ΔJ=2) shown on the left can not accomodate the J=13 state at E_x=10MeV nor do the excitation energies of the J=8-12 states follow the J(J+1) rule, see (1.3). In Fig.1.9 a much more detailed comparison with the rotor values is shown. It is seen that although the excitation energies of the J=0-6 states deviate somewhat from the rotor values, their E2 strengths and g-factors agree quite closely. However for the J=6 level the values Q(shell model)=-5 efm^2 and Q(rotor)=-33 efm^2 deviate strongly. This shows that the K=0 rotor band terminates at J=4.

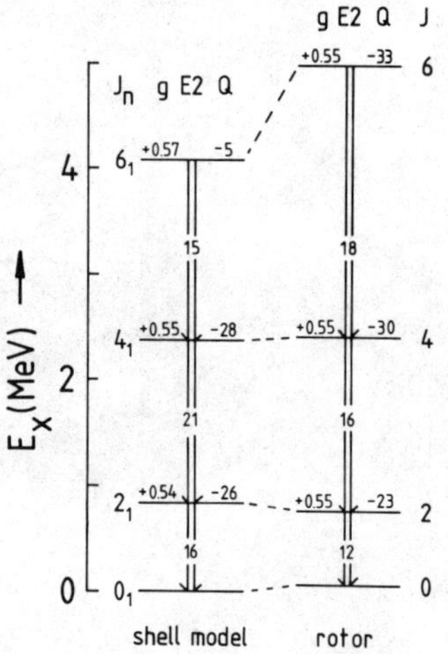

Fig.1.9 Shell-model and rotor values (K=0) in ^{52}Fe for E_x, E2(W.u.) g and Q(efm^2).

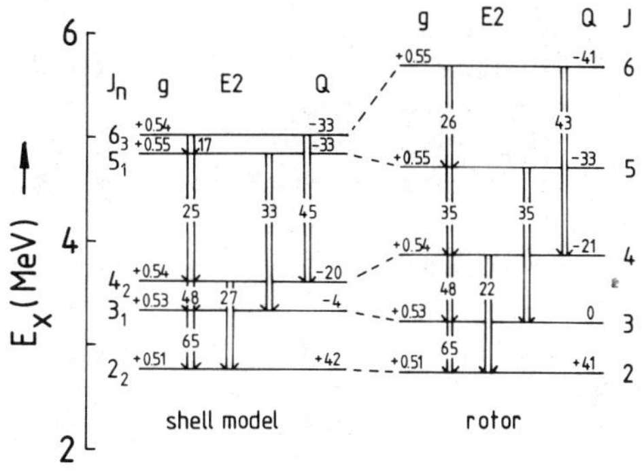

Fig.1.11 Shell-model and rotor values for a K = 3 band in ^{52}Fe.

A more complicated situation holds for the other "bands" shown in Fig.1.8. The strong E2 transitions J = 6→5 and J = 4→3 could well escape experimental detection because of the small energy difference between initial and final states. Hence,one could suggest the existence of three more "bands". Including the g-factors and Q-pole moments, it follows that only the J = 2 to 5(or 6) levels satisfy the properties of a K = 2 band and the J = 3 to 7 levels those of a K = 3 band, see Figs.1.10 and 1.11. The agreement between the two models is seen to be extremely good. It should be realized that for a given K the rotor values of the Q-pole moments and E2 strengths depend on one parameter only, i.e. the value of Q_0, as follows from (1.4) and (1.5).

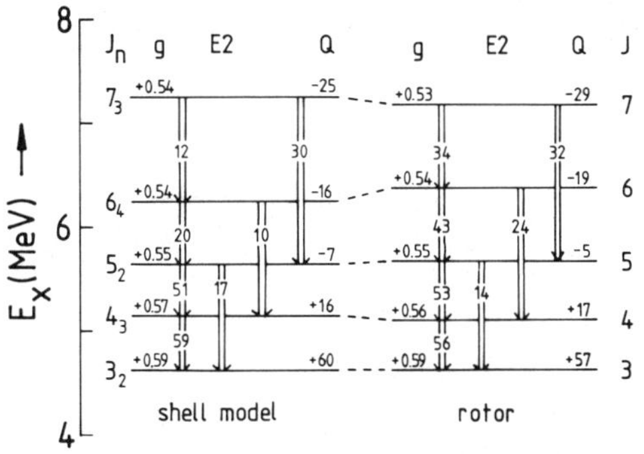

Fig.1.10 Shell-model and rotor values for a K = 2 band in ^{52}Fe.

From the comparisons discussed above it follows that knowledge of the Q-pole moments can be essential for the identification of rotor bands. It should be noted from Fig 1.10 and 1.11 that the "observed" sign change of these moments is correctly reproduced by a rotor, see also (1.4). The slow increase or decrease of the g-factors with E_x is reproduced as well. A naive estimate of the g-factors for a rotor yields $g \simeq Z/A$ which for ^{52}Fe results in $g = 0.50$. The shell-model values $g \simeq 0.55$ can simply be explained also. Because of a vanishing isovector part for a $T_z \doteq 0$ nucleus, only the isoscalar part of the magnetic-dipole operator contributes. The latter part can be approximated by the relation (Brussaard and Glaudemans, ref.5)

$$g = 0.50 \pm \frac{0.38}{2\ell+1} \quad \text{for} \quad j = \ell \pm 1/2 . \tag{1.11}$$

Since in ^{52}Fe the $f^n_{7/2}$ configurations play an important role one finds from (1.11) for $\ell = 3$ and $j = 7/2$ the value $g = +0.55$. It may be remarked that for $T_z \neq 0$ nuclei the values of the g-factor may differ strongly from this simple estimate, as follws also from Fig.1.2.

6. SUMMARY AND CONCLUSION

In this lecture we have shown that the shell model can produce states which exhibit pure collective properties. It turns out that the nuclei in the middle of the fp shell are very good candidates for an investigation of the relation between the microscopic and the collective interpretation.

In the present phenomenological approach the shell model and the axially symmetric rotor are related to each other by a comparison of the numerical values of their observables. It would be very interesting to further study this relation by investigating whether collective states are related to possible cluster structures of the corresponding shell-model wave functions.

II TRANSLATIONAL SYMMETRY

1. INTRODUCTION

Translational invariance is a fundamental symmetry in nuclear structure calculations. This symmetry is often violated in published shell-model studies because of the problems involved in treating it exactly. For calculations in which only particles in one major shell are involved, e.g. the p, sd or fp shell, violation of this symmetry does not affect the calculated observables, however.

The low-lying states in a nucleus can not always be produced in a model space with active particles in one major shell only. This follows clearly from the parity of a many-nucleon state with n particles in a major shell. For a major shell with harmonic-oscillator quantum number N the parity of the n particle system is given by $(-1)^{nN}$. For example for p- and fp-shell nuclei (N=1 and N=3, respectively) one thus obtains states with parity $(-1)^N = (-1)^A$ only, whereas in sd-shell nuclei (N=2) only positive-parity states can be produced. From now on we denote these states as normal-parity states. Experimentally one also may find low-lying states of opposite parity, however. The latter will be denoted as nonnormal-parity states. For example, the p-shell nucleus ^{11}Be has the peculiar property that its ground state has nonnormal parity ($J^\pi = 1/2^+$).

States of nonnormal parity can be formed by exciting one particle (or any odd number of particles) from one major shell into the next higher major shell. For an harmonic oscillator potential this corresponds to a 1 $\hbar\omega$ excitation. For example, a break-up of the ^{4}He core, i.e. lifting a particle from the closed 0s shell into the 0p shell or by exciting an 0p-shell particle into the 1s0d shell. This is illustrated below for an A = 7 p-shell nucleus.

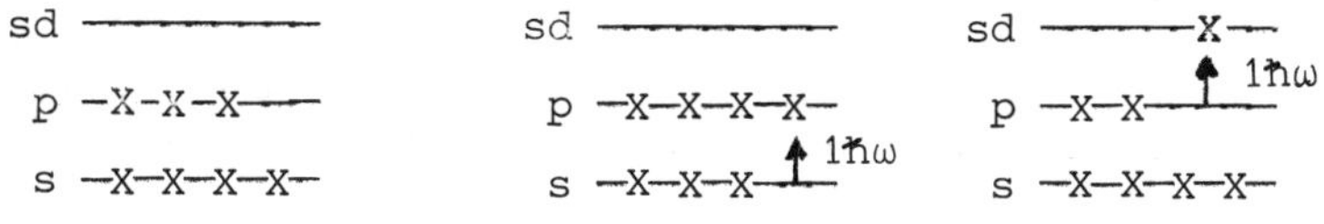

normal parity nonnormal parity

When one includes the excitations of a particle out of a major shell into another shell one encounters a complication, known as the spurious-state problem. In some cases this problem can be solved exactly, however (Elliott,ref.12;Van Hees,ref.13).

2. SPURIOUS STATES

Nuclear spectroscopic properties depend only on the intrinsic structure of the nucleus and not on its center-of-mass motion. A nucleus of mass number A should be described by a translationally invariant Hamiltonian (suppressing spin and isospin dependence)

$$H = \sum_{i=1}^{A} T_i + \sum_{i<j}^{A} V(\vec{r}_i - \vec{r}_j; \vec{v}_i - \vec{v}_j) = \sum_{i=1}^{A} T_i + \sum_{i<j}^{A} V_{ij} , \qquad (2.1)$$

where $T_i = p_i^2/2m$ represents the kinetic energy of a nucleon with mass m. The potential V_{ij} descibing the nucleon-nucleon interaction may show a velocity dependence, but it may contain the difference $\vec{v}_i - \vec{v}_j$ only.

The general many-body problem

$$H\psi(\vec{r}_1 \ldots \vec{r}_A) = E\psi(\vec{r}_1 \ldots \vec{r}_A) \qquad (2.2)$$

can not be solved exactly. In the shell-model approach the nucleon-nucleon potential V is replaced by a sum of single-particle potentials, that are fixed in space and a residual interaction. (A description of the many-particle wave functions in terms of relative coordinates turns out to be unfeasible, because of the antisymmetrization requirements of the Pauli principle). Thus the resulting shell-model Hamiltonian is no longer invariant with respect to translations. The calculated wave functions therefore always contain an unphysical part representing a center-of-mass (c.m.) motion of the nucleus. When this unphysical c.m. motion is the same for all wave functions it does not affect the calculated observables. Otherwise undesired spurious effects are introduced. A method to deal with this problem is presented below.

2.1 The Hamiltonian

The Hamiltonian (2.1) can be split into a part which depends on the c.m. coordinates and a part which depends on the intrinsic coordinates as follows

$$H = \{\sum_{i=1}^{A} p_i\}^2 /2mA + \sum_{i<j}^{A} \{(p_i - p_j)^2/2mA + V_{ij}\} = H_{c.m.} + H_{intr} . \qquad (2.3)$$

Since one is interested in the intrinsic properties of a nucleus only, we are free to modify the c.m. term in the Hamiltonian. We will use the following method to make an exact separation of spurious states possible. Instead of the physical operator

$$H_{c.m.} = P^2/2mA \quad \text{with} \quad P = \sum_{i=1}^{A} p_i \; , \tag{2.4}$$

we introduce, adding a harmonic-oscillator potential,

$$H^{\omega}_{c.m.} = H_{c.m.} + \tfrac{1}{2}Am\omega^2R^2 = P^2/2mA + \tfrac{1}{2}Am\omega^2R^2 \quad \text{with } R = \frac{1}{A}\sum_{i=1}^{A} r_i . \tag{2.5}$$

Here P and R represent the c.m. momentum and position coordinate, respectively, and ω is still a free parameter. With the potential energy $\tfrac{1}{2} Am\omega^2R^2$ in (2.5) the c.m. is bound to the origin of the laboratory system without affecting the intrinsic structure.

The modified Hamiltonian H^{ω} we wish to use is given by, see (2.3) and (2.5)

$$H^{\omega} = H^{\omega}_{c.m.} + H_{intr}$$

$$= \{\sum_{i=1}^{A} p_i\}^2/2mA + \tfrac{1}{2}Am\omega^2R^2 + \sum_{i<j}^{A} \{(p_i-p_j)^2/2mA + V_{ij}\} . \tag{2.6}$$

The operator H^{ω} can finally be rewritten in terms of one-body plus two-body operators as

$$H^{\omega} = \sum_{i=1}^{A} (p_i^2/2m + \tfrac{1}{2}m\omega^2r_i^2) + \sum_{i<j}^{A} \{V_{ij} - \frac{m\omega^2}{2A}(r_i-r_j)^2\} = H_1 + H_2 . \tag{2.7}$$

Here H_1 defines the single-particle basis states and H_2 represents the residual interaction. With the assumption that V_{ij} depends on the intrinsic coordinates only, one obtains the eigenfunctions of H^{ω} as a product of eigenfunctions of $H^{\omega}_{c.m.}$ and H_{intr}, see (2.6). The eigenvalues are found as a sum of the eigenvalues of $H^{\omega}_{c.m.}$ and H_{intr}. For a description of nuclear properties one takes only those eigenstates of H^{ω} with $H^{\omega}_{c.m.}$ in its ground state. Excited states of $H^{\omega}_{c.m.}$ yield spurious states. It should be remarked that the harmonic-oscillator potential for the c.m. motion introduced in (2.5) is the only potential that permits the partition of the Hamiltonian H^{ω} in one- and two-body operators only (Aviles, ref.14). Otherwise many-body operators are introduced.

2.2 The model space

For harmonic-oscillator wave functions it can be shown (Elliott and Skyrme, ref.12; Brussaard and Glaudemans, ref.5) that all wave functions with the lowest energy allowed by the Pauli exclusion principle are completely nonspurious. Excitations of one or more particles into shells 1 $\hbar\omega$, 2 $\hbar\omega$, etc. higher in energy may introduce spuriosity. It thus follows that, if all active particles outside a closed-shell core are confined to one major shell, i.e. a $0\,\hbar\omega$ model space, no spurious states are generated. Examples of such cases are $0\,\hbar\omega$ calculations in the p shell (Cohen and Kurath, ref. 15) with a closed ^{4}He core or in the sd shell (McGrory and Wildenthal, ref.16; Whitehead et al., ref.17) with a closed ^{16}O core.

Spurious states can be identified exactly only when a complete $(0+1)\hbar\omega$ or $(0+1+2)\hbar\omega$, etc. model space is used. For a calculation on nonnormal-parity states at least the complete $(0+1)\hbar\omega$ space must be taken into account. This means that for the A = 4-16 nuclei of the p shell all configurations s^3p^{A-3} and $s^4p^{A-5}(sd)^1$ must be used without any further restriction.

Often one works in a truncated space to keep the calculations feasible. Since in a truncated $(0+1)\hbar\omega$ space only part of the spurious states is incorporated, it is not possible to separate them exactly from the nonspurious states.

2.3 Spins of spurious states

The operator which generates the $1\,\hbar\omega$ c.m. excitations of an A-particle system is a tensor operator of rank 1 (Brussaard and Glaudemans, ref.5; Lawson, ref.6). This operator is equivalent to the isoscalar E1 operator. One finds the $1\,\hbar\omega$ spurious states at an energy of exactly $1\,\hbar\omega$ above the calculated $0\,\hbar\omega$ states and with parity opposite to that of the nonspurious $0\,\hbar\omega$ states. For each state with spin J_0 and isospin T_0 in the $0\,\hbar\omega$ space one obtains spurious states in the $(0+1)\hbar\omega$ space of which spin J_{sp} and isospin T_{sp} are given by

$$J_{sp} = J_0\ ,\ J_0 \pm 1 \text{ and } T_{sp} = T_0 . \tag{2.8}$$

This is illustrated below for $J_0 = 0^+$ and 1^+ states

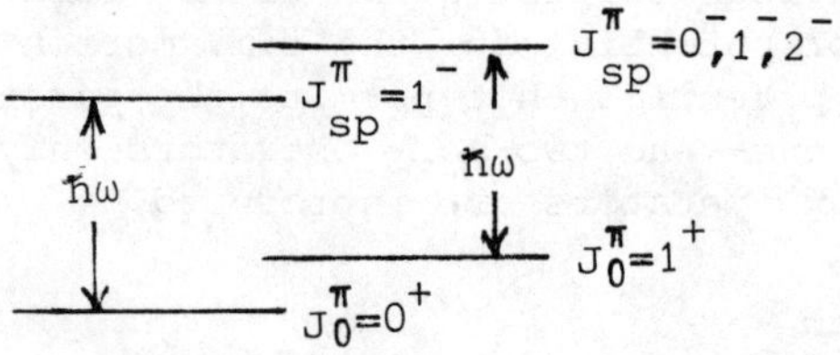

2.4 Effective interactions

Frequently used effective interactions e.g. the Surface Delta Interaction (Brussaard and Glaudemans, ref.5) or several empirically obtained interactions, are not translationally invariant. Suppose one works with j-j or L-S coupled two-body matrix elements. To obtain a translationally invariant interaction one must express the two-particle states in terms of intrinsic and c.m. states. This can be performed (once the two-particle states are given in L-S coupling) with the Brody-Moshinsky transformation coefficients (Brody and Moshinsky,ref.18). This way one can express the two-body matrix elements in terms of relative matrix elements.

In an approach where the interaction is determined empirically from a least-squares fit to the experimental data, the two-body matrix elements are considered as free parameters. The additional requirement of translational invariance may strongly reduce the number of free parameters entering an empirical interaction, since then only the relative matrix elements may be fitted.

The effect of translational invariance on the Hamiltonian parameters is illustrated by a $(0+1)\hbar\omega$ calculation on the p-shell nuclei. For these nuclei one needs in total 140 j-j or L-S coupled two-body matrix elements. Requiring translational invariance these can all be expressed in terms of only 22 relative matrix elements. Since in A = 4-16 nuclei about 150 suitable levels with well assigned J^{π} and T are known experimentally, one could never well determine the j-j or L-S coupled two-body matrix elements from these data. The 22 relative matrix can be well obtained from a least-squares fit, however.

It follows that with the procedure descibed above the spurious states are separated exactly from the nonspurious ones. In other words mixing between spurious and nonspurious states is avoided completely.

For an interaction which violates translational invariance, the effects of spurious states are more severe for light than for heavy nuclei. The mixing between spurious and nonspurious states depends on the ratio $G/\hbar\omega$, where G is the average strength of the residual interaction. This can be understood as follows. When G is large one finds, because of the (strong)effective interaction, that the spurious states at $1\ \hbar\omega$ excitation energy strongly mix with the low-lying nonspurious states. When G is small the (weak) effective interaction does not strongly mix the spurious states at $1\ \hbar\omega$ with the low-lying nonspurious states. Since one has the relations $\hbar\omega \propto A^{-1/3}$ and $G \propto A^{-1}$ (Brussaard and Glaudemans,ref.5) one obtains the ratio $G/\hbar\omega \propto A^{-2/3}$. Thus the mixing of spurious and nonspurious states decreases with mass number A.

3. APPLICATIONS TO P-SHELL NUCLEI

The p-shell nuclei are most easily suitable for a detailed microscopic analysis as follows also from Table 1 in the preceding chapter. Therefore in the past many shell-model studies have been performed on these light nuclei with mass A = 4-16. From these investigations the calculations of Cohen and Kurath, ref.15, are most well-known. Their wave functions and two-body matrix elements are often used in nuclear structure studies.

For a more extensive description of p-shell nuclei one would like to reproduce also states of nonnormal parity, however. A complete $(0+1)\hbar\omega$ calculation inclucing all A = 4-16 nuclei has been performed very recently (Van Hees, ref.13; Van Hees and Glaudemans, ref.19). The main results of these calculations will be presented, since both the energy levels as well as electromagnetic properties show several very interesting features. For example, only for calculations in which both normal- and nonnormal-parity states are taken into account investigations on E1, M2, E3 and M4 transitions are possible.

3.1 Energy levels

With the assumption of a mass-independent Hamiltonian the 22 relative matrix elements (see Sect 2.4) have been determined empirically from a least-squares fit to about 140 experimental energy levels. The value of the parameter ω, see (2.5), has been fixed by fitting the separation between normal- and nonnormal-parity states. Assuming $\hbar\omega$ to be mass independent, one finds $\hbar\omega$ = 9.55 MeV.

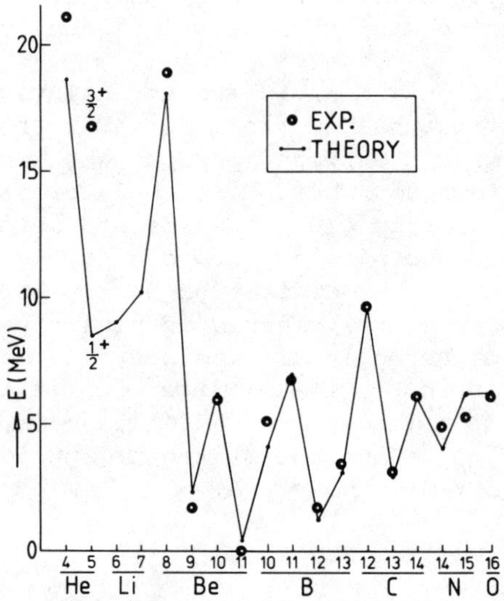

Fig.2.1 Calculated excitation energies of yrast nonnormal-parity states in p-shell nuclei and their corresponding experimental values. The energies are given with respect to experimental ground-state energies.

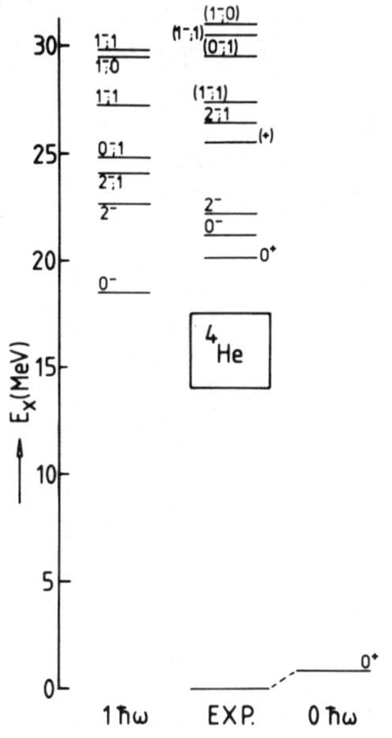

Fig.2.2 Spectrum of ^{4}He.

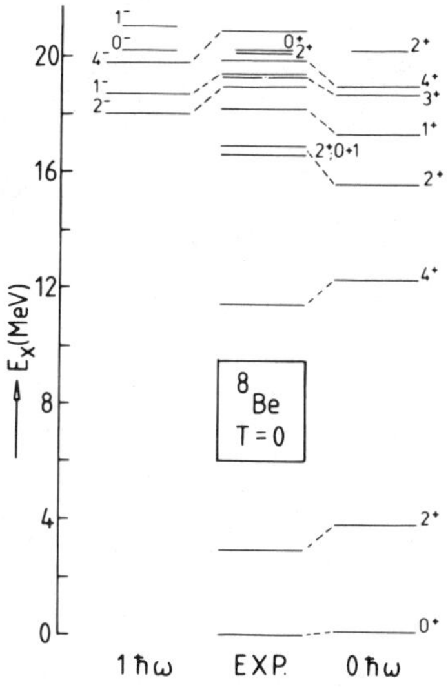

Fig 2.3 Spectrum of ^{8}Be.

As an example of the results the calculated excitation energies of the lowest lying nonnormal-parity states are compared with the avaible experimental data in Fig.2.1. It is seen that the (0+1) ħω model space very well reproduces the observed behaviour of these states. In ^{6,7}Li no nonnormal-parity states have been observed yet. In ^{5}He the lowest positive-parity state is observed at $E_x \simeq 17$ MeV and has $J^\pi = 3/2^+$. From the present shell-model calculation one expects a much lower-lying $J^\pi = 1/2^+$ state at $E_x \simeq 8.5$ MeV, however.

Some typical examples of nuclear spectra resulting from these calculations will now be discussed. An interesting example is the spectrum of ^{4}He given in Fig.2.2. None of the excited states of ^{4}He have been used to fit the interaction parameters. It follows from Fig.2.2, that the shell-model value for the gap between the ground state and the excited states corresponds quite well with the observed value, while the level density for E_x = 20-30 MeV is also reproduced. A closer inspection reveals that the excited $J^\pi = 0^+$ state at $E_x \simeq 20$ MeV is not given by the present calculations, however. This can be easily understood, since one can form a second $J^\pi = 0^+$ state only when two particles make the transition 0s → 0p

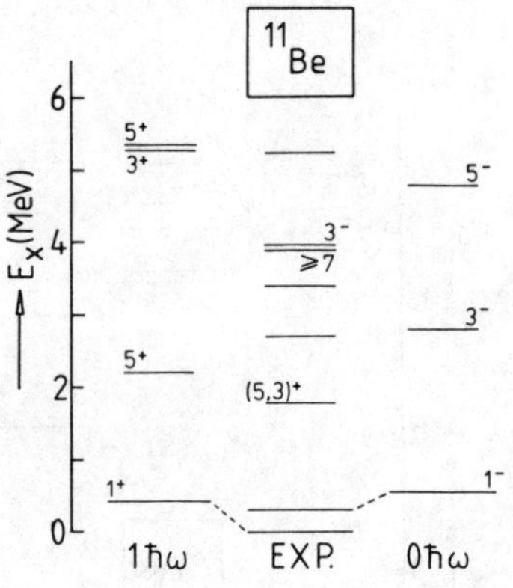

Fig.2.4 Spectrum of ^{11}Be.

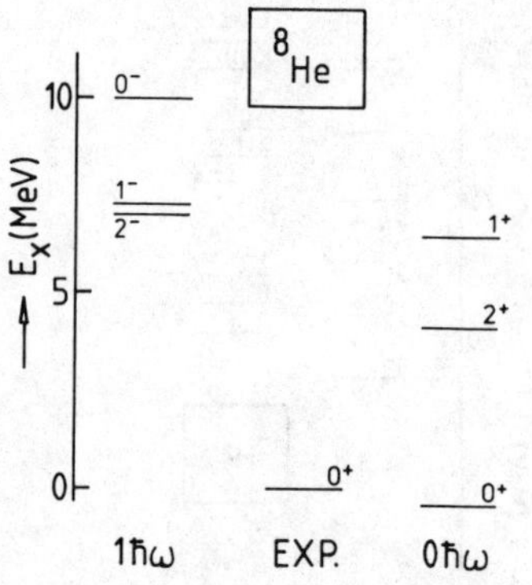

Fig.2.5 Spectrum of ^{8}He.

or by a single-particle 0s → 1s excitation. These transitions involve 2 ħω excitations but these have not been taken into account in the present calculation.

The nucleus ^{8}Be shows another feature. It is seen from Fig.2.3 that the positive- and negative-parity states with T = 0 below $E_x \simeq 20$ MeV are well accounted for. A $J^{\pi} = 2^{+}$, T = 1 state (not shown) is calculated at $E_x = 15.6$ MeV. However, the present theory cannot reproduce the observed isospin mixing of the T = 0 and T = 1 states with $J^{\pi} = 2^{+}$ at $E_x \simeq 16$ MeV. In order to calculate isospin mixing one must take into account the Coulomb interaction between the active protons explicitly.

The spectrum of ^{11}Be is presented in Fig.2.4, since this nucleus has the exceptional property that the ground state has nonnormal parity. Note that this feature is correctly reproduced by the present shell-model calculations.

Finally in Fig.2.5 the calculated spectrum of ^{8}He is shown. For this exotic nucleus (with a ratio of neutrons to protons of three to one)only the ground-state binding energy is known experimentally. The measured value $E^{b} = -31.4$ MeV agrees well with the calculated value $E^{b} = -31.6$ MeV.

3.2 Electromagnetic properties

Since many states in the nuclei studied are unbound with respect to particle emission the amount of information on electromagnetic properties is rather small. Most information is available in the heavier nuclei of the p-shell i.e nuclei with many bound states.

The operators used are specified by Brussaard and Glaudemans, ref.5. The long-wavelength approximation is assumed to be valid. No center-of-mass recoil corrections are included in the operators. For the E1 operator this correction is not necessary since the wave functions do not contain spurious admixtures. For the other electromagnetic transition operators the recoil corrections are small and are ignored in the present work.

In order to specify the matrix elements of the transition operators we take a mass-independent value for the harmonic-oscillator size parameter b = 1.77 fm. According to the relation

$$b = \sqrt{\frac{\hbar}{m\omega}}, \tag{2.9}$$

this corresponds to $\hbar\omega$ = 13 MeV. The value $\hbar\omega$ = 13 MeV has been derived from a fit of calculated charge radii to the experimental values, assuming a mass-independent value of $\hbar\omega$ (Van Hees, ref.13). From the value $\hbar\omega$ = 9.55 MeV as derived from the optimalization of spectra (see Sect.3.1) one would have obtained b = 2.07 fm. We accept this inconsistency since it may be due to the neglect of 2 $\hbar\omega$ and higher excitations. Moreover, the electromagnetic transition strengths can be easily converted into another value of b using the simple relation

$$B(EL) \sim b^{2L} \text{ and } B(ML) \sim b^{2L-2} \tag{2.10}$$

It is clear that only the strength of M1 transitions is independent of b.

For the evaluation of the magnetic transition operators the bare-nucleon g-factors are used. For the evaluation of the E2-transition operators the effective proton and neutron charges used are 1.35e and 0.35e, respectively.

A comparison between theory and experiment for the electromagnetic properties of p-shell nuclei is presented in Figs.2.6-2.12. The x-coordinate of each point denotes the shell-model value, while the y-coordinate represents the measured value. When theory and experiment agree exactly a point on the diagonal (dotted line) is obtained.

Magnetic dipole moments

The g-factors are defined by the relation $\mu = gJ$, with μ denoting the magnetic moment. It is interesting to compare the many-particle shell-model values for odd-A nuclei, with those of the extreme single-particle values. The latter, known as the Schmidt values, can simply be obtained from the expressions (Brussaard and Glaudemans, ref.5).

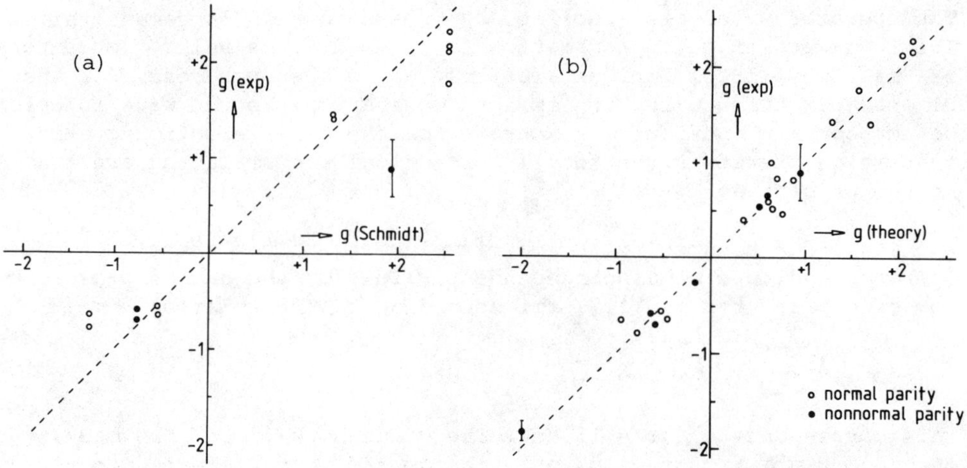

Fig.2.6 Comparison of experimental g-factors for A = 4-16 nuclei to (a) Schmidt values, (b) shell-model values.

$$g^j = \frac{(2j-1)g^\ell + g^s}{2j} \qquad \text{for } j = \ell + 1/2 \tag{2.11}$$

$$g^j = \frac{(2j+3)g^\ell - g^s}{2(j+1)} \qquad \text{for } j = \ell - 1/2 \tag{2.12}$$

with the orbital g-factors $g_p^\ell = 1$, and $g_n^\ell = 0$ and the spin g-factors $g_p^s = +5.59$ and $g_n^s = -3.83$ for the proton and neutron, respectively. This comparison is shown in Fig.2.6a for all measured g-factors in odd-A nuclei. In Fig.2.6b the values of the $(0+1)\hbar\omega$ calculation including also even-A nuclei are presented. From Fig.2.6a it follows that the measured values are usually smaller than the Schmidt values. In other words the single-particle values must be quenched. Although several mechanisms (e.g. meson exchange effects) are proposed to explain this quenching, it follows from Fig.2.6b that the more detailed shell-model calculation employing bare-nucleon g-factors very well reproduces the measured values.

Strengths of M1 transitions

In Fig.2.7 the measured M1 strengths between normal as well as nonnormal parity states are compared with the theory. Not presented in this figure are the very weak pure isoscalar M1 transitions, which are all calculated to be less than 10^{-2} W.u. in agreement with experiment. Not included in the figure are also the measured M1-transition strengths larger than 3 W.u. Also for the latter the

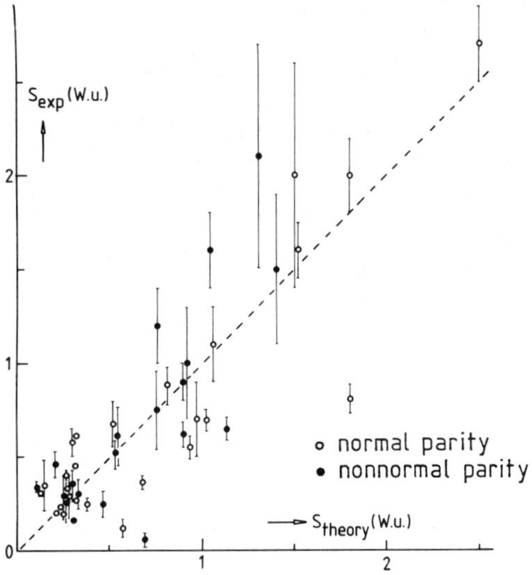

Fig. 2.7 M1 strengths in A = 4-16 nuclei.

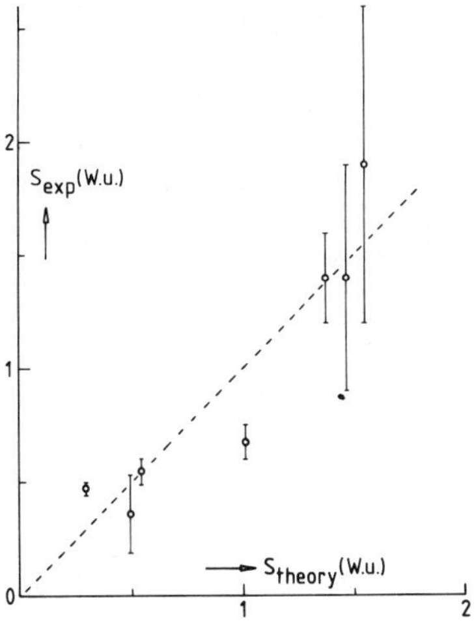

Fig. 2.8 M2 strengths in A = 4-16 nuclei.

agreement between theory and experiment is good. From Fig.2.7 one may conclude that, as found for g-factors, there is no indication for quenching, since except for a few the points are scattered around the dotted diagonal line.

Strengths of M2 transitions

The M2 operator connects states of different parity. In Fig.2.8 the measured strengths of M2 transitions larger than 0.1 W.u. are compared with the theory. Again one sees that there seems to be no need for a renormalisation of the M2 operator in p-shell nuclei.

Strengths of magnetic transitions with multipolarity larger than two

Some M3 and M4 transitions are expected in p-shell nuclei resulting from the single-particle transitions $d_{5/2} \rightarrow s_{1/2}$ and $p_{3/2} \rightarrow d_{5/2}$, respectively. Unfortunately there is no experimental information available so far to check these interesting high-multipolarity operators in some detail.

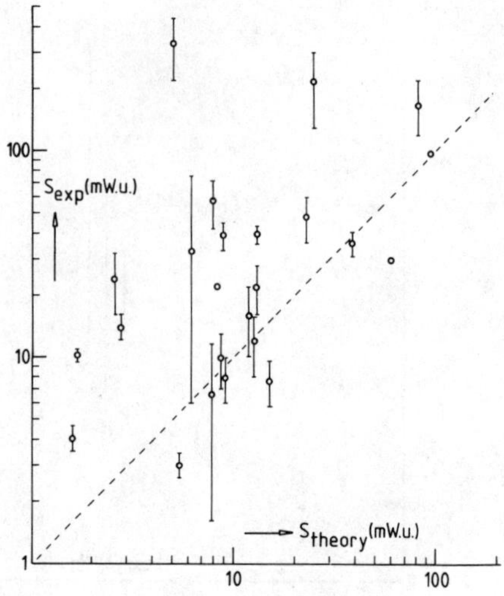

Fig. 2.9 E1 strengths in A = 4-16 nuclei.

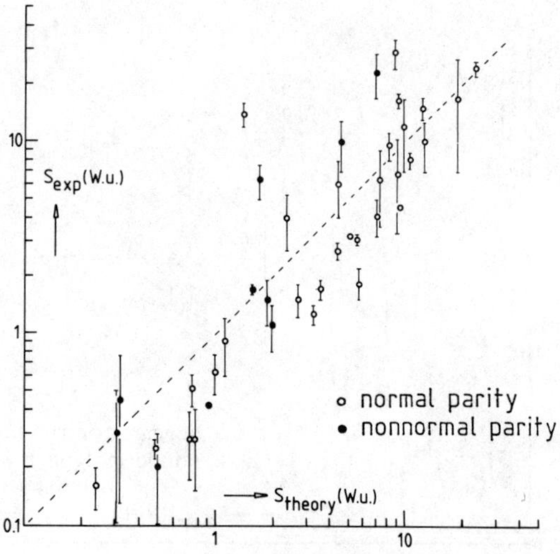

Fig. 2.10 E2 strengths in A = 4-16 nuclei.

Strengths of E1 transitions

The isoscalar part of the E1 operator vanishes exactly for transitions between states which are completely free of spuriosity (Van Hees, ref.13). Therefore only the effective isovector charge $e_p - e_n$ is of importance. We use the bare-nucleon value $e_p - e_n = 1e$. Measured E1 strengths are notoriously hard to reproduce theoretically. This is related to the fact that E1 strengths between low-lying states are usually very weak (typically $\lesssim 10^{-1}$ W.u.). One finds from calculated E1 rates that a large strength must be expected for states at rather high excitation energy. This means that because of the E1 sum-rule limit, not much is left for the strenghts between low-lying states. For an explanation of this feature in terms of favoured quadrupole deformations of the nucleus, see e.g. Lawson, ref.6.

In shell-model language the E1 strengths between low-lying states depend critically on weak components or on cancellations between contributions from strong components of the wave functions. As a result these E1 strengths are difficult to reproduce correctly. This is illustrated in Fig.2.9. Moreover, one sees that the present theory underestimates most measured strengths by roughly a factor of four. Effects of isospin mixing, not well-bound states and the neglect of $2\hbar\omega$ and higher excitations must be investigated first before this difference should be considered serious.

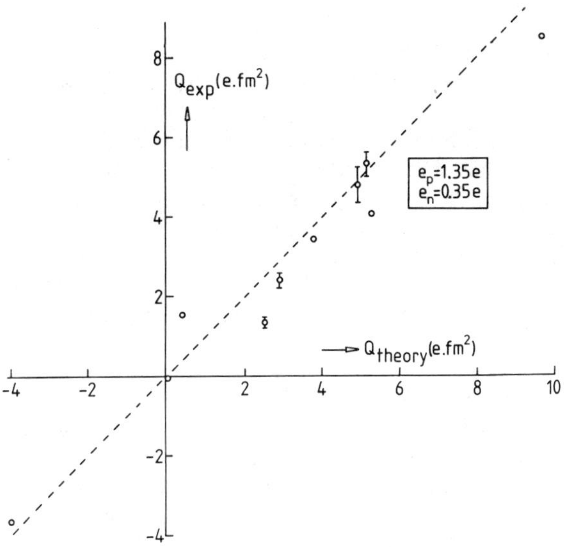

Fig.2.11 Comparison between theory and experiment for Q-pole moments in A=4-16 nuclei.

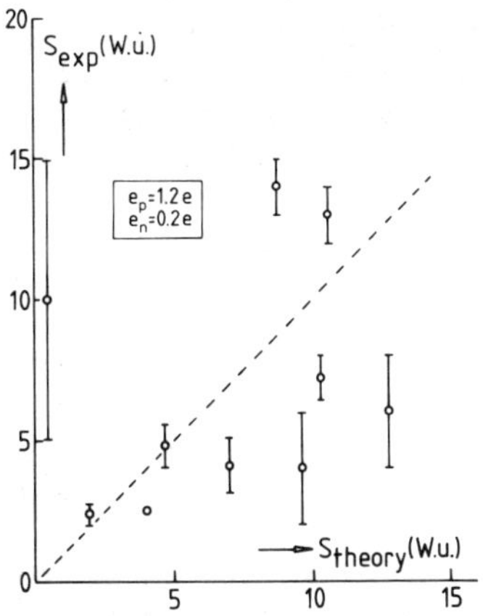

Fig. 2.12 E3 strengths in A = 4-16 nuclei.

Strengths of E2 transitions

In general the measured strengths of E2 transitions in light nuclei can be quite well reproduced by the shell-model. This is clearly illustrated in Fig.2.10, where even the very weak E2 strengths (note the log scale) agree with experiment. It is seen that the effective charges e_p = 1.35e and e_n = 0.35e on the average yield the correct values except for some large E2 rates, where $2\hbar\omega$ excitations might become important.

Quadrupole moments

The calculated and measured Q-pole moments are compared with each other in Fig.2.11. It should be noted that the latter are known for normal-parity states only. It follows that the theory well reproduces most values with the effective charge Δe = 0.35e.

Strengths of E3 transitions

Of the matrix elements that can be compared with experiment, it is that of the E3 operator that depends most critically on the radial wave functions as follows (Brussaard and Glaudemans,ref.5) from the r^6 dependence, see also (2.10). The comparison with present data,taking the effective charge Δe = 0.2e,is shown in Fig.2.12. It is not clear that radial wave functions different from those of the harmonic oscillator should have been taken. The quite accurately measured strengths that deviate most from the theoretical values derive from $3^- \rightarrow 0^+$ transitions in ^{12}C and ^{16}O.

4. SUMMARY AND CONCLUSIONS

Shell-model wave functions always contain an unphysical part, which corresponds to the center-of-mass motion of the nucleus. This may lead to spurious effects when calculating various observables. In this lecture it is shown that under certain conditions the spurious effects can be isolated exactly from the nonspurious ones. The method has been applied to calculate various observables in p-shell nuclei. The translationally-invariant effective interaction has been obtained from a least-squares fit to the data. Because of the requirement of translational invariance the Hamiltonian can be expressed in a rather small number of parameters.

It is interesting to remark that with present day computing techniques the normal-parity states in the p-shell nuclei can even be treated in a full $2\hbar\omega$ model space. The largest Hamiltonian matrix then has order 1897 and the number of relative matrix elements would be 50. A comparison of such a description of the A = 4-16 nuclei with experiment would make a very critical test of the nuclear shell model feasible.

There are some recent speculations that in many-quark systems like nuclei, the quarks could form clusters that differ from the usual protons and neutrons. It is not ruled out that their effects can be detected experimentally. For the analysis of such experiments one may also need reliable nuclear wave functions. In this respect the above mentioned further improvements in the shell-model description may make p-shell nuclei suitable candidates for investigations in this yet unexplored field of nuclear-structure research.

ACKNOWLEDGEMENTS

The literature referred to is by no means complete. On purpose, references are made to books and review papers as much as possible. I am grateful to my collaborators drs. A.G.M. van Hees, B.C. Metsch, R.B.M. Mooy and D. Zwarts for their help in obtaining the material discussed in these lectures and to Mark Glaudemans for typing the manuscript.

REFERENCES

1. A. Bohr, and B. R. Mottelson, Nuclear Structure II, W. A. Benjamin, Inc. Reading, Mass. (1975).
2. F. Iachello, Nuclear Structure, K. Abrahams, K. Allaart, A. E. L. Dieperink, eds., Plenum Press, New York and London (1980).
3. R. B. M. Mooy, P. W. M. Glaudemans, and A. G. M. van Hees, Phys. Lett. 104B:251 (1981).

4. A. deShalit, and I. Talmi, Nuclear Shell Theory, Academic Press, New York, (1963).
5. P. J. Brussaard, and P. W. M. Glaudemans, Shell-model applications in nuclear spectroscopy, North Holland Publ. Co., Amsterdam (1977).
6. R. D. Lawson, Theory of the nuclear shell model, Clarendon Press, Oxford (1980).
7. B. C. Metsch, and P. W. M. Glaudemans, Nucl. Phys. A352:60 (1981).
8. P. M. Endt, Atomic Data and Nucl. Data Tables 23 (1979).
9. A. deShalit, and H. Feshbach, Theoretical Nuclear Physics, John Wiley and Sons, New York (1974).
10. A. Watt, D. Kelvin, and R. R. Whitehead, Nucl. Phys. 6:35 (1980).
11. A. deShalit, and M. Goldhaber, Phys. Rev. 5:1211 (1953).
12. J. P. Elliott, and T. H. R. Skyrme, Proc. Roy. Soc. A232:561 (1955).
13. A. G. M. van Hees, Thesis Utrecht University (1982).
14. J. B. Aviles, Ann. of Phys. 50:393 (1968).
15. S. Cohen, and D. Kurath, Nucl. Phys. 73:1 (1965).
16. J. B. McGrory, and B. H. Wildenthal, Ann. Rev. Nucl. Part. Sci. 30:383 (1980).
17. R. R. Whitehead, A. Watt, B. Cole, and I. Morrison, Advances in Nucl. Phys. 9, New York (1977).
18. T. A. Brody, and M. Moshinsky, Tables of transformation brackets, Monografias del Instituto de Fisica (1960).
19. A. G. M. van Hees, and P.W.M. Glaudemans, Nucl. Phys. to be published (1982).

ENERGY OF A NUCLEON IN A NUCLEUS

Claude Mahaux

Institut de Physique B5
Université de Liège au Sart Tilman
B-4000 Liège 1, Belgium

1. INTRODUCTION

In the <u>independent particle model</u> (IPM), the nuclear ground state wave function ϕ_A is a Slater determinant built with A single-particle wave functions $\varphi_\alpha(r)$, where α generically denotes the orbit quantum numbers n,ℓ,j,m . For simplicity, we restrict the discussion to spherical nuclei and we shall usually omit explicit reference to the quantum number m . A consequence of the IPM is that the configurations

$$\phi_{A-1}^{\alpha} = a_\alpha|\phi_A> \quad , \tag{1.1}$$

$$\phi_{A+1}^{\alpha} = a_\alpha^\dagger|\phi_A> \tag{1.2}$$

are eigenstates of the Hamiltonian H for $A-1$ and $A+1$ nucleons, respectively. Equations (1.1) and (1.2) imply a symmetry property of H . The main purpose of these lectures is to discuss to what extent this symmetry holds in nature. The amount of symmetry breaking depends upon the choice of the basis $\{\varphi_\alpha\}$ and upon the particular nuclei which are considered. We shall limit our discussion to some main features and to a few examples. Further information can be found in review articles, e.g. [1-10].

In Sect. 2, we introduce some concepts in the idealized case of infinite nuclear matter. We then turn to actual nuclei. In Sects. 3-6, we successively consider single-particle states which are at positive energy, or are weakly, deeply or semi-deeply bound. Section 7 describes some recent theoretical progress on the improvement of the independent particle model.

2. NUCLEAR MATTER

We only introduce a few definitions and concepts. For more detail, the reader is referred to [11]. In nuclear matter, translational invariance leads to the choice of plane waves, $\exp(i\vec{k}.\vec{r})$, for the basis states $\varphi_\alpha(\vec{r})$; we omit reference to spin and isospin. The momentum $\vec{k}$ plays the same role as the quantum numbers $\alpha = (n,\ell, j,m)$ in a finite nucleus. In the IPM, the ground state wave function ϕ_o is constructed by filling in all momentum states in keeping with the Pauli principle, up to a maximum value k_F called the <u>Fermi momentum</u>. The density is given by

$$\rho = \frac{2}{3\pi^2} k_F^3 \quad . \tag{2.1}$$

Within the IPM, one still has the possibility of attaching a potential energy $U(k)$ to each nucleon. The nucleon energy is then given by $(\hbar = 1)$

$$e_o(k) = k^2/2m + U(k) \quad . \tag{2.2}$$

The energy of the least bound nucleon is called the <u>Fermi energy</u>. In the IPM, it is given by

$$\varepsilon_F^0 = k_F^2/2m + U(k_F) \quad . \tag{2.3}$$

In reality, nucleons do interact. In the ground state of nuclear matter, momentum states with $k < k_F$ are therefore not fully occupied, and momentum states with $k > k_F$ are not totally empty. This is sketched in Fig. 2.1. Correlations also have the consequence that a nucleon with momentum k does not have a well defined energy. This is described in terms of the <u>spectral function</u> $S(k;\omega)$. The quantity $S(k;\omega)d\omega$ measures the joint probability of being able to remove from or to add to the system a nucleon with momentum k and to find that

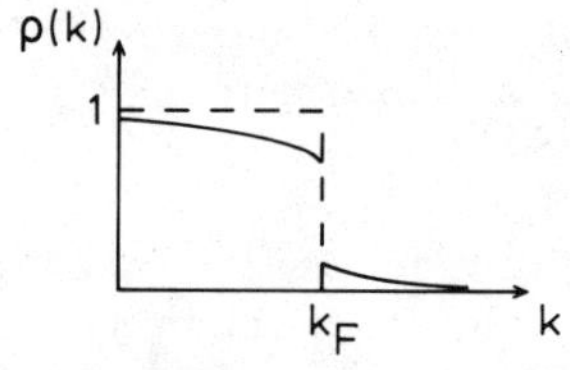

Fig. 2.1. Momentum distribution in the ground state of nuclear matter The dashed line corresponds to the independent particle approximation and the solid line to the actual correlated ground state.

the residual system has an energy contained in the interval $(\omega,\omega+d\omega)$.

In the IPM, eq. (2.2) shows that the spectral function is a δ-function

$$S_o(k;\omega) = \delta(\omega-e_o(k)) \quad . \tag{2.4}$$

This is no longer true when correlations are taken into account. If the symmetry implied by the IPM is not too badly broken, the spectral function $S(k;\omega)$ should have a peak in the vicinity of $e_o(k)$ if $U(k)$ had been suitably chosen. A typical shape of $S(k;\omega)$ is sketched in Fig. 2.2. Note that $S(k;\omega)$ differs from zero for all values of ω . The following sum rule holds

$$\int_{-\infty}^{\infty} d\omega\, S(k;\omega) = 1 \quad . \tag{2.5}$$

The spectral function can be expressed in terms of the mass operator (or self-energy) $M(k;\omega)$. This is a complex quantity

$$M(k;\omega) = V(k;\omega) + i\, W(k;\omega) \quad , \tag{2.6}$$

which fulfills the dispersion relation

$$V(k;\omega) = \pi^{-1} \int_{-\infty}^{\infty} d\omega' \frac{W(k;\omega')}{\omega-\omega'} \quad . \tag{2.7}$$

One has

$$S(k;\omega) = \pi^{-1} \frac{W(k;\omega)}{[\omega-k^2/2m - V(k;\omega)]^2 + [W(k;\omega)]^2} \quad . \tag{2.8}$$

Let us write

$$S(k;\omega) = S^{QP}(k;\omega) + S^{BG}(k;\omega) \quad , \tag{2.9}$$

where

$$S^{QP}(k;\omega) = \pi^{-1} \frac{Z^2(E(k))\; W(E(k))}{[\omega-E(k)]^2 + [Z\; W(E(k))]^2} \tag{2.10}$$

with

$$E(k) = k^2/2m + V(E(k)) \quad , \tag{2.11}$$

$$V(E(k)) = V(k;E(k)) \quad ; \quad W(E(k)) = W(k;E(k)) \quad , \tag{2.12}$$

$$Z(E(k)) = \{1 - \partial V(k;E)/\partial E\}^{-1}_{E=E(k)} \quad . \tag{2.13}$$

If the IPM is a good approximation, the "background" $S^{BG}(k;\omega)$ is

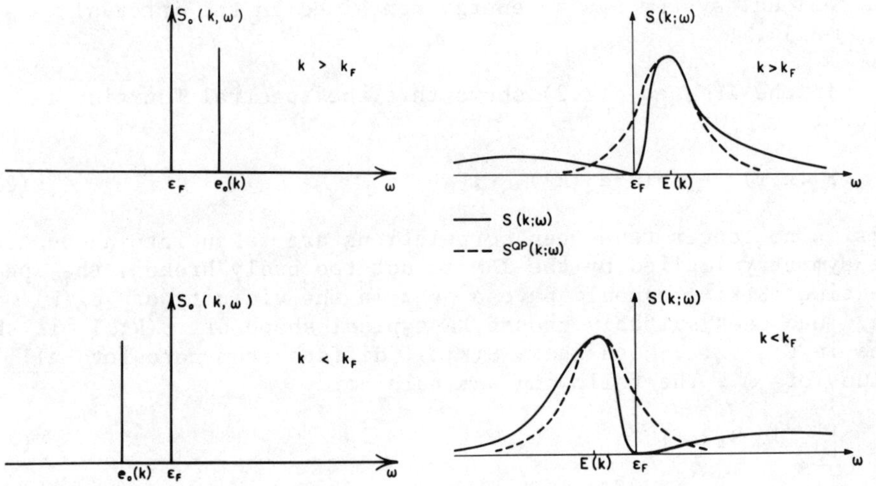

Fig. 2.2. Adapted from [12]. The left-hand side schematically represents the spectral function in the case of the independent particle model. The solid line on the right-hand side indicates the modifications introduced by the nucleon-nucleon collisions, while the dashed curve corresponds to the quasiparticle approximation.

a smooth function of ω . The quasiparticle approximation consists in replacing $S(k;\omega)$ by $S^{QP}(k;\omega)$, see Fig. 2.2. It amounts to assuming that when a nucleon with momentum k is removed from the ground state, the energy of the residual system has a probability distribution given by a Lorentzian centered on the "quasiparticle energy" $E(k)$, with a full width at half-maximum equal to

$$\Gamma^{\downarrow}(E(k)) = 2\, Z(E(k))\, W(E(k)) \quad . \tag{2.14}$$

The comparison of eq. (2.5) with the relation

$$\int_{-\infty}^{\infty} d\omega\; S^{QP}(k;\omega) = Z(E(k)) \quad , \tag{2.15}$$

indicates that the value of Z is a measure of the accuracy of the quasiparticle approximation. We emphasize, however, that the quantities $E(k)$ and $\Gamma^{\downarrow}$ may have a physical meaning even if the background S^{BG} is not small, provided that it is a smooth function of ω : the main criterion for the usefulness of the quasiparticle concept is that $S(k;\omega)$ should have a pronounced peak.[13]

In the IPM it is convenient to choose the potential energy $U(k)$ as follows

$$U(k) = V(k;E(k)) \quad . \tag{2.16}$$

Then indeed, one has $E(k) = e_o(k)$. In addition, the IPM Fermi energy ε_F^0 then becomes identical to the exact Fermi energy ε_F , which is defined by the relation

$$\varepsilon_F = k_F^2/2m + V(k_F;\varepsilon_F) \quad . \tag{2.17}$$

The full curves in Fig. 2.2 show that the exact spectral function vanishes at $\omega=\varepsilon_F$. This is due to the following asymptotic behaviour of $W(k;\omega)$:

$$W(k;\omega) \sim C(\omega-\varepsilon_F)^2 \qquad (C > 0) \quad . \tag{2.18}$$

In the case of nucleon-nucleus scattering the quasiparticle approximation is equivalent to the <u>optical model</u>. The complex quantity

$$M(E) = V(E) + i\,W(E) \tag{2.19}$$

can be identified with the optical-model potential for a nucleon with energy E .[11] Below, we shall indifferently use the expressions "quasiparticle approximation" and "optical model" for negative as well as positive energies. We note that the quantity $M(E)$ only depends upon one variable, while the mass operator (2.6) depends upon the two variables k and ω . The variable k has been eliminated from $M(E)$ by making use of the energy-momentum relation (2.11). It can be shown that this amounts to the construction of a local complex potential $M(E)$ which is "equivalent" to the nonlocal potential $M(k;E)$.[11] We emphasize that this "equivalence" only extends to the quasiparticle properties. Indeed, eq. (2.11) has a physical meaning only as far as the description of the bump in the spectral function is concerned. The description of the full spectral function $S(k;\omega)$ requires the complex mass operator $M(k;\omega)$.

3. SINGLE-PARTICLE STATES AT POSITIVE ENERGY

In an infinite system, $M_\rho(E)$ depends upon the density ρ . In a finite nucleus, the density ρ is a function $\rho(r)$ of the distance r between the nucleon and the centre of the target. One can then tentatively use the "local density approximation"

$$M(r;E) = M_{\rho(r)}(E) \tag{3.1}$$

to construct the nucleon-nucleus optical-model potential.[14] The approximation (3.1) is not accurate at low energy since it does not take into account the dynamical effects associated with surface vibrations. Here we only use eq. (3.1) as a convenient way of visualizing the relationship between the concepts introduced in the case of nuclear matter on the one hand and the more complicated case of finite nuclei on the other hand.

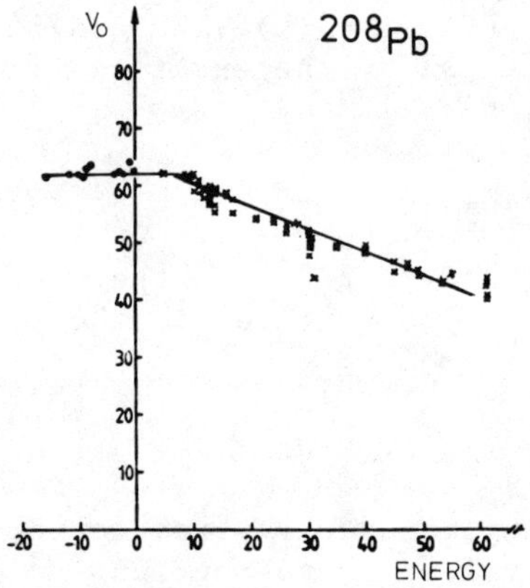

Fig. 3.1. Taken from [15]. *Dependence upon energy of the modulus of the depth of the Woods-Saxon nucleon-nucleus potential as determined from the analysis of p-*^{208}*Pb elastic scattering* $(E > 0$, *crosses) and from the location of weakly bound single-particle level* $(E < 0$, *full dots).*

The analyses of elastic and total cross sections enable one to determine the real part $V(r;E)$ and the imaginary part $W(r;E)$ of the optical-model potential. Excellent fits to the scattering data are obtained by assuming that $V(r;E)$ has a Woods-Saxon shape whose depth V_o is a linear function of E in the domain $10 < E < 60$ MeV . This is illustrated in Fig. 3.1, from which one gets

$$V_o(E) \approx V_o - 0.35\,E \quad , \quad 10\text{ MeV} < E < 60\text{ MeV} \quad , \tag{3.2}$$

with $V_o \approx -65$ MeV ; the latter value depends upon the geometrical properties of the potential well. The linear law (3.2) has been confirmed by analyses of neutron or proton scattering by many targets; the value of the slope is determined within about ten percent.

4. WEAKLY BOUND SINGLE-PARTICLE STATES

In the phenomenological IPM, the ground state wave function ϕ_o is constructed with single-particle wave functions φ_α calculated from a Woods-Saxon (or harmonic oscillator) central potential well supplemented by a spin-orbit component.[16] The corresponding single-particle energies ε_α^0 are represented in Fig. 4.1, which is only schematic in the sense that the ordering of the single-particle energies within one main shell is not fitted to experiment. Let us for definiteness focus on the neutrons in the doubly magic nucleus ^{208}Pb . If the IPM is valid, one should obtain an eigenstate of ^{209}Pb by adding to ^{208}Pb a neutron in a single-particle orbit. Experimentally, this can be achieved by direct stripping reactions. The example of the ^{208}Pb(d,p)^{209}Pb reaction is shown in Fig. 4.2.

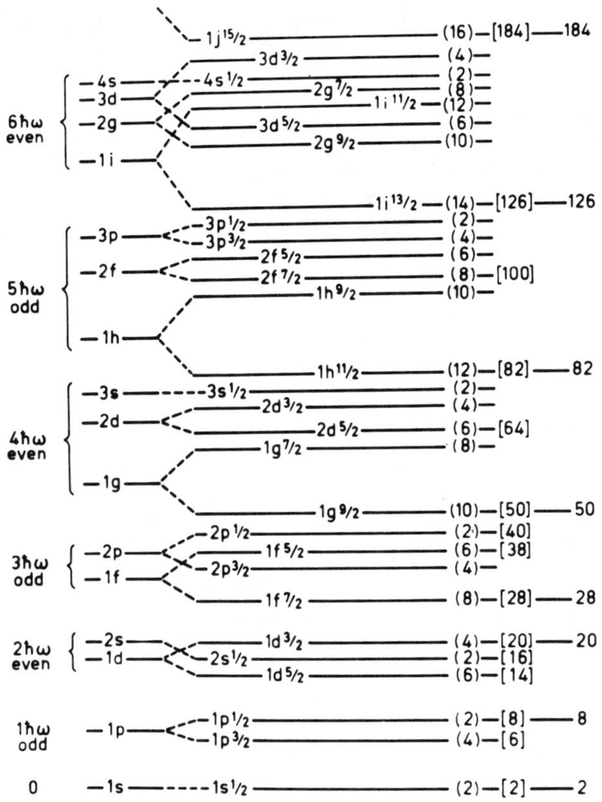

Fig. 4.1. Taken from [16]. Distribution of the single-particle energies among shells in the phenomenological independent particle model.

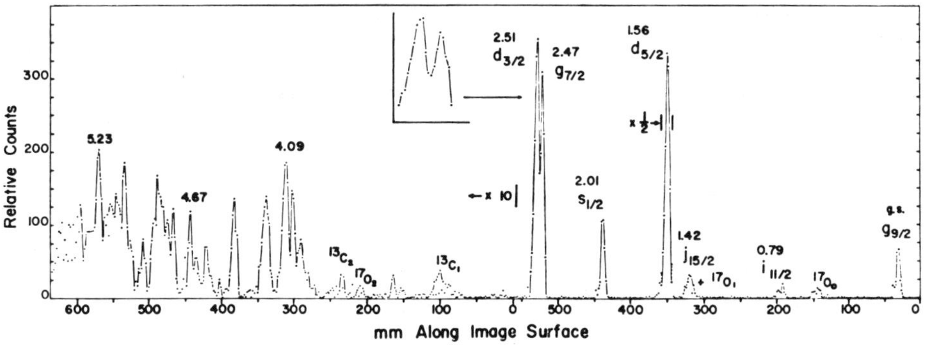

Fig. 4.2. Taken from [17]. Spectrum of the outgoing protons observed in the $^{208}Pb(d,p)^{209}Pb$ stripping reaction, at a deuteron energy of 20.1 MeV and an outgoing angle of 35°.

The spectrum of the outgoing protons is compatible with the assumption that the residual nucleus can be approximated by a configuration in which one would have added a neutron in the $2g_{9/2}$ (ground state of ^{209}Pb), $1i_{11/2}$, $1j_{15/2}$, $3d_{5/2}$, $4s_{1/2}$, $2g_{7/2}$ and $3d_{3/2}$ orbits. The ℓ and j quantum numbers are identified by the angular distribution of the outgoing protons, and the principal quantum number n by the size of the cross section. A check of these assignments is obtained by exciting the isobaric analog resonances which correspond to these single-particle levels, via the ^{208}Pb(p,p') ^{208}Pb* inelastic proton scattering. Within 20 keV, these two types of experiments yield the same values for the single-particle energies.[18]

Single-particle energies which correspond to occupied neutron orbits in the IPM can be measured via direct pickup reactions, like ^{208}Pd(d,t)^{207}Pb, see Fig. 4.3, or ^{208}Pb(^{3}He,α)^{207}Pb.[19-23] Recent

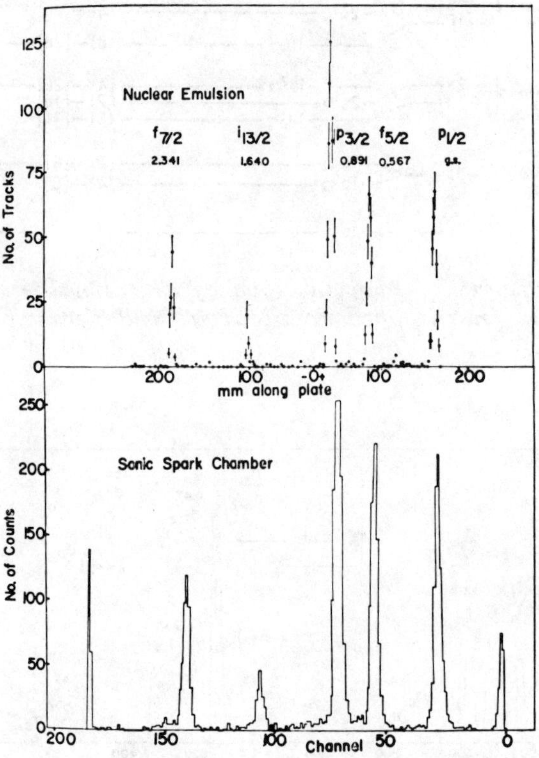

Fig. 4.3. Taken from [17]. Spectrum of outgoing tritons observed with nuclear emulsions (top) or with a sonic spark chamber (bottom) in the ^{208}Pb(d,t)^{207}Pb pickup reaction, at a deuteron energy of 24.8 MeV and an outgoing angle of 30°.

experimental data[21,22] indicate that the narrow peaks in Fig. 4.3 do not exhaust the $i_{13/2}$ and the $h_{9/2}$ strengths; this will be discussed in Sect. 7.

Similarly, the proton single-particle energies can be measured via proton pickup or stripping reactions. In this case too, it appears that in the case of ^{208}Pb the IPM is quite a good approximation for weakly bound single-particle states, i.e. for those states which belong to the main shell which lies either just above or just below the Fermi energy ε_F. The latter will henceforth be identified with the average between the separation energy ($\varepsilon_{p1/2} = \varepsilon_F^-$ in the case of neutrons) and the addition energy ($\varepsilon_{2g9/2} = \varepsilon_F^+$) :

$$\varepsilon_F = \frac{1}{2}(\varepsilon_F^+ + \varepsilon_F^-) \quad . \tag{4.1}$$

One has $\varepsilon_F \approx -6$ MeV in the case of ^{208}Pb .

The experimental values of the single-particle energies for weakly bound neutron states in ^{208}Pb are shown in the third column in Fig. 4.4. They are in good agreement with the values in the fourth column, which are computed from a Woods-Saxon potential whose depth

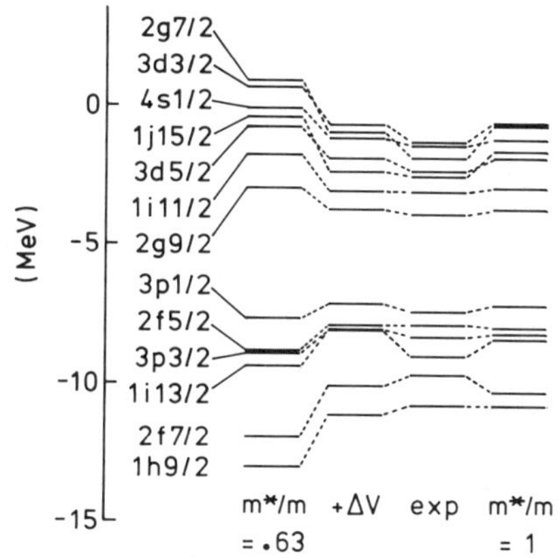

Fig. 4.4. Adapted from [24]. Energies of weakly bound single-neutron states in ^{208}Pb. The first column from the left is calculated from an IPM with a potential well whose depth depends linearly upon energy. The second column includes an estimate of the level shift ΔV due to corrections to the IPM (Sect. 7). The third column shows the experimental values and the fourth column is obtained from an IPM with a potential well whose depth is independent of energy.[25]

is <u>independent of energy</u>. If one uses a potential whose depth has the linear energy dependence found in the analysis of nucleon scattering, see eq. (3.2), one obtains the values shown in the first column of Fig. 4.4. This calculated spectrum is too much spread out as compared to the experimental data. This is illustrated by the difference $\delta = \bar{\varepsilon}_{unoc} - \bar{\varepsilon}_{oc}$ between the (2j+1)-average unoccupied and occupied single-particle energies :

$$\bar{\varepsilon}_{oc} = [\sum_{\ell,j\ oc} (2j+1)\ \varepsilon^0_{\ell j}]\ [\sum_{\ell,j\ oc} (2j+1)]^{-1} \quad . \tag{4.2}$$

The empirical value of δ is 6.50 MeV , while the values calculated from an energy-dependent potential depth would yield δ = 8.34 MeV . A potential with an energy-independent depth would yield δ = 7.07 MeV (fourth column in Fig. 4.4).

Figure 3.1 shows that the same feature also holds in the case of protons : one needs an energy-dependent depth to fit the proton scattering data and an energy-independent depth to reproduce the energy of the weakly bound single-particle proton states. This striking feature was emphasized by Brown, Gunn and Gould,[26] who characterized it in terms of the effective mass m^* . This quantity is defined by

$$m^*(E)/m = 1 - \frac{dV_o(E)}{dE} \quad , \tag{4.3}$$

where $V_o(E)$ is the depth of a Woods-Saxon potential with fixed geometry and which reproduces the single-particle properties of nuclei. The results shown in Figs. 3.1 and 4.4 can be expressed by writing that

$$m^*(E)/m \approx 1 \quad \text{for} \quad -8\ \text{MeV} < E-\varepsilon_F < +8\ \text{MeV} \tag{4.4}$$

$$m^*(E)/m \approx 0.65 \quad \text{for} \quad 15\ \text{MeV} < E-\varepsilon_F < 70\ \text{MeV} \quad . \tag{4.5}$$

Hence, the effective mass $m^*(E)/m$ in ^{208}Pb drops rapidly from unity to 0.65 when $E-\varepsilon_F$ increases from 0 to 15 MeV . While this drop appears quite striking, we note that the phenomenon is not very spectacular in terms of single-particle energies. Indeed, it corresponds to energy shifts of only one or two MeV. This is nevertheless of practical importance, since a 2 MeV difference in the particle-hole gap δ implies a shift of the same magnitude for all the excited states of ^{208}Pb which are built from the lowest particle-hole excitation.[26]

The energy dependence of the effective mass could be exhibited in the case of ^{208}Pb , because the IPM seems to be quite accurate for about fifteen proton and neutron weakly bound states, so that a systematic trend can be found. It is not clear on empirical grounds whether the same phenomenon holds for open-shell of for lighter nuclei

5. DEEPLY BOUND SINGLE-PARTICLE STATES

In order to detect deeply bound single-particle states, one uses direct knockout reactions induced by a high energy projectile, in particular the (p,2p) and (e,e'p) reactions. In principle, both of them enable the determination of the spectral function $S(k;E)$ as a function of the two variables k and E. Figure 5.1 shows the example of ^{16}O. The reason why S can be found as a function of the two variables k and E lies in the existence of three bodies in the outgoing channel. Indeed, the kinematics can then be chosen in such a way that k and E are varied independently. Let $\vec{k}'$ and E' denote the momentum and the energy of the outgoing proton, and $\vec{e}'$ and e' (resp. $\vec{e}$ and e) the momentum and energy of the outgoing (resp. incoming) electron. Momentum and energy conservation yield

$$\vec{k} = \vec{e}' - \vec{e} + \vec{p}' \quad , \tag{5.1}$$

$$E = e' + E' - e \quad , \tag{5.2}$$

where we omitted recoil corrections. Here, $\vec{k}$ is the momentum of the hole state and E its energy.

The analysis is in practice marred by distortion effects, which render the spectral function nondiagonal in momentum $(S(\vec{k},\vec{k}';E))$.[28] Distortion effects are more important in the case of the (p,2p) reaction,[29] but on the other hand the (e,e'p) cross section is four order of magnitude smaller. Pion induced knockout reactions have recently been used.[30,31]

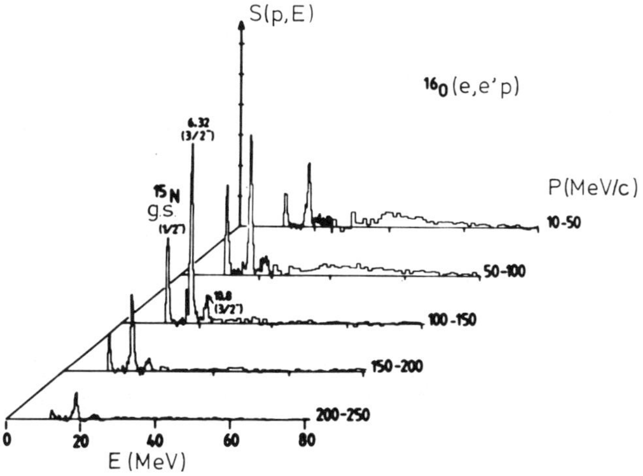

Fig. 5.1. Taken from [27]*. Energy dependence of the spectral function S(p;E) for several momentum bins.*

Let $\varphi_\alpha(p)$ denote the Fourier transform of $\varphi_\alpha(r)$. The expansion

$$S(p;E) \approx \sum_{\alpha\ oc} P_\alpha(E)|\varphi_\alpha(p)|^2 \qquad (5.3)$$

yields an approximation for the energy distribution of the hole state with quantum numbers α . If the quasiparticle approximation is accurate, the quantity $P_\alpha(E)$ should display a peak of width Γ_α near the quasiparticle energy E . This has been observed for practically all the levels α for mass numbers up to about 60.

We shall see in Sect. 7 that the quasiparticle energies E_α are usually calculated in the Hartree-Fock approximation, possibly with some corrections. Figure 5.2 shows that they depend rather smoothly upon the mass number. The quasiparticle width Γ_α is more difficult to evaluate, since it requests a description of the coupling between the single-particle and the other degrees of freedom. Detailed microscopic calculations lead to a complicated spectral function[32] which is not readily comparable to the knockout data. This had led some authors to calculate the spectral function S(k;E) or the width $\Gamma(E)$ in the limiting case of nuclear matter, and to adapt these results to the case of finite nuclei by using some local density and energy approximation. The success of the first attempt of this kind[33] may have been accidental[34,35] but had the merit of stimulating further work along these lines.

On the basis of the dilute hard sphere Fermi gas model, Orland

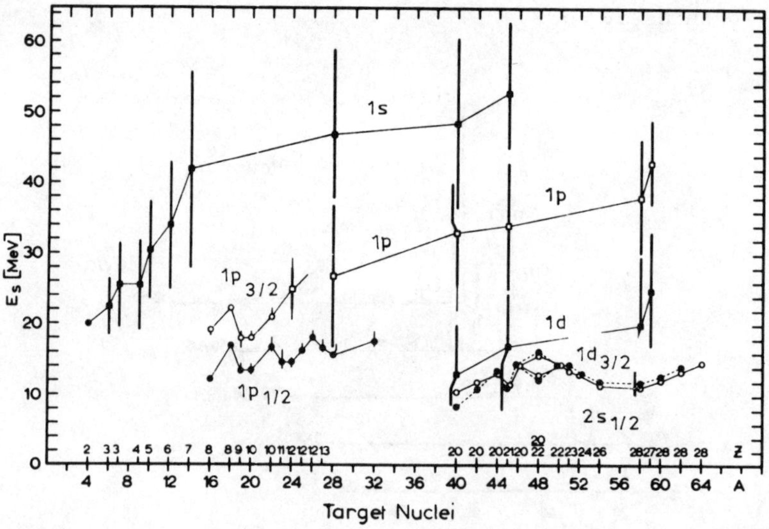

Fig. 5.2. Taken from [13]. Dependence of the quasiparticle energy upon mass number. The thick vertical bars represent the width Γ, and the thin bars the estimated systematical uncertainties.

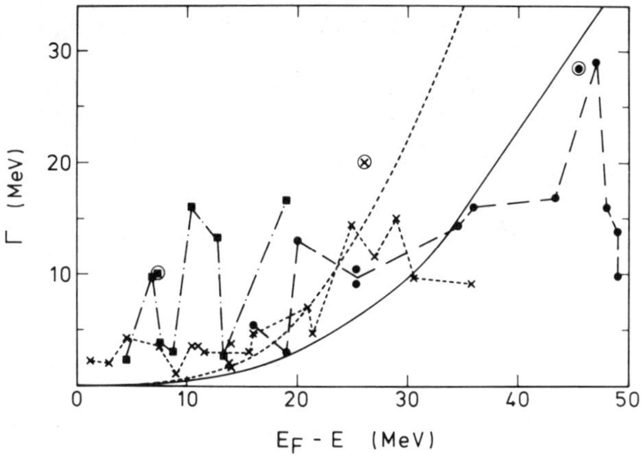

Fig. 5.3. Taken from [37]*. Dependence upon the Fermi energy* (E_F) *and the quasiparticle energy* (E) *of the empirical width,*[3] *for the single-particle states 1s (full dots), 1p (crosses) and 1d (full squares). The solid line is obtained from the Brueckner-Hartree-Fock approximation and the dotted curve from the renormalized Brueckner-Hartree-Fock approximation.*

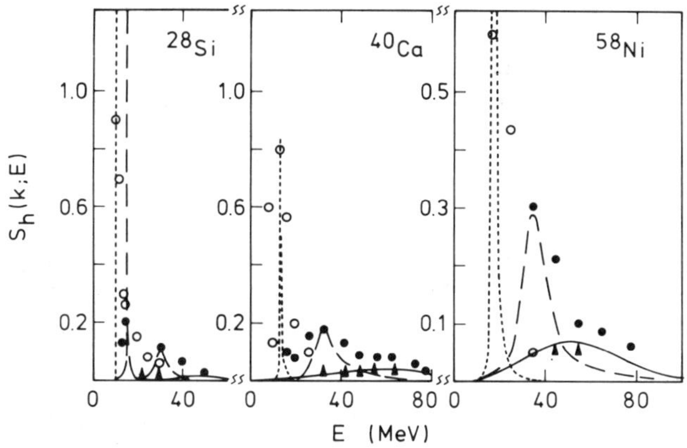

Fig. 5.4. Taken from [37]*. Comparison between the spectral function* $S(k;E) \approx P_\alpha(E)$ *as calculated from eq. (2.8) within the Brueckner-Hartree-Fock approximation on the one hand and the empirical values on the other hand, for the single-particle states 1s (triangles and full curves), 1p (full dots and long dashes) and 1d (open circles and short dashes). In the case of* ^{28}*Si, the two dashed curves correspond to the* $1p_{3/2}$ *and* $1p_{1/2}$ *orbits.*

and Schaeffer[36] suggested that the width Γ_α should mainly depend upon the difference between the quasiparticle energy E_α and the Fermi energy ε_F . Figure 5.3 indicates that this proposal is fairly well fulfilled, except for the weakly bound (d-) levels in light nuclei. Figure 5.3 moreover shows that the measured values are in semi-quantitative agreement with those derived from the Brueckner-Hartree-Fock approximation. This led the authors of ref. [37] to use this approximation in order to calculate the quantity $P_\alpha(E) \approx S(k;E)$ from eq. (2.8). The results are shown in Fig. 5.4, where each energy E_α and the ordinate scale have been fitted to the experimental values.

Antonov et al.[38,39] recently also adopted a nuclear matter approach to calculate the spectral function. Strangely, their "coherent fluctuation model" does not use any nucleon-nucleon interaction. The agreement between their model and the experimental data is about the same as that shown in Fig. 5.4. This is quite intriguing because their model contains no dynamical input other than the experimental density distribution.

6. SEMI-DEEPLY BOUND SINGLE-PARTICLE STATES

The distinction between deeply, weakly and semi-deeply bound states is somewhat arbitrary. Here, we shall call "semi-deeply bound" a single-particle configuration which lies at an excitation energy at which the density of more complicated configurations with the same angular momentum and parity is larger than several levels per MeV. Then, the residual interaction couples the single-particle to the more complicated configurations. If the quasiparticle approximation (2.10) holds, the distribution of the single-particle strength as observed with a direct pickup or a stripping reaction has a peak. The location of its maximum can be identified with the quasiparticle energy, and the full width at half-maximum with the single-particle spreading width. Figure 6.1 illustrates the fact that the weakly bound single-particle states in ^{208}Pb appear as quasiparticles in the open shell nucleus ^{205}Pb , in which case we might call them "semi-deeply bound".

In the case of a semi-deeply bound quasiparticle state, the distribution of the single-particle strength thus remains localized in energy but is spread over several or many complicated states. This spreading can be described by using the same techniques as in the case of intermediate structure. When the density of complicated states is not too large, an explicit calculation can be performed. An example is shown in Fig. 6.2. Although there is no one-to-one agreement between measured and calculated spectroscopic factors, the quasiparticle properties (energy and spreading width) are correctly reproduced. In the theoretical model ten percent of the single-particle strength lies higher than the energy domain covered by Fig. 6.2. If one were to include this tail and to describe the calculated

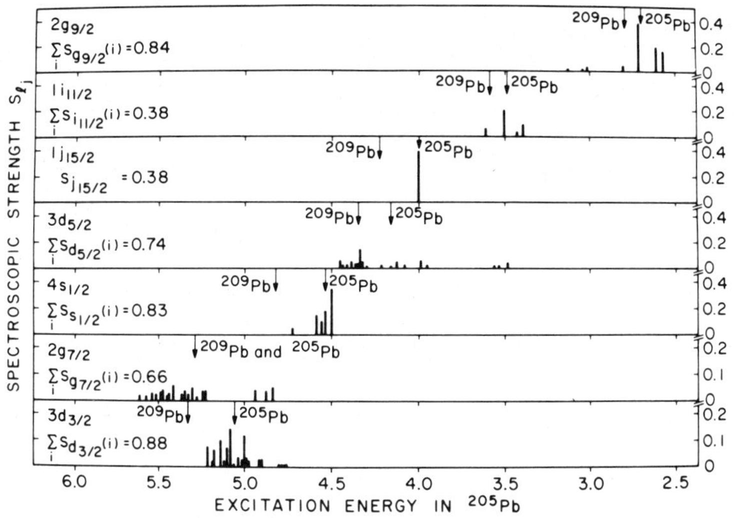

Fig. 6.1. Taken from [40]. *Distribution of the single-particle strength in* ^{205}Pb *as measured from the* $^{204}Pb(d,p)$ *stripping reaction. The location of the centroid is indicated by an arrow, as well as the location of the corresponding single-particle state in* ^{209}Pb *.*

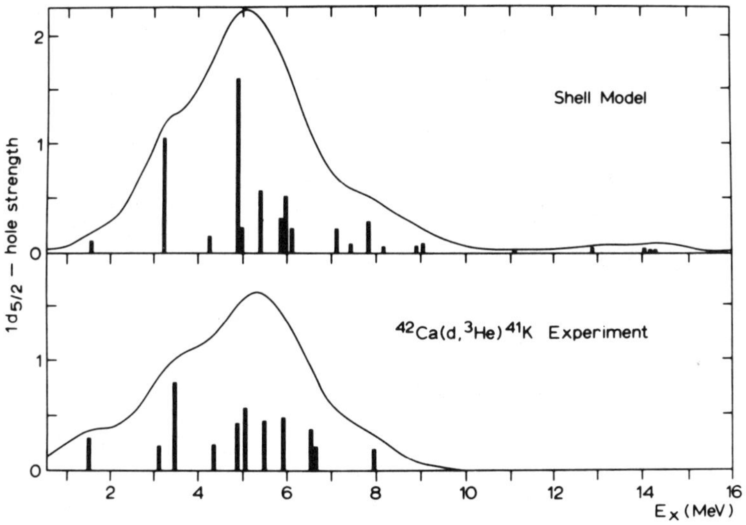

Fig. 6.2. Taken from [4]. *Comparison between the spectroscopic factors of the* $d_{5/2}$ *proton hole in* ^{42}Ca *as measured experimentally (bottom) and as calculated (top, ref.* [41]*) in a two particle-one hole configuration space. The curves are obtained by folding the discrete spectrum with a Gaussian.*

distribution in terms of its first and second moment, one would obtain quite different values for the quasiparticle energy and, mainly, for the spreading width. This illustrates that quasiparticle properties should only be associated with the peak of the strength distribution, and the difficulty of making practical use of the concept of an apparently attractive definition of the single-particle energy in terms of a "mean removal energy".[42]

When the mass number increases, the quasiparticle states which lie one shell away from the Fermi surface (the "next inner shell") overlap, and it becomes difficult to disentangle them experimentally. Much effort has been devoted to this problem during the last few years. In particular, the distributions of the $1g_{9/2}$, $2p_{1/2}$, $1f_{5/2}$ and $2p_{3/2}$ hole strength in the Sn isotopes have very much been investigated both experimentally[43-51] and theoretically.[44,52-55] Let us for instance consider the $1g_{9/2}$ single-particle state. The initial measurements[44,46] indicated that roughly about one-half of the full $g_{9/2}$ strength lies in a quasiparticle peak, represented

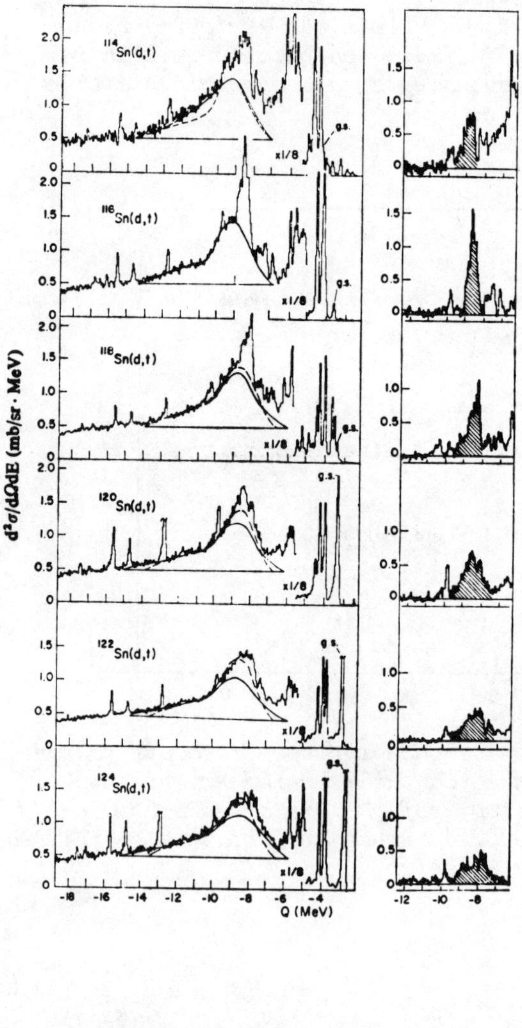

6.3. *Taken from* [46]. *The left-hand side represents the dependence upon Q-value of the triton spectra as observed at 15° in the (d,t) reaction on several Sn isotopes. The shaded area on the right-hand side shows that part of the spectra which has been identified as arising from the pickup of a* $1g_{9/2}$ *neutron.*

by the hatched areas in Fig. 6.3. Recent $(\vec{d},t)$[50,51] and $(\vec{p},d)$[43] measurements with polarized beams show that the "missing" strength had earlier in part been associated with the $1f_{5/2}$ configuration and in part lies at slightly higher excitation energy. This high energy tail contains roughly 30 % of the $g_{9/2}$ strength. It should probably not be viewed as belonging to the quasiparticle peak but should rather be associated with the background contribution S^{BG} to the spectral function, see eq. (2.9). In the present case, the quasiparticle strength Z in eq. (2.15) would then be roughly equal to 0.7 . Microscopic calculations can hardly aim at describing the background, because they necessarily adopt a truncated configuration space. In the spirit of the quasiparticle approximation, the theory should mainly attempt to reproduce the two main features of the bump, namely its energy and its width. We note that most of the $1f_{7/2}$ neutron hole strength, which lies yet half-a-shell deeper, has recently been observed in a $(^3He,\alpha)$ pickup reaction with a 283 MeV ^3He beam.[23]

The selectivity allowed by the use of a polarized beam is confirmed by Fig. 6.4, which exhibits the $1f_{7/2}$ proton quasihole peak in ^{89}Y . The $1f_{7/2}$ neutron quasihole had been detected in the $^{90}Zr(^3He,\alpha)^{89}Zr$ pickup reaction.[19] Neutron pickup reactions have also been performed to locate semi-deeply bound neutron quasiholes in the Pd isotopes,[57] in some Nd isotopes,[58] and in ^{143}Sm.[19,58,59] Proton quasiholes have been observed in some Pm isotopes.[56,60] Recently, the $^{144}Sm(d,t)$ and $^{144}Sm(^3He,d)$ reactions[61] have been able to detect the $h_{9/2}$ and $i_{13/2}$ proton quasiparticles in the next "outer" shells above the Fermi energy.

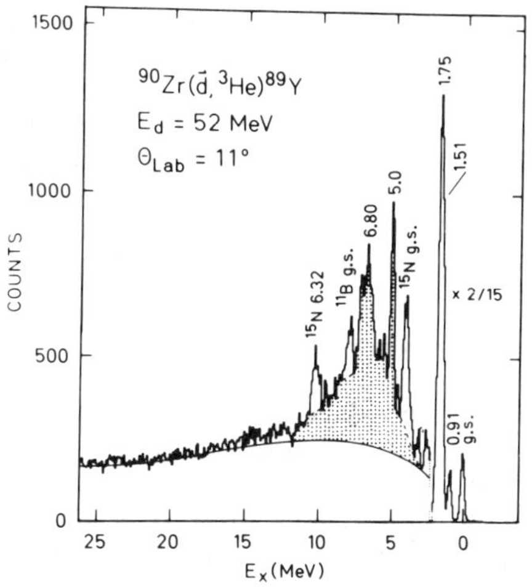

Fig. 6.4. Taken from [56]. *Yield versus excitation energy in* ^{89}Y *in the* $^{90}Zr(d,{}^3He)^{89}Y$ *pickup reaction. The shaded area contains the proton* $f_{7/2}$ *quasihole strength.*

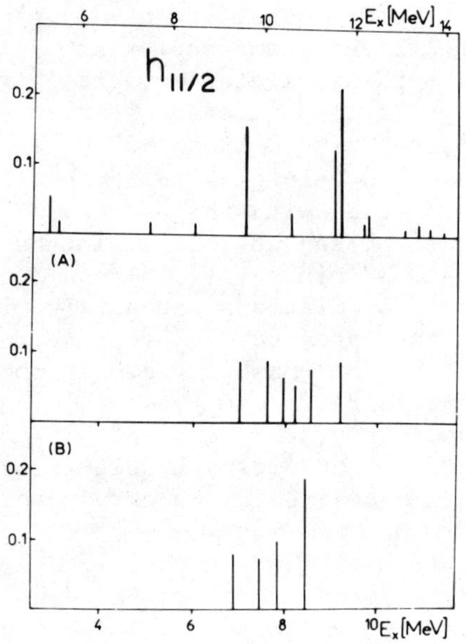

Fig. 6.5. Taken from [9]*. Comparison between the calculated distribution of the* $h_{11/2}$ *neutron strength in* ^{207}Pb *(upper part, ref.* [62]*) and the experimental distribution as given in ref.* [21] *(middle part) and in ref.* [22] *(lower part), respectively.*

The $^{208}Pb(^{3}He,\alpha)$ pickup reaction is particularly successful in locating quasiholes with high angular momentum. It has been able to find part of the strength of the semi-deeply bound $h_{11/2}$ quasihole[19-22] and even of the yet more deeply bound $1g_{7/2}$ and $1g_{9/2}$ configurations. Figure 6.5 shows that the measured quasiparticle properties are in good agreement with a theoretical calculation in which the Hartree-Fock single-particle configurations are coupled to more complicated configurations in the framework of the self-consistent random phase approximation.[62,63]

7. SINGLE-PARTICLE POTENTIAL

In the independent particle model, each nucleon occupies a stationary orbit and the energy dependence of the spectral function reduces to a delta function. The average field in which the nucleons move is real; it can be identified with the Hartree-Fock potential. In practice, the latter includes the effect of virtual (i.e. energy nonconserving) collisions which lead to highly excited states. Indeed, the main effect of these consists in renormalizing the nucleon-nucleon interaction and in replacing it by an "effective" interaction. The latter only smoothly depends upon energy and upon density if only high-energy virtual excitations are included in the renormalization. This effective interaction can be calculated from the free nucleon-nucleon potential.[64] One can also adopt a phenomenological effective interaction whose parameters are adjusted in order to reproduce the main properties of nuclear matter and the static properties of nuclear ground states (density distribution, binding energy). The Skyrme-type effective interactions are popular because they are quite successful[65] while leading to Hartree-Fock equations which are very simple since they only involve the self-consistent nuclear density. More precisely, the radial part of the Hartree-Fock single-particle wave function in a doubly-closed shell nucleus fulfills an equation of the type

$$\frac{1}{2m}\left[-\frac{d^2}{dr^2}+\frac{\ell(\ell+1)}{r^2}+V_{HF}(r;\varepsilon_\alpha)\right]u_\alpha(r;\varepsilon_\alpha) = E_\alpha u_\alpha(r;\varepsilon_\alpha) \quad (7.1)$$

where ε_α is the Hartree-Fock single-particle energy. The Skyrme-Hartree-Fock potential $V_{HF}(r;E)$ depends <u>linearly</u> upon E :

$$V_{HF}(r;E) = V_o(r) + \left[1-\frac{\tilde{m}(r)}{m}\right]E \quad . \quad (7.2)$$

The Hartree-Fock "effective mass" $\tilde{m}(r)$ (see eq. (4.3)) has the radial dependence

$$\tilde{m}(r)/m = [1+B\rho(r)]^{-1} \quad , \quad (7.3)$$

where $\rho(r)$ is the nuclear density while B is a constant determined by the parameters of the Skyrme interaction. For the so-called Skyrme III interaction, $\tilde{m}(r=0)/m \approx 0.7$. Hence, the corresponding Hartree-Fock potential has approximately the same energy dependence as the real part of the empirical potential at <u>positive</u> energy, see eq. (3.2). This leads to the single-particle energies shown in the left-hand column of Fig. 4.4; the corresponding average particle-hole energy gap δ is approximately 2 MeV larger than the empirical value, see Sect. 4.

The practical importance of this discrepancy has led to a renewed theoretical effort to evaluate dynamical corrections to the Hartree-Fock energies. The physical origin of these corrections was understood many years ago,[66,67] but it is only recently that detailed microscopic calculations have been carried out.[63,68-70] Here, we only present the more phenomenological approach of refs.[71,72]. It has the merit of being simple and of yielding an estimate of the radial dependence of the correction to the Hartree-Fock potential well. In contradistinction, the other approaches focus upon the calculation of the single-particle energy shifts; on the other hand, they have the merit of disentangling the various contributions to the energy shifts; they are also able to check the accuracy of the main assumption made in refs. [71,72], namely that the correction to the Hartree-Fock potential is approximately local.

If the latter assumption holds, the average single-particle field can be written in the form

$$V(r;E) = V_{HF}(r;E) + \Delta V(r;E) \quad . \tag{7.4}$$

The correction ΔV is connected to the imaginary part of the optical-model potential by the dispersion relation

$$\Delta V(r;E) = \pi^{-1} \int_{-\infty}^{\infty} \frac{W(r;E')}{E - E'} dE' \quad , \tag{7.5}$$

see eq. (2.7). For positive E', $W(r;E')$ is known from the analysis of nucleon scattering data. In order to calculate the right-hand side of eq. (7.5), one also needs $W(r;E')$ for negative energies E'. In medium-light nuclei, W at negative energy can be estimated from the spreading width $\Gamma^{\downarrow}$, see eq. (2.14); these data suggest that it is reasonable to assume that $W(r;E')$ takes symmetric values about the Fermi energy.[71]

This approach has recently been applied to neutron states in ^{40}Ca and in ^{208}Pb.[72] It appears from these calculations that $\Delta V(r;E)$ is peaked at the nuclear surface and very much depends upon energy in the vicinity of the Fermi energy. These two features correspond to the fact that $\Delta V(r;E)$ mainly originates from the coupling between single-particle and the collective vibrational degrees of freedom.[63,67-70,73,74] Relatedly, it is not justified to interpret the difference between the first and the third column in Fig. 4.4 as reflecting the existence of a plateau in the potential depth, as was made in Fig. 3.1. Rather, the volume integral of $V(r;E)$ has a plateau because the surface geometry of $V(r;E)$ depends upon energy. This is illustrated in Fig. 7.1.

From eqs. (4.3), (7.2) and (7.4), one obtains

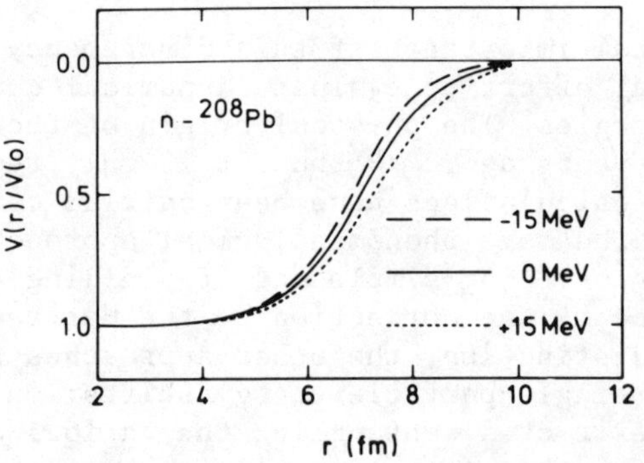

Fig. 7.1. Taken from [72]. *Radial shape of the single-particle field* $V(r;E)$ *for* $E-\varepsilon_F = -15$, 0 *and* 15 *MeV, respectively.*

$$1 - \frac{m^*(r;E)}{m} = \frac{\overline{m}(r;E)}{m} - \frac{\tilde{m}(r)}{m} \quad , \tag{7.6}$$

where the "E-mass" $\overline{m}(r;E)$ is defined by

$$\frac{\overline{m}(r;E)}{m} = 1 - \frac{d}{dE}\,[\Delta V(r;E)] \quad . \tag{7.7}$$

Figure 7.2 shows that at the nuclear surface the energy dependence of the E-mass displays a rather narrow peak centered on the Fermi energy. For fixed $|E-\varepsilon_F| \approx$ several MeV , the radial dependence of $\overline{m}(r;E)$ displays a peak at the nuclear surface.[24] Several implications of these findings have recently been conjectured.[26,70,75-80]

The second column in Fig. 4.4 shows the quantities

$$E_\alpha = \varepsilon_\alpha + \Delta V_\alpha \quad , \tag{7.8}$$

where the level shifts ΔV_α are evaluated from

$$\Delta V_\alpha = \langle\varphi_\alpha|\Delta V(r;\varepsilon_\alpha)|\varphi_\alpha\rangle \quad ; \tag{7.9}$$

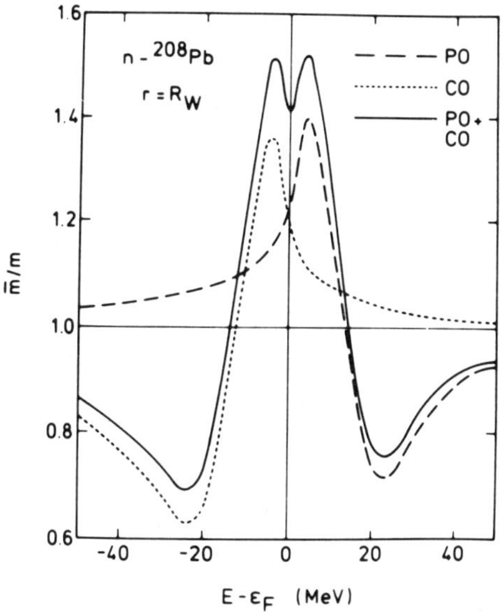

Fig. 7.2. Taken from [72]. The full curve represents the dependence upon $E-\varepsilon_F$ of $\overline{m}(R_W;E)/m$, where R_W approximately corresponds to the nuclear radius. In the domain $E-\varepsilon_F < 0$, the long dashes give the inverse of the probability that normally occupied single-particle states are emptied by the ground state correlations. Conversely, the difference from unity of the short dashes for $E-\varepsilon_F > 0$ gives the probability that a normally unoccupied level is actually occupied because of the ground state correlations.

φ_α denotes the Skyrme-Hartree-Fock single-particle wave function. The good agreement between E_α and the empirical single-particle energies is illustrated by the fact that the calculated average particle-hole energy gap δ is equal to 6.25 MeV and is thus close to the empirical value $\delta = 6.50$ MeV .

The main physical origin of the level shift ΔV_α lies in the coupling between the Hartree-Fock single-particle configurations and core excitations. The latter necessarily lie at higher excitation energy than the Hartree-Fock single-particle state when the latter is close to the Fermi energy. This leads to a widening of the energy separation between the particle + core excited states on the one hand and the level which contains most of the single-particle strength on the other hand. The amount of single-particle strength contained in the empirical quasiparticle state is approximately equal to $Z_\alpha = (\overline{m}/m)^{-1}$, see eq. (2.13). Figure 7.2 shows that $Z_\alpha \approx 0.75$ near the Fermi energy. The rest of the single-particle strength must lie in higher excitation energy ans is thus hard to locate experimentally. Nevertheless, Fig. 7.3 shows that recent

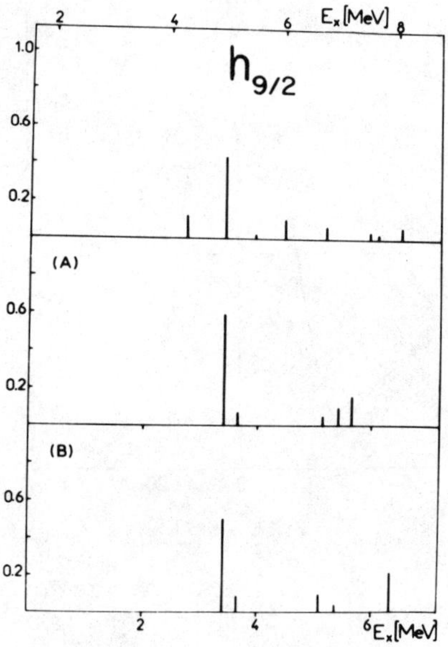

Fig. 7.3. Taken from [9]. Comparison between the calculated distribution of the $h_{9/2}$ single-particle strength (ref. [62], upper part) and the experimental distribution as given in refs. [21] (middle part) and [22] (lower part), respectively.

experiments have been able to establish that part of the $i_{13/2}$ strength indeed lies at higher excitation energy and is not contained in a single state as Fig. 4.3 suggested. The same applies to the other single-particle states of Figs. 4.2 and 4.3.

8. CONCLUSIONS

A better understanding of the validity, of the limitation and of the way of improving the independent particle model has been achieved during the recent years. It now appears feasible to construct a single-particle potential which would include some "dynamical content"[81] and would describe the average properties of deeply, semi-deeply, and weakly bound quasiparticle states as well as of unbound scattering states.

REFERENCES

1. D.F. Jackson, in "Advances in Nuclear Physics", vol. 4, M. Baranger and E. Vogt, eds., Plenum Press (1971).
2. G.J. Wagner, in "Nuclear Structure Physics", U. Smilansky, I. Talmi and H.A. Weidenmüller, eds., Springer Verlag (1973) p. 16.
3. G. Jacob and T.A.J. Maris, Rev.Mod.Phys. 45:6 (1973).
4. G.J. Wagner, in "Proceedings of the International Symposium on Highly Excited States in Nuclei", vol. 2, A. Faessler, C. Mayer-Böricke and P. Turek, eds., (1975) p. 177.
5. A.E.L. Dieperink and T. de Forest, Ann.Rev.Nucl.Sci. 25:1 (1975).
6. P.E. Hodgson, Rep.Prog.Phys. 38:847 (1975).
7. T.A.J. Maris, in "Nuclear and Particle Physics at Intermediate Energies", J.B. Warren, ed., Plenum Publ. Corp. (1976) p. 425.
8. G.M. Crawley, in "Proceedings of the International Conference on the Structure of Medium-Heavy Nuclei", Rhodes, Greece, 1979, Institute of Physics, Bristol and London (1980) p. 127.
9. S. Galès, Nucl.Phys. A354:193c (1981).
10. G.F. Bertsch, P.F. Bortignon and R.A. Broglia, Revs.Mod.Phys. (in press).
11. J.P. Jeukenne, A. Lejeune and C. Mahaux, Phys.Reports 25C:83 (1976).
12. P.F. Bortignon, R.A. Broglia, C.H. Dasso and C. Mahaux, in preparation.
13. H.A. Weidenmüller, in "Nuclear Structure Physics", U. Smilansky, I. Talmi and H.A. Weidenmüller, eds., Springer Verlag (1973).
14. C. Mahaux, in "Microscopic Optical Potentials, H.V. von Geramb, ed., Springer Verlag (1979) p. 1.
15. M. Bauer, E. Hernandez-Saldana, P.E. Hodgson and J. Quintanilla, Journ.Phys. G8:525 (1982).
16. M.G. Mayer and J.H.D. Jensen, "Elementary Theory of Nuclear Shell Structure", Wiley, New York (1955).
17. G. Muehllehner, A.S. Poltorak, W.C. Parkinson and R.H. Bassel, Phys.Rev. 159:1039 (1967).

18. W.R. Wharton, P. von Brentano, W.K. Dawson and P. Richard, Phys.Rev. 176:1424 (1968).
19. E. Gerlic, J. Källne, H. Langevin-Joliot, J. van de Wiele and G. Duhamel, Phys.Lett. 57B:338 (1975).
20. J. van de Wiele, E. Gerlic, H. Langevin-Joliot and G. Duhamel, Nucl.Phys. A297:61 (1978).
21. S. Galès, G.M. Crawley, D. Weber and B. Zwieglinski, Phys.Rev. C18:2475 (1978).
22. J. Guillot, J. van de Wiele, H. Langevin-Joliot, E. Gerlic, J.P. Didelez, G. Duhamel, G. Perrin, M. Buenerd and J. Chauvin, Phys.Rev. C21:879 (1980).
23. H. Langevin-Joliot, E. Gerlic, J. Guillot, M. Sakai, J. van de Wiele, A. Devaux, P. Force and G. Landaud, Phys.Lett. 114B:103 (1982).
24. C. Mahaux and H. Ngô, Physica Scripta (in press).
25. J. Blomqvist and S. Wahlborn, Arkiv Fysik 16:545 (1960).
26. G.E. Brown, J.H. Gunn and P. Gould, Nucl.Phys. 46:598 (1963).
27. J. Mougey, Nucl.Phys. A335:35 (1980).
28. S. Boffi, in "Nuclei to Particles", A. Molinari, ed., North-Holland Publ. Comp., Amsterdam (1981), p. 373, and references therein.
29. D.F. Jackson, Nucl.Phys. A257:221 (1976).
30. E. Piasetzky, A. Altman, J. Lichtenstadt, A.I. Yavin, D. Ashery, W. Bertl, L. Felawka, H.K. Walter, F.W. Schlepütz, R.J. Powers, R.G. Winter and J.v.d. Pluym, Phys.Lett. 114B:414 (1982).
31. N.S. Chant, L. Rees and P.G. Roos, Phys.Rev.Lett. 48:1784 (1982).
32. W. Fritsch, R. Lipperheide and U. Wille, Nucl.Phys. A241:79 (1975).
33. H.S. Köhler, Nucl.Phys. 88:529 (1966).
34. R. Sartor, Nucl.Phys. A267:29 (1976).
35. R. Sartor, Nucl.Phys. A289:329 (1977).
36. H. Orland and R. Schaeffer, Nucl.Phys. A299:442 (1978).
37. R. Sartor and C. Mahaux, Phys.Rev. C21:2613 (1980).
38. A.N. Antonov, V.A. Nikolaev and I.Zh. Petkov, Z.Phys. A304:239 (1982).
39. A.N. Antonov, V.A. Nikolaev and I.Zh. Petkov, Z.Phys. A297:257 (1980).
40. C.F. Maguire, D.G. Kovar, W.D. Callender and C.K. Bockelman, Phys.Rev. C15:170 (1977).
41. S.T. Hsieh, K.T. Knöpfle and G.J. Wagner, Nucl.Phys. A254:141 (1975).
42. D.S. Koltun, Phys.Rev.Lett. 28:182 (1972); Phys.Rev. C9:484 (1974).
43. G.M. Crawley, J. Kasagi, S. Galès, E. Gerlic, D. Friesel and A. Bacher, Phys.Rev. C23:1818 (1981).
44. S.Y. van der Werf, B.R. Kooistra, W.H.A. Hesselink, F. Iachello, L.W. Put and R.H. Siemssen, Phys.Rev.Lett. 33:712 (1974).
45. M. Sakai, M. Sekiguchi, F. Soga, Y. Hirao, K. Yagi and Y. Aoki, Phys.Lett. 51B:51 (1974).

46. S.Y. van der Werf, M.N. Harakeh, L.W. Put, O. Scholten and R.H. Siemssen, Nucl.Phys. A289:141 (1977).
47. G. Berrier-Ronsin, G. Duhamel, E. Gerlic, J. Kalifa, H. Langevin-Joliot, G. Rotbard, M. Vergnes, J. Vernotte and K.K. Seth, Phys.Lett. 67B:16 (1977).
48. M. Tanaka, T. Yamagata, K. Iwamoto, S. Kishimoto, B. Saeki, K. Yuasa, T. Fukuda, I. Miura, K. Okada, M. Inoue and H. Ogata, Phys.Lett. 78B:221 (1978).
49. E. Gerlic, G. Berrier-Ronsin, G. Duhamel, S. Galès, E. Hourani, H. Langevin-Joliot, M. Vergnes and J. van de Wiele, Phys.Rev. C21:124 (1980).
50. R.H. Siemssen, W.P. Jones, W.W. Jacobs, C.C. Foster, D.W. Miller and F. Soga, Phys.Lett. 103B:323 (1981).
51. G. Perrin, G. Duhamel, C. Perrin, E. Gerlic, S. Galès and V. Comparat, Nucl.Phys. A356:61 (1981).
52. T. Koeling and F. Iachello, Nucl.Phys. A295:45 (1978).
53. V.G. Soloviev, Ch. Stoyanov and A.I. Vdovin, Nucl.Phys. A342:261 (1980).
54. P.F. Bortignon and R.A. Broglia, Nucl.Phys. A371:405 (1981).
55. S.P. Klevansky and R.H. Lemmer, Phys.Rev. C25:3137 (1982).
56. A. Stuirbrink, G.J. Wagner, K.T. Knöpfle, Liu Ken Pao, G. Mairle, H. Riedesel, K. Schindler, V. Bechtold, L. Friedrich, Z.Phys. A297:307 (1980).
57. O. Scholten, M.N. Harakeh, J. van der Plicht, L.W. Put, R.H. Siemssen and S.Y. van der Werf, Nucl.Phys. A348:301 (1980); H. Sakai, R.K. Bhowmik, K. van Dijk, A.G. Drentje, M.N. Harakeh, Y. Iwasaki, R.H. Siemssen, S.Y. van der Werf and A. van der Woude, Phys.Lett. 103B:309 (1981).
58. M. Sekiguchi, Y. Shida, F. Soga, T. Hattori, Y. Hirao and M. Sakai, Phys.Rev.Lett. 38:1015 (1977).
59. E. Friedland, M. Goldschmidt, C.A. Wiedner, J.L.C. Ford, Jr. and S.T. Thornton, Nucl.Phys. A256:93 (1976).
60. P. Doll, G.J. Wagner, H. Breuer, K.T. Knöpfle, G. Mairle and H. Riedesel, Phys.Lett. 82B:357 (1979).
61. S. Galès, C.P. Massolo, S. Fortier, E. Gerlic, J. Guillot, E. Hourani, J.M. Maison, J.P. Schapira, B. Zwieglinski, P. Martin and V. Comparat, Phys.Rev.Lett. 48:1593 (1982).
62. Nguyen Van Giai, in "Proceedings of the 1980 Symposium on Highly Excited States in Nuclear Reactions", H. Ikagami and M. Muraoka, eds., Osaka University (1980), p. 682.
63. V. Bernard and Nguyen Van Giai, Nucl.Phys. A348:75 (1980).
64. J.W. Negele, Phys.Rev. C1:1260 (1970).
65. P. Quentin and H. Flocard, Ann.Rev.Nucl.Sci. 28:523 (1978).
66. G.F. Bertsch and T.T.S. Kuo, Nucl.Phys. A112:204 (1968).
67. I. Hamamoto and P. Siemens, Nucl.Phys. A269:199 (1976).
68. P.F. Bortignon, R.A. Broglia, C.H. Dasso and Fu De-ji, Phys.Lett. 108B:247 (1982).
69. H.M. Sommermann, T.T.S. Kuo and K.F. Ratcliff, Phys.Lett. 112B:108 (1982).

70. J. Wambach, V.K. Mishra and Li Chu-hsia, Nucl.Phys. A380:285 (1982).
71. C. Mahaux and H. Ngô, Phys.Lett. 100B:285 (1981).
72. C. Mahaux and H. Ngô, Nucl.Phys. A378:205 (1982).
73. P.K. Bindal, D.H. Youngblood, R.L. Kozub and P.H. Hoffmann-Pinther, Phys.Rev. C12:390 (1975).
74. P. Doll, G.J. Wagner, K.T. Knöpfle and G. Mairle, Nucl.Phys. A263:210 (1976).
75. G.E. Brown and M. Rho, Nucl.Phys. A372:397 (1981).
76. G.E. Brown, J.S. Dehesa and J. Speth, Nucl.Phys. A330:290 (1979).
77. S.O. Bäckman, G.E. Brown and V. Klemt, Nucl.Phys. A345:202 (1980).
78. J. Wambach and R. Fiebig, in "Proceedings of the Conference on Spin Excitations in Nuclei", Telluride, Colorado (March 1982), to be published.
79. Z.Y. Ma and J. Wambach, Stony Brook preprint (1982).
80. X. Campi and S. Stringari, Orsay preprint (May 1982).
81. G.E. Brown, Comments Nucl.Part.Phys. 3:136 (1969).

GENERAL PRINCIPLES OF STATISTICAL SPECTROSCOPY

J. B. French

Department of Physics and Astronomy
University of Rochester
Rochester, N.Y. 14627

I. INTRODUCTION

Rather than repeat the substance of other elementary reviews[1-3] now in print or in press, which pay special attention to techniques and to applications, it seems better to lay more stress on the theorems and structures which underly the subject, and to emphasize their usefulness, for example in fixing parameters in problems which are highly "non-statistical." The domain in which the theorems work is that of the ordinary shell model, say with N single-particle states and m particles. But we may take large values of N and m, thereby generating huge model spaces. Instead of dealing only with the lowest-lying states, we consider the entire spectrum as a single structure, though it may not be at all obvious at the outset that this is a sensible thing to do; for the many-particle spectrum has a span far larger than the single-particle span, so that almost none of the states are directly relevant to physics. This of course is true also in ordinary shell-model calculations, in which usually only the low states are explicitly constructed, the others being hidden in the computer. In the shell-model case there is an implicit argument that, while the high states are not well treated, their effects on the low-lying states are, especially when one takes account of renormalizations of the Hamiltonian. In our case the high-lying states cannot be hidden, their centroid and width being basic parameters of the theory; but if the model spaces are not too large we can rely on a central limit theorem (CLT) to "propagate" downwards from the centroid, and on a theorem[4] of "spectral rigidity" to keep the local fluctuations in check. For larger spaces a method of partitioning, when combined with these theorems, will do the same things.

We stress finally that, whereas ordinary shell-model methods produce numerical values for quantities of interest, it is our aim to produce them as functions of the parameters of the system. In particular we should think of the spectrum as an immensely complicated structure which may display regularities both in its secular behavior and its fluctuations. Its levels move around as we vary the Hamiltonian matrix elements, these motions, under small variations of the elements, then informing us about quantities of interest, expectation values, transition strengths and so forth, in a manner quite analogous to the way in which parametric derivatives on a partition function yield corresponding information.

II. FLUCTUATIONS AND AVERAGES

2.1 The Spectrum Function

The function of central interest will be the Hamiltonian spectrum function, given in terms of the H-eigenvalues by

$$\rho(x) = d^{-1} \sum_{i=1}^{d} \delta(x-E_i) = d^{-1} \langle\langle \delta(H-x) \rangle\rangle^m = \langle \delta(H-x) \rangle^m = d^{-1} I(x). \quad (1)$$

$d \times \rho(x)$ is simply the level density but, by giving it a fancy name, we insist, as mentioned above, that it be given as a function of the Hamiltonian matrix elements. $\langle\langle G \rangle\rangle^m$ stands for the trace of G in the model (or m-particle) space, $\langle G \rangle^m$ then being the average eigenvalue. We introduce also the distribution function $F(z)$, characteristic function $\phi(t)$, and partition function $Z(\beta)$

$$F(z) = \int_{-\infty}^{z} \rho(x)dx; \quad \phi(t) = \langle e^{itH} \rangle = \int \rho(x)e^{itx}dx; \quad Z(\beta) = \phi(i\beta),$$

$$\rho(x) \Longleftrightarrow F(z) \Longleftrightarrow \phi(t) \Longleftrightarrow Z(\beta), \quad (2)$$

where the last equation indicates that all four functions carry the same information. Thus the standard parametric derivatives on $Z(\beta)$ have their counterparts on ρ, F or ϕ, producing thereby the analogs of the standard quantities of statistical mechanics.

We write the moments of ρ as M_p so that the centroid $\mathcal{E}=M_1$, and the central moments $\mathcal{M}_p = \int \rho(x)(x-\mathcal{E})^p dx$, where $\mathcal{M}_2=\sigma^2$, the variance. Then

$$\rho \Longleftrightarrow \{\mathcal{E}, \sigma^2, K_{\nu\geq 3}\} = \{\mathcal{E}, \sigma^2, \sigma^\nu k_{\nu\geq 3}\} \quad (3)$$

with $K_{\nu\geq 3}$ a homogeneous polynomial of order ν in the central moments. Then $\mathcal{E}$ defines the location, σ^2 the scale (translation invariant), and the parameter set $k_{\nu\geq 3}$ the shape (translation and scale invariant).

We mention finally that while $F(x)$ and $\phi(t)$ always exist, and $\rho(x)$ except in pathological cases, the moment series $\phi(t)=\Sigma(it)^p M_p/p!$ need not converge; the moments may not even exist. But none of these difficulties, or related ones, will arise with our finite-dimensional model spaces (nor even in the $N \to \infty$ limit) so that we shall simply ignore them all.

2.2 The Elementary Central Limit Theorem (CLT)

According to Feller[5] *"It is difficult to exaggerate the importance of convolutions in many branches of mathematics"*; including, we might add, communication and information theory, and in physics and other domains as well. Indeed, the essential content of the elementary CLT, which, with extensions, will dominate everything that we do, is that the m-fold convolution of a function $\rho_1(x)$ with itself, $\rho_m = \rho_1 \otimes \rho_1 \otimes \ldots \otimes \rho_1$, becomes Gaussian for large m; the minor restrictions on ρ for the validity of the CLT (Feller[5]) are always satisfied in the model spaces which we shall encounter.

The CLT arises because of the obvious smoothing action of a convolution. Formally we can define in (3) a set of K_ν (the cumulants) which are additive under convolution; $\sigma^2 = K_2$ and the (non-translation invariant) E, are also members of the set. This follows from the fact that the $\phi(t)$ multiply under convolution so that their logarithms add. Then the K_ν defined by $\ell n\, \phi(t) = \Sigma\, (it)^\nu K_\nu/\nu!$ are also additive. But now the shape parameters k_ν behave under m-fold convolution of ρ as

$$k_\nu(m) = k_\nu(1)/m^{\frac{\nu}{2}-1} \xrightarrow[\nu > 2]{m \to \infty} 0 , \tag{4}$$

thus giving for large m, with $\rho_G(x;E;\sigma)$ a Gaussian density with centroid E and variance σ^2,

$$\phi^{(m)}(t) \to \exp[iE(m)t - \tfrac{1}{2}\sigma^2(m)t^2] \Longleftrightarrow \rho_G(x;E(m), \sigma(m)), \tag{5}$$

where of course $E(m) = mE(1)$ and $\sigma^2(m) = m\sigma^2(1)$.

As corrections to the asymptotic Gaussian we have the Gram-Charlier expansion[6] in terms of orthonormal polynomial, $P_\nu = (\nu!)^{-\frac{1}{2}}He_\nu$, excitations of ρ_G,

$$\rho(x) = \rho_G(x;E,\sigma)\left\{1 + \sum_{\nu\geq 3} R_\nu P_\nu\left(\frac{x-E}{\sigma}\right)\right\} ,$$

$$R_\nu = \int \rho(x)\, P_\nu\left(\frac{x-E}{\sigma}\right) dx , \tag{6}$$

which is by no means always convergent[6], but only the first few terms of which are usually taken in practice (see ahead for an exception). With this series, or better its Edgeworth rearrangement[5,6] the terms of which behave uniformly with respect to m, we can study the rate of approach to Gaussian.

Under the summing of independent random variables the probability density convolutes, so that the same CLT applies here also. For the case of non-interacting particles with $H = \sum_{i=1}^{N} \varepsilon_i n_i$ we can take the random variables as the states occupied by the various particles; they are essentially independent when $m \ll N$, but not in general because of the Pauli Principle. Moreover for interacting particles there is usually no convolution which defines the energy. Finally observe that, for the eigenspectrum of an operator G, we have $M_p = \langle G^p \rangle^m$, and that the joint eigenvalue distribution is not defined for non-commuting operators.

2.3 Non-interacting Particles (NIP) with Non-singular Spectra

In an H-diagonal single-particle basis we have, with n_i a number operator,

$$H = \sum_{i=1}^{N} \varepsilon_i n_i \quad ; \quad \sum \varepsilon_i = 0 , \tag{7}$$

the latter equation by a choice of energy zero. If we drop this restriction our results below will be for M_p rather than $\mathcal{M}_p$. Then

(a) $n_i^2 = n_i = n_i^\nu \qquad (\nu \geq 1)$,

(b) $n_i n_j$ $(i \neq j)$ is a 2-body operator; I_ℓ, any k-fold product involving ℓ different n_i, is ℓ-body ,

(c) $\langle\langle I_\ell \rangle\rangle^m = \binom{N-\ell}{N-m} \; ; \quad \langle I_\ell \rangle^m = \binom{m}{\ell}\binom{N}{\ell}^{-1}$

$$\langle\langle F(\ell) \rangle\rangle^m = \binom{N-\ell}{N-m} \langle\langle F(\ell) \rangle\rangle^\ell ; \; \langle F(\ell) \rangle^m = \binom{m}{\ell} \langle F(\ell) \rangle^\ell . \tag{8}$$

The first pair of (c) follows from the fact that $\langle\langle I_\ell\rangle\rangle^m$ counts the number of m-particle states, $\binom{N-\ell}{m-\ell}$, in which a specified set of ℓ single-particle states are already filled. Taking $\ell=1$ we see now that the centroids are additive, $E(m) = mE(1)$, and, since we have chosen $E(1) = 0$ (7), the moments are automatically central, $M_p = \mathcal{M}_p$. The second pair follows from the first when we recognize that, in a direct-product basis, the "diagonal part" of every ℓ-body operator $F(\ell)$ is a linear combination of I_ℓ operators; we have here our first example of the propagation of information throughout a lattice (in this case the line $m=0,1,2,\ldots,N$) of model spaces.

Now, proceeding as in (1), or via a multinomial expansion of H^p, we find easily the exact result that

$$\mathcal{M}_p(m) = \sum_{\pi(p)} d_\pi \langle \varepsilon^{p_1} \wedge \varepsilon^{p_2} \wedge \ldots \wedge \varepsilon^{p_\ell} \rangle N^\ell \binom{m}{\ell}\binom{N}{\ell}^{-1} ; d_\pi = (p:\pi(p))/\pi s_k! \; , \tag{9}$$

where the $\pi(p) \equiv [p_1, p_2 \cdots p_\ell]$ are partitions of p, $(p:\pi(p))$ is the multinomial coefficient, and s_k is the number of p_j's in $\pi(p)$ which are equal to k. For the "augmented symmetric function" in (9) we have, with $\langle\varepsilon^r\rangle = \langle H^r\rangle^1 = N^{-1}\,\Sigma(\varepsilon_i)^r$,

$$\langle \varepsilon^{p_1} \wedge \varepsilon^{p_2} \wedge \ldots \varepsilon^{p_\ell} \rangle =$$
$$N^{-\ell} \sum_{i_1 \neq i_2 \neq \ldots \neq i_\ell} \varepsilon_{i_1}^{p_1} \varepsilon_{i_2}^{p_2} \ldots \varepsilon_{i_\ell}^{p_\ell} \xrightarrow{N\to\infty} \left\{ \langle\varepsilon^{p_1}\rangle \ldots \langle\varepsilon^{p_\ell}\rangle \right\} + O\left(\frac{1}{N}\right), \tag{10}$$

which, as indicated separates, for large N, as a product of single-particle moments. Then

$$\mathcal{M}_p(m) \xrightarrow{N\to\infty} \sum_{\pi(p)} d_\pi \left[\langle\varepsilon^{p_1}\rangle\langle\varepsilon^{p_2}\rangle \ldots \langle\varepsilon^{p_\ell}\rangle \right] \ell! \binom{m}{\ell} \; . \tag{11}$$

Since, for large m, maximum ℓ compatible with $\langle\varepsilon\rangle = 0$ is dominant, we see that the reduced central moments $\mathcal{M}_p = \mathcal{M}_p/\sigma^p$ are, asymptotically

$$\mu_{2\nu}(m) \to (2\nu-1)!! \quad ; \; \mu_{2\nu+1}(m) \to \frac{1}{3}\,\nu(2\nu+1)!! \langle\varepsilon^3\rangle/\langle\varepsilon^2\rangle^{3/2} m^{1/2} , \tag{12}$$

defining thereby the asymptotic Gaussian. Since in the limit (fixed m; $N \to \infty$) all trace of fermion effects (except for the trivial normalization) have disappeared this is simply the classical convolution result. The dominance of binary correlations in the moments, $\pi = [2^\nu]$ for $\mathcal{M}_{2\nu}$, $\pi = [2^{\nu-1},3]$ for $\mathcal{M}_{2\nu+1}$, which gives rise here to a Gaussian distribution, will be seen later to generate more complicated Gaussian phenomena.

Instead of the large-N approximation (11) we have the exact result that

$$\langle \varepsilon^{p_1} \wedge \varepsilon^{p_2} \wedge \dots \wedge \varepsilon^{p_\ell} \rangle = \langle \varepsilon^{p_1} \wedge \varepsilon^{p_2} \wedge \dots \wedge \varepsilon^{p_{\ell-1}} \rangle \langle \varepsilon^{p_\ell} \rangle$$
$$- \frac{1}{N}\left\{ \langle \varepsilon^{p_1+p_\ell} \wedge . \wedge \varepsilon^{p_{\ell-1}} \rangle + \langle \varepsilon^{p_1} \wedge \varepsilon^{p_2+p_\ell} \wedge . \wedge \varepsilon^{p_{\ell-1}} \rangle + . + \langle \varepsilon^{p_1} \wedge . \wedge \varepsilon^{p_{\ell-1}+p_\ell} \rangle \right\} \tag{13}$$

and thus for example

$$\langle \varepsilon^{p_1} \wedge \varepsilon^{p_2} \rangle = \langle \varepsilon^{p_1} \rangle \langle \varepsilon^{p_2} \rangle - \frac{1}{N} \langle \varepsilon^{p_1+p_2} \rangle \ , \tag{14}$$

which with p=2 gives the blocking correction to the variance

$$\sigma^2(m) = m(N-m)(N-1)^{-1}\sigma^2(1) \ . \tag{15}$$

We could have written this at sign by noting that, by (8), $\sigma^2(m)$ is a second-order polynomial in m which moreover, being a variance, must vanish in the one-dimensional spaces with m=0,N. Note the (hole-particle) symmetry corresponding to $m \leftrightarrow (N-m)$; its significance will be clear later.

We find similarly, for the basic shape parameters $k_3=\gamma_1=$ "skewness" and $k_4=\gamma_2=$"excess", that

$$\gamma_1(m)=k_3(m)=\frac{N-2m}{N-2}\left(\frac{N-1}{N-m}\right)^{1/2} m^{-1/2}\gamma_1(1),$$
$$\gamma_2(m)=\frac{(N-1)}{(N-2)(N-3)m(N-m)}\left\{[N(N+1)-6m(N-m)]\gamma_2(1)+6N\left[1-\frac{m(N-m)}{(N-1)}\right]\right\},$$
$$\gamma_2(N/2)=-\frac{2(N-1)}{N(N-3)}\left\{\gamma_2(1)+\frac{3(N-2)}{(N-1)}\right\}, \tag{16}$$

which we can use to convince ourselves that blocking effects are generally of smaller consequence on the shape than on the variance, which for m=N/2 is reduced to half its CLT value.

2.4 Non-interacting Particles with Singular Spectra

"Whenever new coherent states occur far from equilibrium, the application of probability theory, as implied by the counting of the number of complexions, breaks down." Though we are not concerned here with non-equilibrium statistical mechanics, this remark of Prigogine[7] is relevant when we ask about exceptions, even with non-interacting particles, to the Gaussian law derived above.

In approximating the correlated sum (10) we have used an analog of complexion counting by taking for granted that the terms in the sum, or excluded from it, are of the same order with respect to N. But if we take as an example the spectrum $(0^{N-1},1)$ with only a single $\varepsilon_i \neq 0$ (known in the two-particle system as a "pairing spectrum") we see that binary correlations play no role whatever in the correlated sums. The only contributing partition is [p,0,0..] with $\ell = d_\pi = \langle \varepsilon^p \rangle = 1$. Then $M_p(m) = m/N \xrightarrow{m<<N} \mathcal{M}_p(m)$, which of course also defines a non-Gaussian singular spectrum. Thus our approximation has been in error for this case. Since, taking i=1, we have $H = \varepsilon_1 a_1^+ a_1 = \frac{1}{2}\varepsilon_1(a_1^+ + a_1)^2$ we can even recognize a coherent state, generated either by the non-Hermitian a_1 or the Hermitian $\frac{1}{2}(a_1^+ + a_1)$.

Looked at differently we have implicitly taken for granted that $\gamma_2(1) \sim O(1)$. Instead, let us consider large N, m and $\gamma_2(1) = N$. Then with fixed m and $N \to \infty$ we have the canonical result, which of course follows also from (16) that $\gamma_2(m) = N/m$, while, with fixed $m/N = 1/2$ say, we have $\gamma_2(m) \to -2$, both non-Gaussian results. The trivial explanation is that a $\gamma_2(1) \sim O(N)$ spectrum is of such an exotic nature that it would require far more than N convolutions (i.e. particles) to produce a good Gaussian spectrum, this of course being ruled out for fermions. In general these "singular" spectra, as we shall call them, are those in which essentially all the "weight" is carried by a fraction $\sim N^{-1}$ of the levels. They are characterized by $\mu_{2\nu} \sim O(N^{\nu-1})$, whereas, for non-singular spectra such as Gaussian, uniform, χ_r^2 etc, $\mu_{2\nu} \sim O(1)$ (the distinction of course is in the N dependence for large N, not in the ν dependence). The extreme example is the pairing spectrum above, for which obviously $M_p = N^{-1}$ so that $\gamma_2(1) \xrightarrow{N \to \infty} N$, its largest possible value. For this case $\gamma_1(1) \to N^{\frac{1}{2}}$; but of course there are symmetrical singular spectra for which the odd cumulants vanish, an example is $(\pm 1, 0^{N-2})$.

We have mentioned these singular spectra because they have analogs in interacting-particle systems where, as suggested above, they are connected with collective behavior. It is worthwhile to recognize them for the deviations from "normal" behavior which they generate, and even to use them for the prediction of such behavior.

2.5 The CLT for Interacting Particles: Spectral Methods

We have seen that in the NIP case a non-singular single-particle spectrum generates an asymptotic Gaussian density, this result being only mildly affected by the fermion nature of our system. But interaction effects are much more complicated, and it is not obvious that, with interacting fermions, there should be any operative CLT at all. We study the question in two ways, first by analytically tractable examples, (which can always be supplemented by numerical calculations) and secondly via ensembles which give results far more reaching, and significant both for averages and fluctuations.

As a trivial example $H=h^2$, with h non-singular and traceless, is (1+2)-body, the two-body part describing interactions. Its asymptotic spectrum is obviously $\chi_1^2 \sim x^{-\frac{1}{2}} \exp(-x/2)$ which is far from Gaussian, but derives directly from the CLT action generated by h, for which there is a convolution. So convolutions need not act directly to produce the spectrum, which may then not be Gaussian.

For a much more instructive case take $N \to \infty$ in order to eliminate the uninteresting blocking effects, and consider the asymptotic spectrum ($m \to \infty$, $m/N \to 0$) of the sum of squares of traceless one body operators with non-singular spectra,

$$H = \sum_{\alpha=1}^{\ell} h_\alpha^2 \; ; \; h_\alpha = \sum_{i,j=1}^{N} \varepsilon_{ij}(\alpha) U_{ij} \quad ; \quad \varepsilon_{ij}(\alpha) = \varepsilon^*_{ji}(\alpha) \; ;$$

$$\sum_i \varepsilon_{ii}(\alpha) = 0 \; ; \quad U_{ij} = a_i^+ a_j \; . \tag{17}$$

To evaluate the asymptotic moments we may go back to the counting of binary correlations which gave the Gaussian result (12) that $\mu_{2\nu}(m) = (2\nu-1)!!$. It is then proper to say that, in the large-m limit (with $N = \infty$) the different components of H in (17) *effectively commute*. Formally this comes about because, if we re-order the operators in a characteristic term of $H^{2\nu}$, the products containing commutators are of lower particle rank than the simple product (the commutator of two one-body operators is one-body, while the product has a two-body part) and give therefore traces $\propto m^s$ with $s<\nu$. In other terms the effective commutation arises when the particle number is very large because different components almost always generate mutually exclusive transitions, $i \to j$ and $r \to s$ where the sets (i,j) and (r,s) are disjoint. This is clearly a general result for non-singular operators, though not for singular ones.

But now from this effective commutation it follows that

$$M_\nu(m) \xrightarrow[N = \infty]{m \to \infty} \sum_{\pi(\nu)} (\nu:\pi(\nu)) \langle h_1^{2\nu_1} h_2^{2\nu_2} \cdots h_\ell^{2\nu_\ell} \rangle^m \; . \tag{18}$$

Let us now agree that the h_α are uncorrelated in the one-particle space, i.e. that

$$\langle h_\alpha h_\beta \rangle^1 = 0 = \langle h_\alpha h_\beta \rangle^m \qquad \alpha \neq \beta \; . \tag{19}$$

Writing the variances, which we know to be additive for large N, as

$$\langle h_\alpha^2\rangle^m = \hat{\sigma}_\alpha^2(m) = m\langle h_\alpha^2\rangle^1 = m\hat{\sigma}_\alpha^2(1), \qquad (20)$$

we see that the trace in (18) decomposes into an ℓ-fold product of separate traces $\langle h_\alpha^{2\nu_\alpha}\rangle^m$, Gaussian moments of order $2\nu_\alpha$, or equivalently χ_1^2 moments of order ν_α. We then recognize $M_\nu(m)$ as defining a convolution of the ℓ separate h_α^2 densities, which, for large-enough ℓ, (and provided that H is not dominated by a small number of the separate terms) becomes Gaussian. In particular in the important case that the h_α variances are equal, $\hat{\sigma}_\alpha^2 = \hat{\sigma}^2$, the asymptotic spectrum has a $\chi_\ell^{2\alpha}$ form[8]

$$\rho(x) = 2^{\ell/2}\hat{\sigma}^{\ell-2}\{\Gamma(\ell/2)\}^{-1}x^{\ell/2-1}\exp(-x/2\hat{\sigma}^2). \qquad (21)$$

The ℓ=1 case is trivially derivable, as indicated above. The ℓ=3 distribution was derived by Bethe[9] who, in his J-decomposition of the level density, treated J_z as a random variable whose resultant (Gaussian) distribution then gives Maxwellian for J itself. The Q•Q interaction ℓ=5 was first studied by Nomura[10] who derived the skewness and excess but did not give the density. Making a natural extension of (21) we could argue that any H which is decomposable into ℓ non-singular terms of comparable magnitude whose actions are in some appropriate sense "independent" would give a density close to χ_ℓ^2. Since χ_ℓ^2 is the distribution for ℓ independent similarly distributed χ_1^2 variables we see that for large ℓ (say $\ell > 25$) the spectrum for interacting particles with H decomposable as indicated would become Gaussian. This is in fact what happens with realistic nuclear interactions*. The first analytic derivation of this result, done in a different way, is due to Gervois[11].

In coming to (21) and its large-ℓ Gaussian form we have seen a more sophisticated operation of the CLT, giving the asymptotic spectrum of a function of operators which, though linearly uncorrelated, are neither independent nor commuting. In proceeding to the final conclusion we have relied heavily on the fact that the convolution process is described by a powerful and forgiving theorem, the CLT.

The result implied in (19) that one-body operators uncorrelated (orthogonal) in the one-particle space are uncorrelated for arbitrary particle number, follows from the vanishing of $\langle h_\alpha h_\beta\rangle^m$, a

---

*Such H's have a large Q•Q part whose χ_5^2 spectrum is far from Gaussian, but its coefficient is negative so that the major departure of its spectrum from Gaussian which arises from its cut-off at the zero value of the variable occurs in the upper part of the spectrum where it has little effect.

second-order polynomial in m (8), at three values of m, viz m=0,1, N (the latter since d(N)=1) so that $\langle h_\alpha h_\beta \rangle^N = \langle h_\alpha \rangle^N \langle h_\beta \rangle^N) = 0$ for traceless operators. A special property of one-body traceless operators is that they transform as a single irreducible representation of U(N), the group of unitary transformations in the one-particle space. The extended orthogonality theorem is not valid for arbitrary traceless k-body operators. But on the other hand such an operator decomposes into k irreducible U(N) tensors (which are themselves orthogonal) and the extension implied in (19) applies to these, i.e. to unitarily irreducible operators either belonging to the same irrep or not. We have therefore a considerable extension of the CLT to particles acting via k-body interactions. Finally we mention that the notion of operator independence is not very useful to us; such operators, U_{ij} and U_{rs} with $(i,j) \neq (r,s)$ for example, must operate in different subspaces.

2.6 The CLT for Interacting Particles: Ensemble Methods

The connection of convolutions with random variables, Prigogine's remark quoted at the beginning of (2.4) and our speculation about the kinds of H which might lead to Gaussian spectra, all remind us that "real" probability considerations (as opposed to the calculation of traces and the use of a CLT, a theorem of non-probabilistic mathematics) have a natural role when we consider general problems about spectra, for example not whether the spectrum of a given H has a certain property but whether the spectra of "almost all" H's have it. The purpose of an ensemble, in our case {H} an ensemble of H's, is to produce a realization of the probability laws involved, while eliminating the inessential features (of H) which complicate everything. Since our present interest is in the origin of the CLT, the arguments above lead us to suspect that the essential features are that {H} should describe interactions of particle ranks k, much smaller than the particle number m, and that a negligible fraction of its members should display a singular spectrum. The question then is whether an ensemble of this kind generates mostly Gaussian spectra. We proceed in three stages:

1) By showing that the ensemble average is Gaussian in a natural limit.
2) By showing that large deviations of individual spectra from the average are improbable, the probability vanishing in the limit. Thus for "almost all" H's in {H} we find Gaussian.
3) By realizing that if we evaluate measures for the deviations of (2), instead of merely demonstrating that they vanish in some limit, we have a theory for the "ensemble fluctuations," i.e. measures for the fluctuations from member to member of {H} ("across the

ensemble") in quantities of interest evaluated in the same part of the spectra. If the ensemble is ergodic these measures will coincide in a natural limit with measures for the "spectral fluctuations" (along the spectrum of a given H) for "almost all" members of {H}. Thus we produce a theory for the energy-level fluctuations, a subject which has been studied experimentally for many decades.

The basic starting ensemble[12] is of real symmetric matrices of large dimensionality d, with distinct elements chosen independently in accordance with zero-centered Gaussian laws with variance v^2 for off-diagonal and $2v^2$ for diagonal. The joint probability distribution for the eigenvalues is known for this ensemble[12], so that we can verify that the relative probability of finding a singular spectrum in it is small, vanishing in the large-d limit. If we realize this "Gaussian orthogonal ensemble (GOE)" in an m-particle system we find only m-body interactions; for the number of independent matrix elements is $\sim \frac{1}{2}\binom{N}{m}^2$ while that for a k-body operator (defined in a k-particle space) $\sim \frac{1}{2}\binom{N}{k}^2$. It is not the fact that m-body interactions are unexpected in nuclei which disqualifies them here, but rather the fact that they generate no CLT action and no Gaussian spectra (which when properly understood, see ahead, do agree with experiment). There is in fact a binary correlation structure in the GOE moments since the $W_{\alpha\beta}$ must be correlated in pairs (higher-order correlations being down in number by some power of N and first order ruled out by the ensemble average $\overline{W}_{\alpha\beta}=0$). But these correlations are inhibited by the matrix structure since for example a sub-sequence $W_{\alpha\beta}W_{\beta\beta}W_{\beta\alpha}$, in which the first and third are correlated, forces the second to be diagonal (one free summation variable), as opposed to $W_{\alpha\beta}W_{\beta\alpha}W_{\alpha\gamma}$ (two free variables). The result is that only correlations across fully correlated products contribute in the large-d limit. In the moment $M_{2\nu}$, with normalization $dv^2=1$, we find, instead of the number, $(2\nu-1)!!$, of correlation patterns which gives Gaussian, the Catalan number $t_\nu=(\nu+1)^{-1}\binom{2\nu}{\nu}$ which defines the (ensemble averaged) GOE density[13]. Since, for large ν, $t_{\nu+1} \rightarrow 4t_\nu$ we see that, unlike the Gaussian, this distribution is bounded, $\rho(x)$ vanishing when $|x|>2$. The moments in fact are those of Wigner's semicircle, $\rho_o(x)=\pi^{-1}[4-x^2]^{\frac{1}{2}}$, which, as weight function, defines the orthonormal polynomials

$$v_\zeta(x) = (-1)^\zeta \sin(\zeta+1)\psi(x)/\sin\psi(x) = \sum_s (-1)^s \binom{\zeta-s}{s} x^{\zeta-2s}, \quad (22)$$

where $\psi(x) = \pi - \cos^{-1}(x/2)$.

We have given these details because, while the GOE is not an appropriate model for the density, it is so for the fluctuations, as we shall see. For the density we should define the GOE in a k-particle space $(d(k)=\binom{N}{k})$ but allow it to operate in an m-particle space $(d=\binom{N}{m})$. But now when m>>k it is immediately obvious by our "effective commuting" theorem that all H pairings are allowed; thus as m increases from k the binary inhibitions are gradually relaxed and the density changes from semicircular to Gaussian.

It is worthwhile to proceed further[14] by writing H in terms of k-body state (creation) operators $\psi_\alpha(k)$ where $\Psi_\alpha(k)=\psi_\alpha(k)|0\rangle$. Then

$$H = W_{\alpha\beta}\psi_\alpha(k)\psi_\alpha^+(k) = W_{\alpha\beta}\ \alpha\beta^+ \tag{23}$$

where summation is understood and we use an abbreviated notation. The moments are still dominated by binary correlations and can be evaluated for low order by using a simple theorem[14], that for "almost all" operators O and a correlated pair of H's, $\overline{H(k)O(t)H(k)} = \binom{n-t}{k}O(t)$. Thus $M_2(m)=\binom{m}{k}$ and easily $M_4(m)=\binom{m}{k}\{2\binom{m}{k} + \binom{m-k}{k}\}$; observe then that μ_4=2,3 respectively for m=k (GOE) and m>>k (asymptotic EGOE), just as is expected. In the latter case $\gamma_2 \rightarrow -k^2/m$ behaving therefore with (large) particle number in the classical way. The general moment has not been evaluated but the moments are known to order 8 and from them one finds that $k_6(m) \sim m^{-2}$ and $k_8(m) \sim m^{-3}$ so that we can probably assume that for large enough m the approach to Gaussian is characteristic of a convolution.

2.7 Spectral Fluctuations

It is not enough to show that the ensemble-averaged density is Gaussian. Before we can argue that we have a true CLT action we must show that the ensemble variance of the density is small over the entire spectrum. If, instead of merely demonstrating smallness, we actually evaluate the variance function, or rather its extension to the covariance function (or two-point or autocorrelation function, the latter name being sometimes reserved for a renormalized version)

$$S^\rho(x,y) = \overline{\rho(x)\rho(y)} - \overline{\rho(x)}\ \overline{\rho(y)}\ , \tag{24}$$

we not only demonstrate the CLT action but we produce a theory for the two-point fluctuations as well. With this we can reassure ourselves about the accuracy of spectral averaging results, but can also make direct comparison with experimental fluctuations, as in slow-neutron resonance spectra on heavy nuclei and proton spectra with lighter ones, and in the fluctuations of other quantities as well[14]. There are of course (t>2)-point functions which follow from higher-order functions than (24) but the experimental evaluation of measures for these would require longer runs of data than are available.

The exact evaluation of $S^\rho(x,y)$ is formidable[15], but we again rely on moment evaluations with binary correlations, which gives excellent results. With some forethought (or hindsight) we might expect to find a representation of the ensemble of densities in the form of orthonormal-polynomial expansions of the asymptotic density, as in (6) for the EGOE case with the R_ν uncorrelated random variables; for the GOE we might similarly expect Chebyshev-polynomial excitations of the semicircle. We shall see that these expectations are close to the truth.

We sketch for GOE the evaluation[14,3] of the $S^\rho(x,y)$ moments (which incidentally are the covariances of the density moments). With ρ_0 Wigner's semicircle and v_ζ given by (22),

$$\int S^\rho(x,y)x^p y^q dxdy = \Sigma^2_{p,q} = \overline{\langle H^p\rangle\langle H^q\rangle} = \sum_{\zeta>0} \overline{\underbrace{\langle H^p\rangle\langle H^q\rangle}_{\zeta}}$$

$$= \sum_{\zeta>0} \mu^p_\zeta \mu^q_\zeta \overline{\underbrace{\langle H^\zeta\rangle\langle H^\zeta\rangle}_{\zeta}} = \frac{2}{d^2}\sum_{\zeta>0} \zeta\mu^p_\zeta\mu^q_\zeta \Longrightarrow S^\rho(x,y)$$

$$= \frac{2}{d^2}\frac{\partial}{\partial x}\frac{\partial}{\partial y}\left\{\rho_0(x)\rho_0(y) \times \sum_{\zeta>0} \zeta^{-1} v_{\zeta-1}(x)v_{\zeta-1}(y)\right\} . \qquad (25)$$

Here the ζ expansion recognizes that, by the same argument used above for the density,we count only those patterns in which the (p+q) H's are *pairwise* correlated. Such correlations can involve two H's in the same trace ("internally correlated") or one in each trace ("cross-correlated"); ζ is simply the number of pairs of the second type. The $\zeta=0$ term does not enter since it is eliminated by the second term of S^ρ (24). For fixed ζ we could have generated the cross-correlated product by starting with $p=q=\zeta$. For GOE this product is simple to calculate and gives $2\zeta/d^2$. We have now to insert into the first factor $(p-\zeta)/2$ correlated pairs in such a way that each pair contracts only over internally correlated pairs. For $\zeta=0$, which we do not encounter here, the number of ways of doing this would be given by a Catalan number which we have met in (2.6); for $\zeta\neq 0$, by a rather tricky combinatorial argument, we find its value to be $\mu^p_\zeta=\binom{p}{(p-\zeta)/2}$; we have a similar factor for the other trace, and thus a ζ-decomposition of the two-point correlation moments. This moment expansion is inverted by using a standard elementary result and gives $S^\rho(x,y)$ in terms of a series of correlated Chebyshev-polynomial excitations of the densities at the two points. That the excitation should be Chebyshev is natural since, as mentioned above, these are the polynomials with the average density as weight function. But our "prediction" of the S^ρ form is not quite correct; because of the double derivative occurring in the last form of (25) it is the two-point distribution function

$$S^F(x,y) = \int_{-\infty}^{x} dx' \int_{-\infty}^{y} dy' \; S^\rho(x',y') \,, \tag{26}$$

which has the expected expansion (just drop $\frac{\partial}{\partial x}\frac{\partial}{\partial y}$ in (25)).

Using the trigonometrical form (22) for v_ζ, and noting that the number of orthonormal polynomials is bounded ($\zeta \leqslant d$) because the spectra are discrete, the evaluation of S^F is now simple. With $\alpha \sim d^{-1}$, its precise value being irrelevant as long as x,y are separated by a few levels, we have, as an excellent approximation[16]

$$\begin{aligned} S^F(x,y) &= \frac{2}{\pi^2 d^2} \sum_{\zeta \geqslant 1} \zeta^{-1} \sin(\zeta\psi(x))\sin(\zeta\psi(y)) \\ &= \frac{1}{2\pi^2 d^2}\, \ell n \left\{ \frac{1 + e^{-2\alpha} - 2e^{-\alpha}\cos(\psi(x) + \psi(y))}{1 + e^{-2\alpha} - 2e^{-\alpha}\cos(\psi(x) - \psi(y))} \right\} , \end{aligned} \tag{27}$$

from which we can evaluate all the two point measures. In particular we find that the individual level motions (deviations from the ensemble average) are very small, $\sim \pi^{-1}(\ell n\ d)^{\frac{1}{2}}$ in units of the local spacing; for the relative motion of two levels, r spacings apart, the result is $\sim \pi^{-1}(\ell n\ r)^{\frac{1}{2}}$.

These results, and refinements of them all deriving from S^F, which are originally due to Dyson and Mehta[4] constitute a huge extension of the von Neumann-Wigner level repulsion ("no crossing") theorem. The ergodicity of the GOE ensemble insofar as the density is concerned follows immediately[17,14] from the variance function $S^F(x,x)$, or from (28) below, so that for large d almost all the spectra are closely semicircular. That the ensemble is ergodic much more generally has been shown by Pandey[18]. That the GOE is a good model for the experimental fluctuations has been convincingly demonstrated only very recently in an important paper of Haq, Pandey and Bohigas[19], who analyze the high-quality data in terms of new spectrally averaged measures (which by their structure already take explicit advantage of the ergodicity). That an essentially featureless ensemble should give results in agreement with data is more or less well understood[20,2], but we shall not discuss that. It implies of course that the fluctuation laws are of a general nature and that the fluctuations carry little specific information.

Knowledge of the two-point function (quadratic in the density) enables us to express the density itself as a random function $\hat{\rho}(x)$ with known second-order statistical properties. We find easily[14] (note that in this reference $\zeta\rho_o G_\zeta = -(\rho_o v_{\zeta-1})'$) for the "ensemble density" that

$$\hat{\rho}(x) = \rho_o(x)\left\{1 - \rho_o^{-1}(x)\,\frac{d}{dx}\sum_{\zeta\geq 1} R_\zeta \rho_o(x) v_{\zeta-1}(x)\right\}$$

$$\bar{R}_\zeta = 0 \; ; \; \overline{R_\zeta R_{\zeta'}} = \frac{2\zeta}{d^2}\,\delta_{\zeta\zeta'} \; , \qquad (28)$$

thereby giving an explicit second-order representation of the complicated structure described at the end of the first section.

Returning now to our question, whether or not there is an operative CLT for "almost all" members of an ensemble of k-body interactions, we should repeat the S^F calculation for EGOE in the limit (large m; $N \to \infty$). This is quite simple[14], the result being a Gram-Charlier (6) expansion for the ensemble density

$$\hat{\rho}(x) = \rho_G(x)\left\{1 + \sum_{\zeta\geq 3} \zeta!^{-1} S_\zeta He_\zeta\left(\frac{x-E}{\sigma}\right)\right\} ,$$

$$\bar{S}_\zeta = 0 \; ; \; \overline{S_\zeta S_{\zeta'}} = 2\zeta\delta_{\zeta\zeta'}\binom{N}{k}^{-2}\binom{m}{k}^{2-\zeta} , \qquad (29)$$

but valid only for excitations of long and intermediate wavelengths, not for excitations with $\zeta \sim d$. The very rapid decrease of the amplitudes with increasing ζ defines the operation of a very strong CLT which, for increasing m, rapidly fixes the density (smoothed to within fluctuations) as very close to Gaussian. If we care to incorporate the very long-wave-length excitations (say those with $\zeta \leq 5$ for a $(ds)^{12}$ shell-model matrix with $d \sim 1000$) into the shape for each individual spectrum, we can assert that for almost all EGOE spectra (and in strong contrast with GOE results) there are only very short-wave-length fluctuations. This fact, which is very favorable for spectral averaging, has been well verified by Monte-Carlo calculations[21], it would be good to find some kind of experimental confirmation.

If (29) were valid for all excitations the asymptotic EGOE spectra would be uniform without fluctuations. It is not valid because the short-wave-length fluctuations escape the CLT action. As we add more and more particles to the system the spectrum does indeed become smoother (via the effective convolutions); but this happens on a fixed energy scale, not on the scale of the local spacings which, on the CLT scale, rapidly decreases. In other words things become smoother but we examine them more closely. It would appear from this that, when we embed the k-body GOE in the m-particle space the "medium" should have no effect on the shortest-range fluctuations, which should then be identical with GOE and, a fortiori, with experiment. These results are well verified and the technical reason for the high-frequency failure of (29) is known. But a rigorous proof that GOE and EGOE fluctuations are identical has not been found.

We have said that the fluctuations, being described by an almost featureless ensemble, carry little specific information. However the close agreement with experiment[19] does impose limits on mechanisms, characteristic of H, which would give different fluctuation patterns. Broadly speaking the only such mechanism known is the failure of an almost exactly good symmetry, in particular the failure of time reversal invariance[12,22] (TRI). This would show up in the Hamiltonian as a departure from real symmetry in the matrices, and in the fluctuations as an increased rigidity (consider the no-crossing theorem for d=2 real symmetric vs. general Hermitian matrices[2]).

For the ensemble we have

$$\{H_\alpha\} = \{H(GOE) + i\alpha\{H(A)\} , \qquad (30)$$

where the antisymmetric H(A) matrices are constructed by the same process as those of the GOE. For d=2 the change from GOE fluctuations (α=0) to those of another classical ensemble, the Gaussian unitary ensemble (GUE) of general Hermitian matrices defined by α=1, is very slow. But, by the same argument used to describe the escape of the short-range fluctuations from the CLT action, while the number of levels increases by a factor d the spectrum span increases only as $d^{\frac{1}{2}}$, so that (on a scale of fixed matrix elements) the levels are pushed together. Then a perturbation argument for small α, gives the effective transition parameter as $d^{\frac{1}{2}}\alpha$ so that the GOE $\rightarrow$ GUE transition speeds up greatly and in the limit corresponds to a discontinuous phase change in the chaos represented by the fluctuations.

Formally[22] the result of the binary-correlation calculation is to insert a factor $\frac{1}{2}\{1+[(1-\alpha^2)/(1+\alpha^2)]^\zeta\}$ in the mode summation of (27), and then everything follows easily for arbitrary α. By comparison with the data[19], upper limits on αv, the RMS non-TRI matrix element, have been determined[22] but not yet a limit for α itself. We mention finally that an exact analytic treatment (very formidable indeed!) for the interpolating ensemble (30) has been recently given by Pandey and Mehta[23].

III. SYMMETRIES AND INFORMATION

3.1 Preliminaries

All the m-particle states together form an irreducible representation of the group U(N) of unitary transformations in the single-particle space; thus, in an almost trivial sense, we have generated above a relationship between symmetry and statistical averaging; signs of this are to be seen in the propagation law for m-particle traces (8), a particle-hole symmetry (invariance under $m \leftrightarrow N-m$) in (15), and something about U(N) tensors and the CLT at the end of (2.5). As soon as a group is recognized, one is led almost automatically to consider the symmetries defined by its subgroups. An almost frivolous reason for doing that in the present case comes from the fact that, without in some way partitioning the model space, we cannot indefinitely increase its dimensionality without losing all accuracy in the ground-state region (which we must always consider, if only to fix the zero of energy).

The only useful partitionings are by means of symmetries. Only in that case can the distribution be expected to be close to a characteristic form, calculable via a few moments; and only then will propagation methods emerge for calculating the moments. And besides that of course the study of symmetries, which then becomes feasible, is in itself a matter of the greatest interest.

An obvious form of partitioning, consistent with the strong shell-structure effects, is via shell-model configurations in which the m particles are distributed among the orbits (spherical or otherwise) which make up the single-particle space. The group here is defined by an additive decomposition of the U(N) algebra, the transformations being products of the transformations in the subspaces. Other obvious partitionings are via isospin, very useful for light nuclei because of the large isospin splittings (quite apart from our direct interest in isospin itself), and angular momentum (which is technically much harder to deal with).

Going beyond these obvious things, many different situations arise depending on the groups involved; there are many different chains and lattices of subgroups; they may define good symmetries* or broken ones; the groups may involve free parameters to be chosen in some optimal way; all states for a given symmetry may be localized or spread out in energy; all the states of a given symmetry may form together an irreducible representation of some group (as with isospin) or they may not (as with angular momentum). Moreover, in contrast to the U(N) case above, there seems often

*In our notation a symmetry is good if every H-eigenfunction belongs to an irreducible representation. Nothing is implied about degeneracy.

no natural way to proceed to an asymptotic limit which would define characteristic forms although such forms are encountered. Not enough study has been made of these things. It is clear though, from available calculations on the one hand, and on the other hand from general theorems which are known about probability on algebraic structures, and in particular about the corresponding limit theorems[17], that the whole subject is a rich one waiting to be studied. We must content ourselves however with a few practical aspects of it, and will not for example discuss the origin of CLT's in the fixed-symmetry subspaces,nor attempt to demonstrate that in such spaces we still have GOE fluctuations.

Symmetries enter when instead of studying the density for *all* m-particle states in our "universe" defined by N we consider the distribution of the summed intensity of a subset of the states as they are found in the Hamiltonian eigenstates. If the symmetry which labels the states is a good one we can regard the density as an intensity distribution as above, or alternatively as an eigenvalue distribution; but when the symmetry is not a good one only the first of these choices is open to us, for in that case a subspace cannot in general be spanned by a set of eigenstates. If this point is not appreciated an apparent paradox arises when the energy-level fluctuations are considered; to the decomposition of the m-particle space, there is a corresponding decomposition of the density. If we think of the partial densities as those of eigenvalues we know that, in a region where several spectra overlap, the energy level fluctuations are drastically modified, the nearest-neighbor spacing distribution going towards Poisson instead of the very different Wigner form which displays the level repulsion. If in fact the defining symmetry is good this is precisely what is known to happen; but if it is very badly broken it has no effect on the Wigner form and we have a conflict. The resolution is that the correspondence of the partial density with eigenvalues is lost in the case of a broken symmetry. In the same way it must be understood that the density being a superposition of partial densities does not in any sense imply that one set of states has no effect on the other; when the symmetry is broken the distribution of one irrep is affected by all of the others which are connected to it by H, this showing up in the "partial widths" of the representation which go to intermediate states of different symmetries, and in higher moments as well. Thus each irrep accomodates itself to the others. We stress these matters, perhaps unnecessarily, because confusion has in the past arisen on both counts.

If (m) labels the space and (m,Γ) the subspaces we have the linear decompositions

$$(m) = \sum_{\Gamma} (m,\Gamma) \ ; \ \langle\langle G\rangle\rangle^{m} = \sum_{\Gamma} \langle\langle G\rangle\rangle^{m,\Gamma} \ ; \ I_{m}(x) = \sum_{\Gamma} I_{m\Gamma}(x) \qquad (31)$$

so that the formal problem is the evaluation of traces over the irrep subspaces and the construction of the (interacting) partial densities. In many cases, as often with configurations, it will be adequate to deal with lower-order moments than for fairly large unpartitioned spaces; the Gaussian assumption for the subspace densities is often adequate, except perhaps for those which are centered near the ground-state region.

3.2 Propagation of Information

If we change a single matrix element in the two-body interaction Hamiltonian the effect of this change propagates in some complicated way throughout the m-particle spaces and symmetry subspaces defined by the same set of single-particle states. But this really should be of little interest to us. It is true that the many-particle spaces may be immensely more complicated than the two-particle one in which H is defined, but the CLT filtering away of most of the information, and the fact that the fluctuations carry little if any, generate such simplicities that we should not have to worry about such microscopic things as the propagation of single matrix elements. The real things to be concerned with are traces over U(N) irreps (the m-particle spaces) and irreps of U(N) subgroups (occasionally of other groups which we shall not discuss). The simplicity shows up in the way in which these traces propagate throughout the set of subspaces. For example, with several important symmetry decompositions, the many-particle traces are simply linear combinations of those in the defining space, with coefficients which are calculable in terms of the eigenvalues of Casimir operators.

Observe now that this simplicity of trace propagation, coupled with CLT action in the irrep spaces, gives us a greatly expanded view of the spectrum discussed at the end of the first section. For we have now an extension of that structure to cover systems of different particle numbers, and of decompositions according to a very large array of symmetries. That this kind of description is appropriate, and not simply an artifice, derives from the relationships between the quantities (traces) which define everything, and the way in which they are exhibited as functions of the Hamiltonian matrix elements. There is no counterpart for example in conventional shell-model calculations.

Since propagation is discussed in many places[1-3,8] we restrict ourselves to comments about some special features. For U(N) propagation (of m-particle traces) (8) gives the m-particle trace of a k-body operator as a simple multiple of its defining trace $\langle F(k)\rangle^k$ (observe that we describe both $\langle\langle G\rangle\rangle^m$ and $\langle G\rangle^m$ as traces). Then for F, an operator of mixed particle ranks $\leq u$, $\langle F\rangle^m$ is a polynomial of order u in m, representable therefore as a linear combination of any set $\langle F\rangle^{t_i}$ with i=0,1,...u. In practice we would choose the set $\{t_i\}$ with care since many-particle traces may not be easy to calculate. If F is of fixed particle rank ν and, to within a sign (which obviously must be $(-1)^{\nu(2\nu-1)}$) invariant under hole $\leftrightarrow$ particle transformations, $a_i^+ \leftrightarrow a_i$, i=1...N, then it should be clear that $\langle F(\nu)\rangle^m = 0$ unless $\nu = 0$, and

$$\langle F^2(\nu)\rangle^m = \binom{m}{\nu}\binom{N-m}{\nu}\binom{N-\nu}{\nu}^{-1}\langle F^2(\nu)\rangle^\nu \xrightarrow{m \ll N} \binom{m}{\nu}\langle F^2(\nu)\rangle^\nu, \quad (32)$$

so that in this case the trace of the square of an operator propagates in the simplest possible way.

There is something quite special about such an operator; it is an irreducible U(N) tensor[24,25], belonging to the irrep $[2\nu,1^{N-2\nu}]$ which we prefer to label by its column structure as $[N-\nu,\nu]$. From the fact that a general k-body operator has a structure $\psi(k)\times\psi^+(k)$, being defined by transformations in the k-particle space, while $\psi(k) \sim [k]$ it follows that F(k) transforms as $\sum_{\nu=0}^{k} F^\nu(k)$ where necessarily $F^\nu(k)$ contains as a factor $\binom{n-\nu}{k-\nu}$ (but no polynomial of higher order in the number operator n). Thus

$$F(k) = \sum_{\nu=0}^{k} F^\nu(k) = \sum_{\nu=0}^{k} \binom{n-\nu}{k-\nu} F^\nu(\nu), \quad (33)$$

which considerably simplifies the U(N) propagation and greatly simplifies the propagation for configurations, to which there is an immediate extension.

It should be clear that $F^\nu(k)$ is a k-body operator whose algebraic complexity is only that of $(\nu\leq k)$-body; i.e. transformations which it induces in a given m-particle space could be equally well induced by a ν-body operator. For a two-body interaction Hamiltonian, $H^0(2)=\langle H\rangle^2\binom{n}{2}$ is responsible for the average energy, $H^1(2)=(n-1)h(1)$ renormalizes the single-particle part of the total H, and the residual $H^2(2)$ is unitarily the most complex part of H. The decomposition defined by (33) is of course orthogonal with respect to the (unitary) norm defined by the trace of the Hermitian square $\langle F^+F\rangle^m$, i.e. with respect to a geometry whose relevance to the problems at hand is assured by the CLT.

Intuitively we might guess for the U(N) average that

$$\langle F(k)\rangle^m = \langle F(k)\rangle^k \times \{\text{weight with which (k) is found in (m)}\}, \quad (34)$$

in which the natural definition of the weight is

$$\langle \sum_{\alpha} \psi_{\alpha}(k)\psi_{\alpha}^{+}(k)\rangle^{m} = \langle e(k)\rangle^{m} = \langle \binom{n}{k}\rangle^{m} = \binom{m}{k} , \tag{35}$$

where the identification of the quadratic state sum follows, for example, from the fact that it is a k-body operator which gives unity on every k-particle state. We must ask whether this very pleasing picture, which gives of course the correct result, extends to more complicated averages.

Let us write the "hole-particle adjoint" ($a_i \leftrightarrow a_i^{+}$) of F as $F^{\times}$, and the (N-m)-particle state which is complementary to $\psi_{\alpha}(m)$ as $\psi_{\alpha_c}(N-m)$ (then by definition $\psi(N)=\psi_{\alpha_c}(N-m)\psi_{\alpha}(m)$). Since holes are fermions we have now

$$\langle\langle F\rangle\rangle^{m,\Lambda} = \langle\langle F^{\times}\rangle\rangle^{N-m,\Lambda_c} , \tag{36}$$

where (m,Λ) denotes an arbitrary subset of the m-particle states. By introducing "trace operators" $e(m,\Lambda)$ we find compact explicit forms for the trace $\langle F\rangle^{m,\Lambda}$, in terms of traces over the spaces in which F is *defined*. Let

$$\begin{aligned} e(m,\Lambda) &= e^{+}(m,\Lambda) = \sum_{\alpha\varepsilon\Lambda} \psi_{\alpha}(m)\psi_{\alpha}^{+}(m) , \\ e^{\times}(m,\Lambda) &= e^{+\times}(m,\Lambda) = \sum_{\alpha\varepsilon\Lambda} \psi_{\alpha}^{+}(m)\psi_{\alpha}(m) . \end{aligned} \tag{37}$$

Consider the case of a k-body operator

$$F(k) = \sum_{\beta'\beta} \langle k\beta'|F(k)|k\beta\rangle \psi_{\beta'}(k)\psi_{\beta}^{+}(k) , \tag{38}$$

where the $\langle k\beta'|F(k)|k\beta\rangle = \langle\psi_{\beta'}^{+}(k)F(k)\psi_{\beta}(k)\rangle^{0} \equiv F_{\beta'\beta}(k)$ are the defining matrix elements of F(k). Then

$$\begin{aligned} \langle\langle F(k)\rangle\rangle^{m\Lambda} &= \langle\langle F(k)^{\times}\rangle\rangle^{N-m,\Lambda_c} \\ &= \sum_{\substack{\beta\beta' \\ \alpha\varepsilon\Lambda}} F_{\beta'\beta}(k)\langle\psi_{\alpha_c}^{+}(N-m)\psi_{\beta'}^{\times}(k)\psi_{\beta}^{+\times}(k)\psi_{\alpha_c}(N-m)\rangle^{0} \\ &= \sum_{\substack{\beta'\beta \\ \alpha\varepsilon\Lambda}} F_{\beta'\beta}(k)\langle\psi_{\beta'}^{\times}(k)\psi_{\alpha_c}^{+}(N-m)\psi_{\alpha_c}(N-m)\psi_{\beta}^{+\times}\rangle^{0} \\ &= \sum_{\beta'\beta} F_{\beta'\beta}(k)\langle\psi_{\beta'}^{+}(k)e^{\times}(N-m:\Lambda_c)\psi_{\beta}(k)\rangle^{0} \end{aligned} \tag{39}$$

and thus

$$\langle\langle F(k)\rangle\rangle^{m\Lambda} = \langle\langle e^{\times}(N-m,\Lambda_c)F(k)\rangle\rangle^{k} = \langle\langle e(N-m,\Lambda_c)F^{\times}(k)\rangle\rangle^{N-k}. \quad (40)$$

We have used the essential equivalence of $\psi^{\times}$ and ψ^{+} to carry out commutations in (39) (it is to be able to do this that we have made the first transformation $F\to F^{\times}$) and again in the last form where we have used the hermiticity of $e^{\times}$. Note that the final forms in (40) involve a trace over the entire defining space. On the other hand since $e(k,\Gamma)$ acts, in the k-particle space, as a projection operator we have

$$\langle\langle Ve(k,\Gamma)\rangle\rangle^{k} = \langle\langle V\rangle\rangle^{k\Gamma}, \quad (41)$$

so that these traces are in fact restricted. Moreover, from (40),

$$\langle\langle e(k,\Gamma)\rangle\rangle^{m\Lambda} = \langle\langle e^{\times}(N-m,\Lambda_c)\rangle\rangle^{k\Gamma} = \langle\langle e(N-m,\Lambda_c)\rangle\rangle^{N-k,\Gamma_c}, \quad (42)$$

where $\langle e(k,\Gamma)\rangle^{m\Lambda}$ is the natural measure (the extension of (35)) for the weight with which the (k,Γ) space is "contained in" (m,Λ), the average $\langle\ \rangle$ rather than the trace $\langle\langle\ \rangle\rangle$ being used in order to give unity when the spaces coincide. Then, *inter alia* (42) gives the indicated relationship between the weight with which one space is contained in another and that with which the complement of the second is contained in that of the first. Note too that $e(k,\Gamma)$ represents the trace of a density-matrix operator.

These elementary manipulations and results are of consequence only when (m,Λ) defines an irreducible representation space of a U(N) subgroup, say U(N) ⊃ G. For then by construction the density operators $e(m,\Lambda)$ are G-scalars constructed from the U(N) generators (and similarly for more complicated subgroup chains). It is known however[26] that all such scalars are representable as polynomials in a finite set of independent scalars, which form the so-called "integrity basis"; and of course Casimir operators of G, and of any group higher up in the subgroup chain, are also G scalars. If these form the complete integrity basis the construction of the density operators is immediate and we have (C describing "Casimir" propagation)

$$\langle\langle F(k)\rangle\rangle^{m\Lambda} \xrightarrow{(C)} \sum_{\Gamma} \langle e^{\times}_{\Gamma_c}(N-m)\rangle^{k\Gamma}\langle\langle F(k)\rangle\rangle^{k\Gamma}, \quad (43)$$

which displays beautifully how the trace information propagates throughout all the subspaces, the propagation coefficient depending only on the pair of irreps involved, in fact, exactly as in (34), on the weight with which one irrep space is found in the other. This kind of propagation, which occurs in many important cases, including U(N) ⊃ U(N/t) × U(t) for isospin (for which the integrity basis is n, T^2), or spin-isospin SU(4), must not be thought of in terms of technicalities; instead, when coupled with the CLT, it gives quite direct information, expressed in terms of the Hamiltonian parameters, about the location, spreading and admixing of the symmetries involved, quite in consonance with our search for simplicity of behavior in many-particle systems.

The general features of propagation should now be clear. Things are more complicated in detail when the integrity basis is not entirely generated by Casimir operators[8,27], or when we need the propagation of non-scalar information (represented by "double-barred" traces), for which an extended integrity basis is needed. There are also dilute-system expansions in terms of the occupancies which promise to clarify the way in which symmetries are effective even with very complicated subspace decompositions. We cannot now discuss any of these things, nor the ways in which the goodness and significance of a given symmetry can be estimated. We do remark however that the simple application of configuration or configuration-isospin symmetries, in which one calculates and combines even a huge number of partial densities, appears to give a highly accurate theory of the level density for interacting particles.

IV. EXPECTATION VALUES AND SUM RULES

The things involved here are really applications of what we have been doing so far and since, our stress is on principles, we need discuss them only very briefly, extremely interesting though they are. Since sum rules determine expectation values we have really only one thing to talk about, $K(x)$, the locally averaged expectation values $\langle x|K|x\rangle$ of an operator K in H-eigenstates at energy x. We should discuss also its fluctuations. Since the strength sums are moments of the transition strength $|(x'|T|x)|^2$ distribution we should for completeness discuss that also, but we shall not[28].

For $K(x)$ we have immediately that

$$\rho(x)K(x) = \langle K\delta(H-x)\rangle^m \tag{44}$$

and expanding $\delta(H-x)$ in the polynomials $P_\mu(x)$ defined by $\rho(x)$ as weight function

$$K(x) = \sum_\mu \langle KP_\mu(H)\rangle^m P_\mu(x) , \tag{45}$$

where then truncating the series corresponds to a local smoothing. As a practical matter we can deal only with small values of μ so that in any case we average over the level-to-level spectrum fluctuations. If then our arguments following (29), that the power spectrum of the density expansion divides sharply into a secular and a short-wave-length domain, applies here also, there is a natural low limit to the expansion order in (45) which should give an excellent account of expectation values, at least when combined with a calculation of the variance or covariance function.

But where in (45) do we see any CLT convergence and how does (45) exemplify our argument following (2) that everything follows from the density $\rho(x)$? For the answers we write K(x) in a "linear response" form. With $H_\alpha = H + \alpha K$ we have, as an adult version of the trivial result $(x|K|x) = \left(\frac{\partial x_\alpha}{\partial \alpha}\right)_{\alpha=0}$, that

$$\rho(x)K(x) = -\left(\frac{\partial F_\alpha(x)}{\partial \alpha}\right)_{\alpha=0} . \tag{46}$$

But now with $\rho \Longleftrightarrow \{\mathcal{E}, \sigma^2, \sigma^\nu S_{\nu \geq 3}\}$ as in (3), where S_ν are any set of shape parameters, we have

$$\frac{\partial \rho_\alpha(x)}{\partial \alpha} = \frac{\partial \rho_\alpha(x)}{\partial \mathcal{E}} \frac{\partial \mathcal{E}(\alpha)}{\partial \alpha} + \frac{\partial \rho_\alpha(x)}{\partial \sigma^2} \frac{\partial \sigma^2(\alpha)}{\partial \alpha} + \sum_{\nu \geq 3} \frac{\partial \rho_\alpha(x)}{\partial S_\nu} \frac{\partial S_\nu(\alpha)}{\partial \alpha} ,$$

$$\frac{\partial \mathcal{E}(\alpha)}{\partial \alpha} = \int \rho(x)K(x)dx = \langle K\rangle^m ; \quad \left(\frac{\partial \sigma^2(\alpha)}{\partial \alpha}\right)_{\alpha=0} = 2\langle K(H-\mathcal{E})\rangle^m \tag{47}$$

and then

$$\begin{aligned} K(x) &= \langle K\rangle^m + \langle K(H-\mathcal{E})\rangle^m \frac{(x-\mathcal{E})}{\sigma^2} - \rho^{-1}(x) \sum_{\nu \geq 3} \left(\frac{\partial S_\nu}{\partial \alpha}\right)_{\alpha=0} \frac{\partial F(x)}{\partial S_\nu} \\ &= K^{CLT}(x) + \Delta K(x) , \end{aligned} \tag{48}$$

where $K^{CLT}(x)$ coincides with the first two terms of (45) and the other terms can be made to coincide by an appropriate choice of the shape parameters. Thus we have once again a rather sophisticated operation of the CLT. Relying on the CLT convergence of $\rho(x)$, we have a general assurance that the simple polynomial expansion (45) is also rapidly convergent. Note also that, since they are polynomial rather than exponential functions, the calculated values of K(x) are far less sensitive to errors than is the density itself. Finally observe that the CLT limit result for K(x), corresponding as it does only to centroid and scale deformations, should follow by the most casual consideration.

Once again polynomial expansions are of no value for fluctuations. Instead one uses the GOE, along with various results from multidimensional geometry, in order for example to calculate the variance of the expectation values. We mention only two things about this, the first that a sum rule quantity, e.g. $\langle x|T^{+}T|x\rangle$ will fluctuate very weakly (strongly) according as the number of intermediate states (final states for the transition) is large or small. This can be used to predict collectivities[28]. In a similar way the fitting of a GOE ensemble to a given nucleus, i.e. the determination of its parameter (which is what we need to complete the TRI problem; see the end of (2.7)) will follow from a calculation of the variance of the expectation value of the two-body interaction. We mention these things to make clear that, though separated in the power spectrum, fluctuations and averages are connected in different ways.

Finally, we have probably not stressed sufficiently the way in which the model-space geometry determines much of the behavior, consistent with our remark at the beginning that the methods used are by no means restricted to "statistical" problems.

ACKNOWLEDGMENTS

I am indebted to Mrs. Edna Hughes for the preparation of the manuscript.

This work was supported in part by the U. S. Department of Energy.

REFERENCES

1. B. J. Dalton, S. M. Grimes, J. P. Vary and S. A. Williams,eds. "Moment Methods in Many-Fermion Systems," Plenum, N.Y. (1980).
2. J. B. French and V, K. B. Kota, in Annual Reviews of Nuclear and Particle Science, Vol. 32, J. D. Jackson, ed., Annual Reviews Inc., Palo Alto (1982), in press.
3. J. B. French, in "Nuclear Spectroscopy," G. F. Bertsch and D. Kurath, eds., Springer, Berlin (1980).
4. F. J. Dyson and M. L. Mehta, J. Math. Phys. 4:701 (1963).
5. W. Feller, "An Introduction to Probability Theory and its Applications," Vol. 2, Wiley, N.Y. (1971).
6. H. Cramér, "Mathematical Methods of Statistics," Princeton University Press, Princeton, N.J. (1946).
7. I. Prigogine, "From Being to Becoming," Freeman, San Francisco (1980) p.89.
8. J. B. French and J. P. Draayer, in "Group Theoretical Methods in Physics" eds. W. Beiglböck, A. Böhm and E. Takasugi, Springer, Berlin (1979).
9. H. A. Bethe, Phys. Rev. 50:332 (1936).
10. M. Nomura, Prog. Theor. Phys. 48:442 (1972).
11. A. Gervois, Nucl. Phys. A184:507 (1972).
12. E. P. Wigner, SIAM Review 9:1 (1967).
13. E. P. Wigner, Ann. Math. 62:548 (1955).

14. K. K. Mon and J. B. French. Ann. Phys. (N.Y.) 95:90 (1975).
15. M. L. Mehta, "Random Matrices and the Statistical Theory of Energy Levels," Academic, N.Y., (1967).
16. J. B. French, P. A. Mello and A. Pandey, Ann. Phys. (N.Y.) 113:277 (1978).
17. U. Grenander, "Probabilities on Algebraic Structures," Wiley, N.Y. (1963).
18. A. Pandey, Ann. Phys. (N.Y.) 119:170 (1979).
19. R. U. Haq, A. Pandey and O. Bohigas, Phys. Rev. Lett. 48:1086 (1982).
20. A. Pandey, Ann. Phys. (N.Y.) 134:110 (1981).
21. S. S. M. Wong, private communication.
22. J. B. French, V. K. B. Kota, and A. Pandey, to be published.
23. A. Pandey and M. L. Mehta, Comm. Math. Phys., in press.
24. C. M. Vincent, Phys. Rev. 163:1044 (1967).
25. F. S. Chang, J. B. French, and T. H. Thio, Ann. Phys. (N.Y.) 66:137 (1971).
26. H. Weyl, "The Classical Groups," Princeton University Press, Princeton (1946).
27. C. Quesne, in reference 1.
28. J. P. Draayer, J. B. French and S. S. M. Wong, Ann. Phys. (N.Y.) 106:472,503 (1977).

SEARCH FOR NEUTRINO MASSES AND OSCILLATIONS

Petr Vogel

California Institute of Technology
Pasadena, CA 91125, U.S.A.

INTRODUCTION

In these lectures we shall discuss the various aspects of the neutrino mass problem. This has been an interesting story for several reasons. First, if neutrinos are massive one has to modify the "standard" (Weinberg-Salam-Glashow) theory of weak interactions. A nonvanishing neutrino mass would represent a strong argument for grand unification with its tremendous mass scale. Second, the search for neutrino mass involves an interplay of very different, and often isolated, subfields of physics. Thus we shall mention astrophysics, particle physics, nuclear physics, and even geophysics. It is quite unusual that results of so many disciplines contribute to a common goal.

Only a few selected references will be quoted here. There are several review articles[1-3] where details can be found. We shall mostly describe how one looks for a proof of a nonvanishing neutrino mass. However, we begin with a few purely theoretical remarks.

Let us begin with a brief description of the difference between the Dirac and Majorana neutrinos. The Dirac equation is obtained from the Lagrangian

$$L_D = \bar{\psi}(i\not{\partial})\psi + m\bar{\psi}\psi. \qquad (1)$$

The second term, the so called mass term, obviously describes the mass energy. The corresponding antiparticle is described by the wave function

$$\psi^c = i\gamma_2\psi^*. \tag{2}$$

Besides the mass term $\bar{\psi}\psi$ there are other similar Lorentz invariant expressions, namely

$$\bar{\psi}^c\psi^c = \bar{\psi}\psi \tag{3a}$$

and

$$m_1\,\bar{\psi}^c\psi + m_1^*\ \ \bar{\psi}\psi^c. \tag{3b}$$

The term (3b) would transform an electron into a positron and thus it would violate the conservation of electric charge. However, one cannot a priori exclude such a term for neutral fermions and thus we are led to the general mass hamiltonian

$$H_{mass} = d(\bar{\psi}\psi + \bar{\psi}^c\psi^c) + M\ (\bar{\psi}^c\psi + \bar{\psi}\psi^c). \tag{4}$$

The first term above (Dirac mass) conserves lepton number, while the second one (Majorana mass) violates it. We can diagonalize H_{mass} and obtain

$$\chi = \frac{\psi + \psi^c}{\sqrt{2}} = \chi^c, \qquad \phi = i\,\frac{\psi - \psi^c}{\sqrt{2}} = \phi^c, \tag{5}$$

where χ and ϕ are self-conjugate and, therefore, two component objects. The mass term is now

$$H_{mass} = (d + M)\ \bar{\chi}\chi + (d - M)\ \bar{\phi}\phi, \tag{6}$$

and the kinetic energy is also diagonal in χ and ϕ.

Thus we see that a massive fermion can be always described by a pair of two-component spinors. For a neutral fermion, therefore, we quite naturally have two possibilities: either $d \neq 0$ (Dirac neutrino) and the lepton number is conserved, or $M \neq 0$ (Majorana neutrino) and the lepton number is not conserved.

There is another relation between the two-component and four-component descriptions. The Dirac equation can be written in the form

$$E\psi_+ = (\vec{\sigma}\cdot\vec{p})\ \psi_+ + m\psi_- \tag{7}$$

$$E\psi_- = -\ (\vec{\sigma}\cdot\vec{p})\ \psi_- + m\psi_-,$$

where ψ_+ and ψ_- are two-component spinors. In the ultrarelativistic limit $m/E \to 0$ these two spinors become independent and the helicity is conserved. Further, we know that the particle and the corresponding antiparticle have opposite helicities. Now we see that even if the neutrino Lagrangian contains the Majorana mass term, the transformation of a left-handed neutrino into a right-handed antineutrino will be suppressed by $(m/E)^2$.

Hence our present confusion. In most cases we are dealing with ultrarelativistic neutrinos. Even if we can somehow establish that their mass does not vanish, it will be difficult to decide whether we are dealing with Dirac or Majorana particles. In the limit of $m \to 0$ and standard theory of weak interactions there is no distinction between Dirac and Majorana neutrinos. The distinction becomes important if either the neutrinos have a nonvanishing mass or if both left-handed and right-handed currents participate in the neutrino weak interactions.

If we want to introduce massive neutrinos into the theory we have to go beyond the standard Weinberg-Salam-Glashow model. That is so because in the standard model there are only left-handed neutrinos (and right-handed antineutrinos). We see from Eq. (7) that this is possible only when $m = 0$.

Once we accept $m \neq 0$ we have to understand why the mass is so small. That is relatively easy in grand unified theories where the smallness of m_ν is related to the large value of the unification scale. Very crudely one obtains

$$m \sim \frac{\lambda^2}{M_X} , \tag{8}$$

where $M_X \sim 10^{14}$ GeV is the unification scale and $\lambda \lesssim 250$ GeV is the "ordinary" scale (vacuum expectation value giving mass to familiar particles). Thus we have a hint that $m_\nu \sim 10^{-1}$ eV is not unreasonable (albeit uncertain by several orders of magnitude). Let us note that the grand unified theories usually also predict a highly nondegenerate neutrino mass spectrum (related to the highly nondegenerate charged lepton or quark spectrum), and that Majorana neutrinos are prefered.

To conclude: It appears that for the time being almost anything is possible and it is worthwhile to look everywhere for manifestations of the neutrino mass. There is, however, a slight prejudice for light ($m_\nu \lesssim 10$ eV) nondegenerate neutrinos. The following sections describe the various forms of the search for neutrino mass.

COSMOLOGICAL AND ASTROPHYSICAL CONSIDERATIONS

Although it is beyond the immediate interests of most nuclear physicists, it is worthwhile to mention the relation of the neutrino mass problem to cosmology and astrophysics[4].

Cosmology gives us an important upper limit on the neutrino mass. The arguments go as follows: From the age and expansion rate of the universe one concludes that the present average mass density in the universe is limited by

$$\rho \leq 2 \times 10^4 \text{ eV/cm}^3. \tag{9}$$

The blackbody 2.7^0 K background radiation has a number density $n_\gamma \sim 400$ photons cm^{-3}. If one believes in the Big Bang cosmology there is around us a similar neutrino background of a somewhat lower temperature $T_\nu \sim 1.9^0$K and correspondingly somewhat lower number density $n_\nu \sim 100 \text{ cm}^{-3}$ for each two-component neutrino species. Therefore, the heaviest stable neutrino cannot weigh more than

$$m_\nu \leq 2 \times 10^4/100 = 200 \text{ eV}. \tag{10}$$

(One often finds smaller limits in the literature, but I prefer to be rather conservative). Thus our prejudice for light neutrinos is supported.

It is often argued that astrophysics gives a clue for the actual value of the neutrino mass. Let us describe the arguments which deal with the large scale objects known as galactic halos. (Because these objects are so large it is, perhaps, not surprising that this neutrino mass estimate is not very far from the upper limit of Eq. (10)).

Observations of radiation of hydrogen atoms suggest that at distances $R \sim 3 \times 10^5$ light years from the galactic center the orbital velocity is $v_H \sim 300$ km/sec. From these values, and from the virial theorem, the enclosed mass is found to be $M \sim 4 \times 10^{45}$ g, exceeding by an order of magnitude the aggregate mass of visible stars. Could one relate this "dark" mass to massive neutrinos? [5]

Let us assume that the galactic halo is filled with neutrinos of mass $m_\nu = \chi$eV, that these neutrinos form a degenerate Fermi gas, and that the mean neutrino velocity is equal to v_H. Then the density is related to v_H by the relation

$$\rho_\nu = \frac{4\pi}{3} \frac{(m\, v_H)^3}{(2\pi h)^3} = 2200\, \chi^3 \text{ cm}^{-3}. \tag{11}$$

On the other hand, the average density is

$$\bar{\rho} = \frac{3M}{4\pi R^3 m_\nu} = \frac{2 \times 10^7}{\chi} \text{ cm}^{-3} \tag{12}$$

Assuming $\rho_\nu \sim \bar{\rho}$ we obtain $m_\nu = 10$ eV. This is, naturally, not a proof that the neutrino mass is indeed 10 eV, but shows that neutrinos of such a mass would have important astrophysical consequences.

NEUTRINO MASSES FROM MEASUREMENT OF ENERGY AND MOMEMTUM OF CHARGED PARTICLES

Neutrinos are produced in weak decays, usually together with other particles. Any possible neutrino mass will affect the energy and momentum of all outgoing particles, and one can deduce its value from measurement of their spectra.

The two-body decays $M \rightarrow \ell + \nu$ are the simplest to analyze. The neutrino mass is related to the lepton momentum by the formula

$$m_\nu^2 = m_M^2 + m_\ell^2 - 2m_M \, (p_\ell^2 + m_\ell^2)^{\frac{1}{2}}. \qquad (13)$$

This is very straightforward, but not very practical experimentally (quadratic dependence). For example, in order to determine the neutrino mass of 10 eV in pion decay one would have to know the pion mass, the muon mass, and the muon momentum to 14 digits each.

If the neutrino produced in a weak decay represents a superposition of several mass eigenstates (neutrino mixing), the lepton ℓ will have several discrete values of momentum, each related to the corresponding m_ν value by Eq. (13). The probability of each momentum will depend on the corresponding mixing amplitude. (The dependence of decay rate on neutrino mass is rather complicated. The detailed theory for massive mixed neutrinos was developed by Schrock[6]). If, further, some of these m_ν values are relatively large, there is a hope of seeing additional peaks in the lepton momentum spectrum. Figure 1 shows the present status of the search for heavy weakly admixed neutrinos. No evidence has been seen so far, and rather stringent limits on the mixing amplitude have been established.

Three-body decays offer a better opportunity to find small neutrino masses. In such decays there is always a region of the momentum space in which neutrinos have very little energy. If they are massive, they can become nonrelativistic, and their energy will depend linearly on mass. Of course, there is a price to pay. The corresponding corner of the phase space is not much populated, with only $(m_\nu/Q)^3$ of the total number of decays there (Q is the usual Q value of the decay). Hence one would like to have a strong source with a small Q value for such experiments. In nuclear beta decay the finite neutrino mass will cause a deviation from the straight Kurie plot $(N/E{\cdot}p{\cdot}F)$ at distances m_ν from the corresponding endpoint

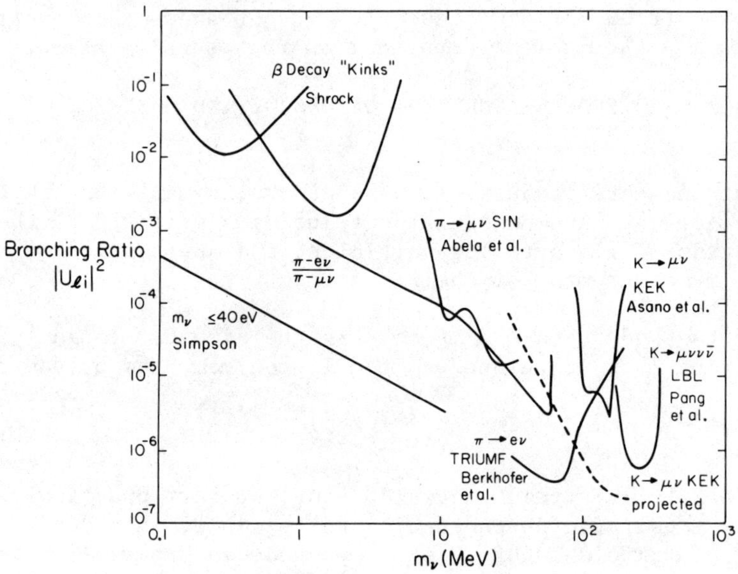

Fig. 1. Limits for branching ratio for decay into heavy neutrino based on variety of experiments (Refs. 6-11). (Reproduced from Ref. 12)

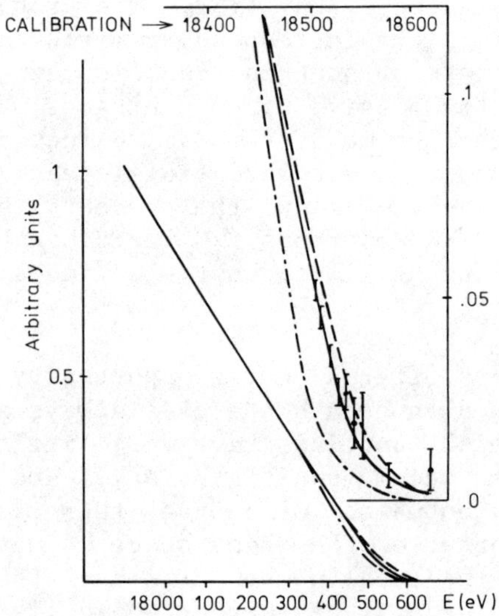

Fig. 2. The Kurie plot of tritium beta decay[13]. The χ^2 fit for different m_ν values are also shown (solid line m_ν = 37 eV, E_0 = 18578 eV; dashed line m_ν = 0, E_0 = 18574 eV; dot-and-dashed line, m_ν = 80 eV, E_0 = 18586 eV).

$$E_{Max} = [M(Z)^2 + m_e^2 - (M(Z+1) + m_\nu)^2] \Big/ 2M(Z), \qquad (14)$$

where M(Z) (M(Z+1)) are the initial (final) nuclear masses.

The ^3H beta decay is the most popular, and clearly the most suitable, case. In the famous experiment of the ITEP Moscow group[13] evidence was found for a finite neutrino mass $14 \leq m_\nu \leq 46$ eV. The Kurie plot of that experiment is shown in Fig. 2. It was a very difficult experiment and it would be very important if it could be independently tested. Unfortunately, in the two years since the announcement no one has been able to challenge it. This also means that one cannot expect a substantial improvement of sensitivity of similar experiments anytime soon.

There are other possible three-body decays. The lowest Q value seriously considered is the electron capture of ^{163}Ho with Q = 2.3 ± 1.0 keV[14]. This is not really a three-body decay, but a bremsstrahlung photon spectrum can be used in a way similar to the electron spectrum in beta decay[15]. The practical problems are, however, formidable.

We can thus conclude that measurement of charged particles (or possibly photons) offers a very direct method of neutrino mass determination. Unfortunately, it appears difficult to achieve sensitivity below ~ 10 eV, which is just in the most promising region. On the other hand, the technique is very well suited for study of heavy, weakly coupled neutrinos.

DOUBLE BETA DECAY

Double beta decay is a second order semileptonic weak process in which the initial nucleus A,Z makes a direct transition into the final nucleus A,Z + 2*. The process is observable if the intermediate nucleus A,Z + 1 has larger mass than A,Z because the initial nucleus is then stable with respect to ordinary beta decay. There are several suitable pairs of nuclei in nature (~ 25), all even-even ones. The total transition energy (Q value for the $Z \rightarrow Z+2$ transition) varies between a fraction of an MeV (for example $\varepsilon_0 = 0.872$ MeV for ^{128}Te) to several MeV (for example $\varepsilon_0 = 2.543$ MeV for ^{130}Te, $\varepsilon_0 = 2.045$ MeV for ^{76}Ge, and $\varepsilon_0 = 4.267$ MeV for ^{48}Ca).

The process can proceed in two ways as illustrated schematically in Fig. 3. The standard second order process (2ν $\beta\beta$ decay) $[A,Z] \rightarrow [A,Z+2] + 2e^- + 2\bar{\nu}$ is always possible. The interesting process for us is the neutrinoless transition (0ν $\beta\beta$ decay) which, if it exists, would violate lepton number conservation and possibly signal an existence of a Majorana neutrino mass.

*Recent reviews[16] contain original references on $\beta\beta$ decay.

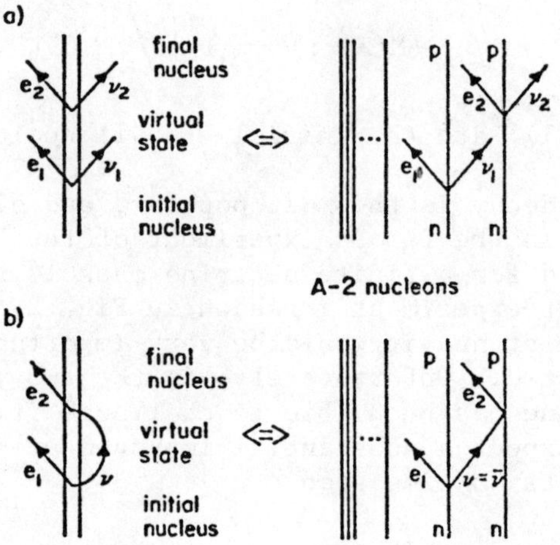

Fig. 3. Two-nucleon mechanism for (a) two-neutrino and (b) no-neutrino ββ decay.

The 2ν ββ decay is for our purpose an unwanted background. We shall discuss it nevertheless because it allows us to appreciate the time scales and nuclear physics problems involved. The rate is governed by the second order perturbation matrix element

$$\sum_m \frac{\langle f|H_W|m\rangle \langle m|H_W|i\rangle}{E_m - E_i} . \tag{15}$$

It is generally assumed that Gamow-Teller transitions dominate the sum over intermediate states and that the strength is concentrated so that one can use the closure approximation to evaluate (15). The rate for the 2ν ββ decay is then given by the formula

$$\Gamma^{2\nu} = 2\times 10^{-21}\, F_{2\nu}(\varepsilon_0)\, C(\varepsilon_0, Z)\, \frac{|M_{GT}|^2}{\langle \Delta E_N + \varepsilon_0/2 + \rangle^2}\ \text{years}^{-1} \tag{16}$$

Here $F_{2\nu}(\varepsilon) \sim \varepsilon^{11} + 22\varepsilon^{10} + \ldots$ is the 4-fermion phase-space factor, $C(\varepsilon,Z)$ is the Coulomb correction, M_{GT} is the matrix element of the operator $(\vec{\sigma}\cdot\vec{\tau})_1\ (\vec{\sigma}\cdot\vec{\tau})_2$, ε_0 is the maximum kinetic energy of the electrons, and ΔE_N is the average excitation energy in the intermediate nucleus (here all energies are in units of $m_e c^2$). A rough estimate of the 2ν halflife is $T^{2\nu} \simeq 10^{22\pm 2}$ years; the main source of uncertainty is the nuclear matrix element M_{GT}. The electron spectrum is a continuous, bell shaped function of the total electron kinetic energy. Experimentally, if one wants to identify the 2ν decay unambiguously, one has to distinguish this continuous spectrum from the ever present background.

The 0ν $\beta\beta$ decay proceeds by a virtual neutrino exchange. It is possible only if an antineutrino can somehow turn itself into a neutrino (Majorana particle) and at the same time the emitted (absorbed) particle is not completely right handed (left handed). Two mechanisms can achieve this: (a) finite neutrino Majorana mass m_ν (amplitude proportional to m_ν/E due to helicity conservation); or (b) explicit admixture (amplitude η) of the right handed lepton current in the weak hamiltonian. The analog of Eq. (16) for the 0ν $\beta\beta$ decay is

$$\Gamma^{0\nu} = 5\times10^{-16}\, F_{0\nu}(\varepsilon_0)\, C(\varepsilon_0, Z)\left|\frac{M_{GT}}{< r_{ij} > m_p}\right|^2 \text{years}^{-1}. \qquad (17)$$

The phase-space factor above reflects the two possible mechanisms

$$F_{0\nu}(\varepsilon) = m_\nu^{\ 2}\, f_m(\varepsilon) + m\eta f_{m\eta}(\varepsilon) + \eta^2 f_\eta(\varepsilon), \qquad (18)$$

and the functions reflect the 2-fermion phase-space nature, e.g. $f_m(\varepsilon) \sim \varepsilon^5 + 10\varepsilon^4 + \ldots$ (for details see Rosen[16]). The $< r_{ij} >$ in Eq. (17) is the average internucleon distance and m_p is the nucleon mass. (Actually, the matrix elements in Eqs. (16) and (17) are only approximately the same).

The factor in front of Eq. (17) is $\sim 10^5$ times larger than the similar factor in Eq. (16) and correspondingly one gets roughly $T^{0\nu} \simeq 10^{15\pm2}\, m_\nu^{-2}$ years. This increase in rate is partially explained by the large momentum of the virtual neutrinos $p \simeq \hbar/R_{Nucl} \sim 35$ MeV/c. The signature of the 0ν $\beta\beta$ decay is obvious, a single peak at $\varepsilon_1 + \varepsilon_2 = \varepsilon_0$ in the summed electron spectrum. This feature makes an experimental search for 0ν $\beta\beta$ decay very attractive.

There are two ways to study double beta decay experimentally. In the geochemical method one uses old ores containing parent nuclei and counts the daughter nuclei in them. In this way one can establish the total lifetime ($0\nu + 2\nu$) and this was done with varying degrees of precision for noble gas (Xe, Kr) daughters.

The counter experiments, on the other hand, can distinguish the two possible modes of the $\beta\beta$ decay. They lose, however, the $\sim 10^9$ advantage of counting time in comparison with the geochemical methods. Nevertheless, good upper limits were obtained for a number of 0ν $\beta\beta$ decays. For example, Fiorini et al.[17] obtained the limit $T^{0\nu} > 5\times10^{21}$ y for $^{76}Ge \rightarrow {}^{76}Se$. In order to interpret this result as an upper limit for the Majorana neutrino mass, we have to know the matrix element M_{GT} in Eq. (17). Can one calculate such a matrix element reliably ?

The 2ν decay of $^{82}Se \to {}^{82}Kr$ illustrates the problems encountered. Two geochemical measurements[18,19] give essentially consistent lifetimes $T \sim 1.5 \times 10^{20}$ y while the laboratory experiment[20] (the only laboratory $\beta\beta$ decay with a positive result) gives $T^{2\nu} = 1.0 \pm 0.4 \times 10^{19}$ y. Shell model calculations of Haxton et al.[21] agree with the shorter laboratory lifetime, and consequently underestimate the geochemical lifetime. If we ignore these problems and blindly use the shell model matrix element[21] for ^{76}Ge we obtain $m_\nu < 20$ eV showing that the $\beta\beta$ decay experiments have reached the same (or better) sensitivity as the ^3H beta decay.

Is there then some positive evidence for neutrino mass in $\beta\beta$ decay? The answer lies in the $^{128}Te/^{130}Te$ puzzle. Both these isotopes can undergo $\beta\beta$ decay, but with very different Q values (0.872 MeV and 2.543 MeV). Assuming that the nuclear matrix elements are similar, one can predict the 2ν lifetime ratio

$$T^{2\nu,128}/T^{2\nu,130} \simeq F_{2\nu}^{130}/F_{2\nu}^{128} = 5600 .$$

The 0ν lifetime ratio contains very different phase-space factors and one obtains, for example

$$f_m^{130}/f_m^{128} = 31,$$

and only slightly larger ratios for f_η and $f_{m\eta}$. Thus, if the experimental ratio of lifetimes is significantly smaller than 5600 we can argue that the deviation signifies presence of the 0ν mode.

The geochemical evidence is summarized in Fig. 4 and is clearly contradictory. The experiment of Hennecke would imply $m_\nu \sim 10$ eV, but the experiment of Kirsten is compatible with $m_\nu = 0$.

There is yet another problem, the absolute lifetime of ^{130}Te. The various geochemical determinations all basically agree and give $T \simeq (1\text{-}3) \times 10^{21}$ y. However, shell model calculations of Haxton et al.[22] give ~ 100 times shorter lifetimes, and thus the disagreement is even worse than for ^{82}Se.

So, what is the moral of the double beta decay story? This is an important and potentially powerful source of information. However improvement of experimental sensitivity of 0ν decay, resolution of contradiction between the different geochemical results, and solution of the theoretical problems with the matrix elements are the necessary ingredients of further progress.

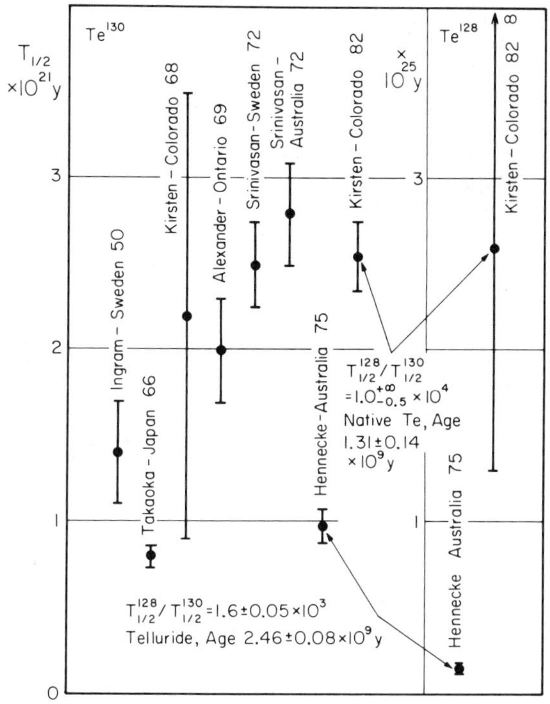

Fig. 4. Lifetime data for ^{130}Te and ^{128}Te. The references are M.G. Inghram and J.H. Reynolds. Phys. Rev. 78, 822 (1950); N. Takaoka and K. Ogata, Z. Naturf. 219, 84 (1966); T. Kirsten et al., Phys. Rev. Lett. 20, 1300 (1968); E. Alexander et al., Earth Plan. Sci. Lett. 5, 478 (1969); B. Srinivasan et al., J. Inorg. Nucl. Chem. 34, 2381 (1972); E.W. Hennecke et al., Phys. Rev. C11, 1378 (1975); T. Kirsten et al., to be published, (1982).

NEUTRINO OSCILLATIONS

The methods discussed so far are limited in sensitivity to neutrino masses $m_\nu \geq 10$ eV. It appears that the only practical way to study smaller masses is through the dynamic effect of neutrino oscillations. The idea of oscillations is well known[1,2], and we shall explain it only briefly.

The basic assumption is that the "physical" neutrinos ν_ℓ, that is the particles emerging from weak decays and associated each with its charged lepton partner ℓ, do not have a definite value of mass. Instead, they are superpositions of "mass-eigenstate" neutrinos ν_i, i.e.,

$$\nu_\ell = \sum_i U_{\ell i} \nu_i , \tag{19}$$

where the summation is over all mass eigenstates i. It is assumed that the unitary matrix U has at least one nonvanishing nondiagonal matrix element. The time development of the mass-eigenstates is described by a single phase ($p \gg m_i$)

$$\nu_i(t) = \nu_i(0) e^{-iE_i t} \simeq \nu_i(0) e^{-i(p + \frac{m_i^2}{2p}) t} . \tag{20}$$

Consequently, if one creates at $t = 0$ the physical neutrino ν_ℓ (for example in a weak decay), and if the initial state is properly selected so that the coherence conditions are obeyed[2], the time development of the ν_ℓ state is described by the superposition

$$\nu_\ell(t) = \sum U_{\ell i} \nu_i e^{-i(p + \frac{m_i^2}{2p}) t} , \tag{21}$$

and we encounter typical interference effects.

The only practical way to detect neutrinos is through their weak interaction. Thus one does not detect the individual terms in Eq. (21), but their combination corresponding to a certain weak interaction eigenstate, Eq. (19). In particular, the amplitude to detect a neutrino of flavor ℓ' at the time t using a source of neutrinos of flavor ℓ emitted at $t = 0$ is

$$A_{\ell\ell'} = \sum e^{-it \frac{m_i^2 - m_j^2}{2p}} U_{\ell i} U^*_{\ell' i} , \tag{22}$$

and the corresponding probability is the square of this amplitude

$$P_t(\ell \to \ell') = \sum_i |U_{\ell i}|^2 |U_{\ell' i}|^2 + \sum_{i \neq k} U_{\ell i} U_{\ell' k} U_{\ell' i} U_{\ell' k} \cdot e^{-it \frac{m_i^2 - m_k^2}{2p}} \tag{23}$$

This is an oscillating function of time t or distance $L = ct$, hence neutrino oscillations. The phase relation is governed by the factor (in practical units)

$$\Delta_{ik} = \frac{2.54(m_i^2 - m_k^2)(eV)^2 \cdot L(m)}{E(MeV)}, \tag{24}$$

and the corresponding wave length $\Delta = 2\pi$ is

$$L_{osc}(m) = \frac{2.5\ E(MeV)}{|m_i^2 - m_k^2|\,(eV)^2}\ . \tag{25}$$

The oscillations do not continue for ever. Far away from the source the coherence is lost and only the first constant term in Eq. (23) survives; the signal then does not depend on the neutrino masses, or on the distance from the detector.

There are two classes of possible experimental arrangements: Diagonal (disappearance) tests involve the probability $P(\ell \rightarrow \ell)$. To prove oscillations one has to show that the probability varies with distance or at least that it is smaller than unity at a certain distance. The second type of test involves $\ell \neq \ell'$ (nondiagonal or appearance). Typically one starts with a beam of neutrinos of flavor ℓ and looks for the "wrong" flavor ℓ'. These tests depend only on certain part of the mixing matrix U.

Generally one should note that the phase Δ, Eq. (24) depends on the distance L and on the energy E and the two quantities are interchangeable; the real parameter governing oscillations is L/E.

Typically the distance L between the neutrino source and detector is given by the experimental conditions. The sensitivity to mass squared difference is then approximately

$$\Delta m^2 \equiv |m_1^2 - m_2^2| \simeq \frac{2.5E}{L} \times f, \tag{26}$$

where f depends on the apparatus, and usually $f \sim 0.1$. The farther the detector from the source, the smaller Δ values it can detect and lower energies are better than higher energies. This is why nuclear reactors play such a prominent role.

It is often assumed that only two neutrino flavors mix with each other. The matrix U is then characterized by a single parameter $U_{11} = U_{22} = \cos\theta$, $U_{12} = -U_{21} = \sin\theta$. The probability of appearance $P(\ell \rightarrow \ell')$ and disappearance $P(\ell \rightarrow \ell)$ are then given by

$$P(\ell \rightarrow \ell') = \frac{\sin^2 2\theta}{2}\,[1 - \cos(2.53\Delta m^2 L/E_\nu)],$$

$$P(\ell \rightarrow \ell) = 1 - P(\ell - \ell')\ . \tag{27}$$

The wave-length of the oscillations determines the mass parameter Δm^2 and the amplitude determines the mixing angle θ.

Accelerators make mainly μ-neutrinos and therefore many accelerator oscillation experiments examine the $\nu_\mu \to \nu_e$ channel. No evidence for oscillations has been seen so far, and the limits are $\Delta m^2 \geq 0.6$ eV^2 for $\theta = 45^0$ (maximal mixing) and $\sin^2 2\theta \lesssim 5 \times 10^{-3}$ for large Δm^2. The limits for $\nu_\mu \to \nu_\tau$ are several times worse. In the near future experiments at Brookhaven and LAMPF, among others, will improve the sensitivity to small Δm^2 in the $\nu_\mu \to \nu_e$ channel by ~10 times. The review talk of Baltay[23] summarizes the results of the accelerator neutrino oscillation studies.

The beam dump experiments should be mentioned separately. In them the beam is stopped in a dense block of material; the detectors downstream are able to select neutrinos produced either directly, or by the decay of short-lived ($T < 10^{-11}$ s) particles. It is assumed that the bulk of the prompt leptons is due to charm production and that the resulting beam should contain nearly equal numbers of the muon and electron neutrinos. Yet the first series of CERN experiments[24] imply a 1:2 ratio between the electron and muon neutrinos. The effect remains unexplained, but it is generally agreed that it is not caused by neutrino oscillations. The oscillations would demand values of Δm^2 and θ excluded by other experiments. Further work on this problem is in progress at CERN and at Fermilab.

Another topic worth mentioning is that of atmospheric neutrinos observed with detector deep underground. These neutrinos are produced by pions decaying in the atmosphere, and are, therefore, essentially only muon neutrinos. By measuring the neutrino flux for different zenith angles, one can vary the effective pathlength L between a few and 12000 kilometers. In 1978 Crouch et al.[25] measured the horizontal neutrino flux and got only 0.62 ± 0.17 of the expected signal. Could this slight discrepancy imply neutrino oscillations? Apparently not, because in the more recent measurement at Baksan[26] dealing with "upward" neutrinos one observes just the right amount, namely 0.98 ± 0.20. Nevertheless, with a further development of the experimental techniques one can expect that studies of atmospheric neutrinos will play an important role.

At the present time the reactor neutrino studies represent the most sensitive probe of neutrino oscillations. Nuclear reactors produce copious amounts of electron antineutrinos, with an energy spectrum extending to ~ 8 MeV. Due to this small energy one can study only the "disappearance" of $\bar{\nu}_e$. Table 1 shows the four reactions which have been observed and which can be used as antineutrino detectors.

Table 1. Reactor Antineutrino Induced Reactions

Reaction	Symbol	σ^{a}_{tot} 10^{-44} cm^2/fiss	Threshold MeV	Max. Yield at MeV
$\bar{\nu} + p \to n + e^{+}$	ccp	63.4	1.8	3.8
$\bar{\nu} + d \to n + n + e^{+}$	ccd	1.13	4.0	5.2
$\bar{\nu} + d \to n + p + \bar{\nu}$	ncd	3.10	2.3	4.5
$\bar{\nu} + e \to \bar{\nu} + e^{-}$		1.37[a]	-	-

[a]Integrated over the final electron energies 1-6 MeV.

The charged current reactions (ccp, ccd) are sensitive to neutrino oscillations. (The $\bar{\nu}_e$ scattering is a combination of charged and neutral current reactions; it is also somewhat sensitive to neutrino oscillations). In the ccp reaction one can measure the positron energy which is simply related to the $\bar{\nu}$ energy, $T(e^{+}) = E(\bar{\nu}) - 1.8$ MeV. The kinematics of the ccd reaction is more complex and only the integrated cross section can be measured.

The positron yield in the ccp reaction is given by (neglecting the effect of the reactor and detector sizes, and the energy resolution)

$$Y(E_{e^+}, L) = \frac{n_p \cdot \varepsilon}{4\pi L^2} \sigma(E_{\bar{\nu}}) \, S(E_{\bar{\nu}}) \, P_L(\ell \to \ell). \tag{28}$$

Here n_p is the number of protons in the detector, ε is the efficiency, $P(\ell \to \ell)$ is the function (27) depending on the unknown parameters Δm^2 and θ, $\sigma(E)$ is the known cross section for monoenergetic antineutrinos, and $S(E)$ is the spectrum of $\bar{\nu}$ produced in the reactor core at $L = 0$. Thus one can either measure Y at two or more distances (movable detector, essentially no knowledge of S(E) and ε necessary), or at one distance absolutely (ε and S(E) needed).

All four "modern" ccp experiments [27-30] are of the "one distance" type and rely on the knowledge of S(E), even though the most recent one[30] is really the first stage of the "two distances" experiment. (Two other similar experiments are running or are about to run). Thus we should discuss the status of our knowledge of the reactor $\bar{\nu}$ spectrum.

There are two ways to determine this spectrum, independent of detecting antineutrinos. In the summation method one adds the contributions of all branches of all beta decaying fission fragments (see e.g. Ref. 31 where other related papers are quoted). The difficulty is that beta decay of many short lived, high Q value fission fragments has not been studied and one has to use some theoretical model for them. There is no general consensus on how to construct such a model, and even less agreement about the associated uncertainties. The other difficulty is related to the "known" nuclei. It is sometimes hard to judge how complete (and uncertain) the individual decay schemes are, and the possibility of introduction of systematic errors is serious. According to my own work[31] the resulting uncertainty of the spectrum gradually increases from ~ 6% at 1-2 MeV to ~ 30% at 7-8 MeV. Others are much more optimistic. In any case, the summation method gives us a good idea about the general character of the spectrum. It also allows us to calculate quite reliably the relative change of the $\bar{\nu}$ spectrum associated with different fuels (^{235}U, ^{239}Pu, ^{238}U, ^{241}Pu, etc.)

The other method of spectrum determination relies on the one-to-one correspondence between the electron and antineutrino spectra. First one has to measure accurately the electron spectrum. In the next step this spectrum is "converted" into the antineutrino spectrum[31,32]. In this way the $\bar{\nu}$ spectra of ^{235}U and ^{239}Pu have been obtained to ~ 5% accuracy[32]. As an added benefit the electron spectrum measurement constitutes a severe test of the summation method calculations. The agreement is often quite good and the uncertainties quoted above appear too conservative. Thus, taking all this together it appears that we have a reliable, ~ 5% accurate spectrum S(E). It is, therefore, possible to analyze the "one distance" ccp experiments in terms of the oscillation parameters Δm^2 and θ.

Figure 5 shows the analysis of the most recent ccp experiment[30]. No evidence for oscillations is seen and limits on the values of the oscillation parameters are obtained. The other experiment by the same group[29], performed at ILL at L = 8.7 m, also did not show any evidence of oscillations. The 6 m experiment[27] is difficult to interpret because the background level is not known.

The data for L = 11.2 m need a special discussion. F. Reines and his collaborators performed two experiments at this distance[28]. The one widely known is the deuteron target experiment. In it the neutral current reaction ncd was used as a monitor of the neutrino flux. The experimental ratio of the ccd and ncd cross-sections was determined to be only 0.4 ± 0.2 of the expected value. When taken as an evidence for neutrino oscillations it would imply $\Delta m^2 \simeq 0.4 - 0.8$ eV^2 and $\sin^2(2\theta) \geq 0.5$ at the one standard deviation level. This is clearly excluded by the analysis shown in Fig. 5.

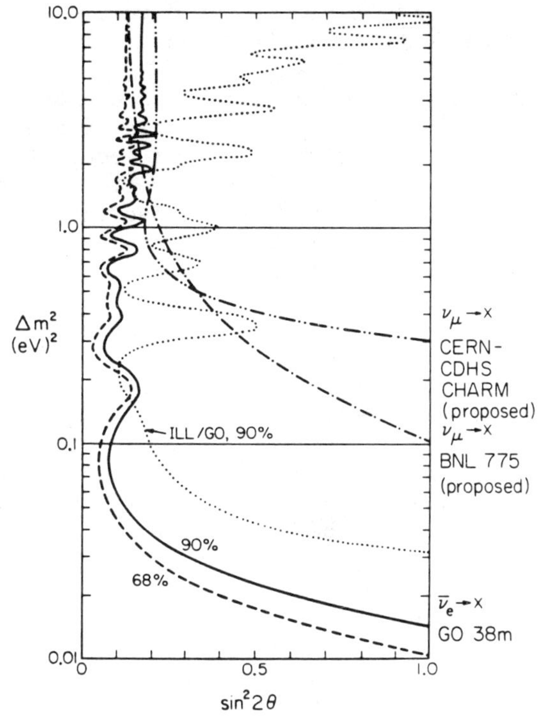

Fig. 5. Limits for neutrino oscillations $\nu_e \to X$. The solid (dashed) curves represent current experimental limits[30]. Limits obtained from the ratio of the ILL (8.7 m) and Gosgen (38 m) data are shown by the dotted curve. Forthcoming experiments on $\nu_\mu \to X$ are shown by broken lines (Reproduced from Ref. 12)

The same group also measured the ccp reaction at L = 11.2 m. The positron spectrum is softer than expected for energies above ~ 4 MeV and also implies oscillations, although with smaller mixing angles $\sin^2(2\theta) \simeq 0.3$. There is no explanation of the discrepancy between the deuteron results, the 11.2 m ccp result, and the other ccp results. However, one clearly cannot consider this evidence a proof of neutrino oscillations.

THE SOLAR NEUTRINO PUZZLE

The Brookhaven solar neutrino experiment has been operating since 1967. It is based on the $^{37}C\ell(\nu,e^-)\ ^{37}Ar$ reaction (threshold 0.814 MeV). The average result since 1970 is 0.39 ± 0.06 atoms/day[33] (background subtracted). It is customary to express the rates in SNU (solar neutrino unit) and the experimental value corresponds to 2.1 ± 0.3 SNU. On the other hand the predicted rate[34],

7.6 ± 1.1 SNU, is 3-4 times larger. Could this discrepancy have something to do with neutrino oscillations?

Suppose that the electron neutrinos indeed oscillate during the journey from the sun into some other neutrino flavors, which do not transform the Cℓ nuclei into Ar. The source-detector distance is so large that already $\Delta m^2 \geq 3 \times 10^{-12}$ $(eV)^2$ would be sufficient. Such a small neutrino mass differences are unobservable in experiments with terrestial neutrino sources. What about the magnitude of the effect? Assuming that the oscillating term in Eq. (27) is averaged out, the experiment would imply the value ~1/3 for the constant term. This is just the minimal value for mixing among three neutrino types. Thus it is theoretically possible that the apparent lack of solar neutrinos is caused by the full mixing between three neutrino flavors with very small values of Δm^2.

However, before making such a conclusion it is instructive to consider the difficulties encountered in deriving the predicted rate. Table 2 illustrates the solar proton-proton cycle and the contribution of the individual reactions to the expected ^{37}Cℓ rate. One can see that the bulk of the rate comes from a reaction with only 1.5×10^{-4} branching ratio. The rate of this reaction is quite sensitive to the various nuclear cross-sections and extremely sensitive to the temperature in the solar center. We can understand better all the effort devoted to the development of the ^{71}Ga based detector (threshold 0.23 MeV), which will be sensitive to the pp neutrinos and rather insensitive to the solar model. Until the Ga or some similar experiment is performed, it is difficult to attribute the lack of detectable solar neutrinos to neutrino oscillations.

Table 2. The Proton-Proton Chain in the Sun

Reaction	%	Max. ν energy (MeV)	^{37}Cℓ capt. rate SNU
$p + p \to {}^2H + e^+ + \nu$	99.75	0.42	
or			
$p + e^- + p \to {}^2H + \nu$	0.25	1.44	0.23
${}^2H + p \to {}^3He + \gamma$	100.00		
${}^3He + {}^3He \to {}^4He + 2p$	86.00		
or			
${}^3He + {}^4He \to {}^7Be + \gamma$			
${}^7Be + e^- \to {}^7Li + \nu$	14.00	0.86(90%), 0.38(10%)	1.02
${}^7Li + p \to {}^4He + {}^4He$			
or			
${}^7Be + p \to {}^8B + \gamma$			
${}^8B \to Be^* + e^+ + \nu$	0.015	14.1	6.05
${}^8Be^* \to {}^4He + {}^4He$			

CONCLUSIONS

The easiest way to summarize the preceding discussion is to put all the "evidence" together and consider whether it has been challenged and how it withstood the challenge. This is done in Table 3.

We see that, first of all, the different pieces of evidence have each something to do with neutrino mass, but they certainly do not form a coherent picture. Secondly, we see that all challenged evidence is contradicted by the newer results. The last two entries remain untested, but as explained before they represent at best a circumstantial evidence. Only the ^3H beta decay remains, awaiting challenge.

Table 3. Status of Evidence on Neutrino Mass

Experiment	Possible Interpretation	Challenge
Reactor $\bar{\nu}_e \to X$	Oscillations[28] $\Delta m^2 \sim 1(eV)^2$	no oscillations[29,30] $\Delta m^2 < 0.02(eV)^2$
^3H spectrum	mass[13] $14 < m_\nu < 46$ eV	no challenge
Beam dump	oscillations[24] Δm^2 large	no oscillations (CERN, NAL)
$\beta\beta$ decay in $T\ell$	mass (Hennecke et al) $m \sim 10$ eV	no mass (Kirsten) $m < 3$ eV
Deep mine	oscillations[25] $\Delta m^2 > 10^{-3}(eV)^2$	no oscillations[26]
Solar neutrinos	oscillations $\Delta m^2 > 10^{-11}(eV)^2$	no challenge but other explanations not excluded
Galactic halos	mass $m \sim 10$ eV	

REFERENCES

1. S. M. Bilenky and B. Pontecorvo, Phys. Rep. 41, 225 (1978).
2. P. H. Frampton and P. Vogel, Phys. Rep. 82, 339 (1982).
3. Ya. B. Zeldovich and M. Yu Khlopov, Usp. Fiz. Nauk 135, 45 (1981).
4. A. D. Dolgov and Ya. B. Zeldovich, Rev. Mod. Phys. 53, 1 (1981).
5. S. Tremaine and J. E. Gunn, Phys. Rev. Lett. 42, 407 (1979).

6. R. E. Schrock, Phys. Lett. 96B, 159 (1980); Phys. Rev. D24, 1232 and 1275 (1981).
7. R. Abela et al., Phys. Lett. 105B, 263 (1981).
8. Y. Asano et al., Phys. Lett. 104B, 84 (1981).
9. C. Y. Pang et al., Phys. Rev. D8, 1989 (1973).
10. D. Berghofer et al., Neutrino 81, Vol. II, p.67, Univ. of Hawaii, 1982.
11. J. J. Simpson, Phys. Lett. 102B, 35 (1981).
12. F. Boehm, Talk at the Int. Conf. on Novel Results in Particle Physics, Vanderbilt Univ., Nashville, Tenn., May 1982.
13. V. S. Kozik et al., Yad. Fiz 32, 301 (1980); V. A. Lubimov et al., Phys. Lett. 94B, 266 (1980), Soviet Phys. JETP 54, 616 (1981).
14. J. U. Andersen et al., Phys. Lett. 113B, 72 (1982).
15. A. De Rújula, Nucl. Phys. 188B, 414 (1981).
16. D. Bryman and C. Picciotto, Rev. Mod. Phys. 50, 11 (1978), Y. G. Zdesenko, Sov. J. Part. Nucl. 11, 582 (1981), and S. P. Rosen, Neutrino 81, Vol. II, p.76, Univ. of Hawaii, 1982.
17. E. Fiorini et al., Nuovo Cim. A13, 747 (1973).
18. T. Kirsten and H. Muller, Earth and Plan. Sci. Lett. 6. 271 (1969).
19. B. Srinivasan et al., Econ. Geol. 68, 252 (1973).
20. M. Moe and D. Loventhal, Phys. Rev. C22, 2186 (1980).
21. W. C. Haxton, G. J. Stephenson, and D. Strottman, Phys. Rev. Lett. 47. 153 (1981).
22. W. C. Haxton, G. J. Stephenson, and D. Strottman, Phys. Rev. D, to be published.
23. C. Baltay, Neutrino 81, Vol. II, p.295, Univ. of Hawaii, 1982.
24. M. Jonker et al., Phys. Lett. 96B, 435 (1980); P. Fritze et al., Phys. Lett. 96B, 427 (1980); F. Dydak, Neutrino-80, Erice, 1980 (Plenum Press New York, 1981).
25. M. F. Crouch et al., Phys. Rev. D18, 2239 (1978).
26. M. M. Boliev, Yad. Fiz 34, 1418 (1981).
27. F. A. Nezrick and F. Reines, Phys. Rev. 142, 852 (1966).
28. F. Reines, H. W. Sobel, and E. Pasierb, Phys. Rev. Lett. 45, 1307 (1980).
29. F. Boehm et al., Phys. Lett. 97B, 310 (1980); H. Kwon et al., Phys. Rev. D24, 1097 (1981).
30. J.-L. Vuilleumier et al., Phys. Lett., to be published.
31. B. R. Davis et al., Phys. Rev. C19, 2259 (1979); P. Vogel et al., Phys. Rev. C24, 1543 (1981).
32. K. Schreckenbach et al., Phys. Lett. 99B, 251 (1981); and to be published.
33. R. Davis Jr., in Neutrino Mass Miniconference, Telemark, Wisconsin, Oct. 80, Univ. of Wis. Report 186, p.38.
34. J. N. Bahcall et al., Rev. Mod. Phys., 54, 767 (1982).

ELECTRON SCATTERING

Ingo Sick

Dept. of Physics, University of Basel

4056 Basel, CH-Switzerland

EXPERIMENT

The experimental equipment used for electron scattering experiments obviously is of great diversity. Here, we want to describe a "typical" set by discussing some of the apparatus used at the three highest-energy facilities (Amsterdam, Bates and Saclay).

The existing electron accelerators suitable for nuclear physics investigations produce electron beams of up to 700 MeV energy. These linear accelerators deliver beams of an intensity of several 100 μA (average) with a relative energy resolution $\Delta E/E$ of a few times 10^{-3}, an emittance of 0.1/mm mrad and a duty cycle of up to a few percent. The beam intensity is often marginal, the relative energy resolution is more than an order of magnitude too bad for most nuclear physics experiments, while the duty cycle for the non-coincidence experiments of interest here is of no concern.

The principles determining the experimental set-up for beam preparation before the target and detection of the scattered electrons are governed by two considerations:

1) Due to the weakness of the electromagnetic interaction-the main asset of the electron as a probe of nuclei - the cross sections are very small; typically they are α^2 times smaller than the ones for hadronic probes. This smallness can be compensated experimentally to a very limited degree only through the use of thicker

targets or spectrometers of larger solid angle. In order to obtain sensible counting rates, the experimental set-up therefore must be designed to allow for the maximum beam intensity on target. (While many targets a priori could be destroyed by beams in the 10 μA-range, it is in the hands of the experimentalist to design appropriate target cooling systems, such that the maximum intensity available can be used).

2) The momentum transfer q to be reached in an electron scattering experiment is imposed by the nucleus which shows features of the size of a few tenths of fermis. The spatial resolution of the electron microscope, which amounts to roughly 1.5/q, has to be adapted to allow observation of the desired detail. This requires momentum transfers of the order of $4 fm^{-1}$, i.e. electron energies of the order of 500 MeV. The energy resolution, on the other hand, is imposed by the spacing of nuclear levels. In order to observe the lowest states of deformed nuclei (let alone higher lying levels), an energy resolution of the order of 50 KeV is required. The resulting relative energy resolution $\Delta E/E$ of 10^{-4} in the above example represents a formidable challenge.

The need to perform experiments at excellent $\Delta E/E$ and with the full beam intensity delivered by the accelerator has led to the general use of socalled energy loss spectrometer systems. In order to use the full beam despite its wide energy band, the beam is prepared such that electrons differing in energy by ΔE hit the target at points separated by a distance $x = D \cdot \Delta E$. The magnetic spectrometer used to energy-analyse the scattered electron is designed to, in principle at least, measure this coordinate as well.

For most energy-loss systems in use, the coordinate x actually is not measured. The spectrometer is designed to have, for a given point in the focal plane, an energy dispersion dx/dE identical to - D. In this case a measurement of the coordinate of the electron trajectory in the focal plane yields directly the quantity of interest, the electron energy loss, which amounts the sum of nuclear excitation plus recoil energy.

For obvious reasons, the magnetic system that transports the beam to the different experimental halls deflects the beam in the horizontal plane. An energy dispersion in the horizontal plane is thus easily available, but does not fit the vertical dispersion required by most spectrometers, which are built to deflect in the vertical plane such as to achieve the largest range in scattering angles. A special device to move the dispersion from the horizontal to the vertical plane is thus required.

Spectrometers with the energy dispersion in the plane orthogonal to the scattering plane are advantageous for a second reason. For experiments with good energy resolution the scattering angle has to be measured with a resolution superior to the one given by the solid-angle defining spectrometer slits. Only then can the variation of the recoil nucleus kinetic energy be corrected for, and the excitation energy determined with the desired resolution. The decoupling of scattering angle and energy dispersion is achieved easily if they occur in orthogonal planes.

With these general considerations, we can discuss the various elements of the apparatus used.

Beam deflection system

The beam deflection systems used at different facilities are quite different, and specific of the site used. For this particular subject, I therefore will discuss one particular set-up only. For reasons of familiarity I chose it to be the one of the Saclay ALS accelerator and the HEI experimental hall (Fig. 1).

At the end of the accelerator, the beam is brought to a focus in the horizontal direction. At this point the position of the beam is constantly measured by a Ferrite monitor. The beam passes a fast switching magnet (E2) that allows, on a time-basis of several 10 ms, to switch from the main to a parasitic user. The magnet B4 deflects the beam by $\sim 45^{o}$, and focusses electrons of a given energy to a vertical line at the location of the energy slit. These slits are set to select from the beam a given band ΔE in energy; for the energy-loss mode operation $\Delta E/E$ is typically $1 \div 2 \cdot 10^{-3}$. Electrons outside this band, up to a maximum intensity of 100 μA, are energy degraded by the slit and ultimately removed.

The quadrupole lens QD has a double function. In the conventional achromatic beam transport system it refocusses electrons of different energies, separated spatially at the location of QD, to a single beam at the exit of magnet B5. In the energy loss mode, QD defocusses the beam in order to increase the energy dispersion produced by B4 by a factor of ~ 2. In this mode, QD allows to tune the dispersion to the desired value. The magnet B5 bends the beam by another 45° and sends it towards the experimental hall HEI. The main function

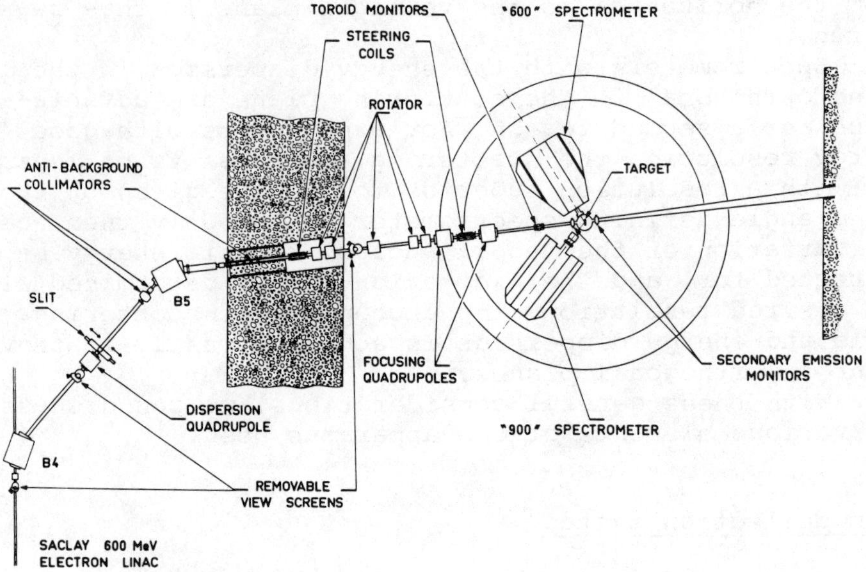

Fig. 1 HE1 Switchyard and experimental hall

of B5 is to clean up the beam by removing all lower-energy electrons created by showers originating in the energy slits. Various collimators before and after B5 absorb these low energy electrons.

At the entrance of the experimental hall, the energy dispersion of the incident beam is moved from the horizontal to the vertical plane. The rotator R1÷R5 consists of 5 (tilted) quadrupole magnets; by appropriate choice of strength and distance the exchange of horizontal and vertical coordinates is obtained without any further change of the optical properties of the beam. The set of quadrupole lenses QF1, QF2 produces, for a monochromatic beam, a horizontal line focus at the target. For an energy-dispersed beam of $\Delta E/E \sim 10^{-3}$ the vertical dispersion amounts to ~10mm.

A screen at the end of the accelerator allows to visually check beam position and the focussing to a vertical line required by beam transport optics. Another screen near B4 ensures alignment of the beam along the optical axis of the transport system. Screens near the center of the rotator, and at the location of the target determine the alignment of the beam along the 0°-axis of the spectrometer. Various horizontal and vertical steering magnets near S30, B4, R and QF allow to simultaneously fulfill all the conditions desired.

After passing the target, the beam is stopped in a Faraday cup. In order to avoid background, this beam-stop is located in a well-shielded separate hall. The particular cup used at Saclay has a useful aperture of 60 cm and is able to accept up to 100 kW of beam power.

The Faraday cup also serves as the primary reference for the integrated beam current. Despite the cooling with (very pure) water, leak currents are of the order of 1nA, and can be compensated by injecting a current of opposite polarity. Unless the electrons are refocussed after the target (as done at Amsterdam) a few percent of the electrons miss the cup due to multiple scattering in the target. Using a non intercepting toroid beam intensity monitor placed up-stream of the target, the losses of the Faraday cup can be determined from toroid to Faraday ratios with and without target. The overall precision on the absolute integrated charge derived from the Faraday cup is better than 1%.

Spectrometers

For an energy loss set-up, the magnetic systems for beam transport and scattered electron analysis form a single unit. They only differ in the emittance of the beam they have to deal with. We therefore will discuss next the spectrometer designs used.

The choice of the magnetic spectrometer is governed by a number of considerations. The primary criteria concern the energy resolution, the solid angle and, for certain applications, the momentum acceptance. Large angular range, the smallness of background or the shape of the focal plane are often treated as secondary.

The traditional type of spectrometer, which features a deflection of the scattered electron in an inhomogeneous field by ~180°, was introduced to electron scattering with the pioneering experiments performed at Stanford. In its modern version, as used for example at the Saclay ALS, this spectrometer features curved entrance and exit pole faces and achieves an intrinsic resolution of $\Delta E/E \sim 10^{-4}$ with a 5 msr solid angle and a 10% momentum accpetance. This type of double focussing spectrometer is very compact and provides, in combination with its large deflection angle and narrow focal plane, for excellent background supression.

The split-pole spectrometer developed for the

Bates accelerator differs in several respects. Two 45° bending magnets with a homogeneous field are used to energy-analyze the scattered electron. Focussing and corrections of higher order aberrations is done using the four curved pole faces; in particular, aberrations due to finite source (target) extension can be controlled. This spectrometer features properties similar to the ones of the Saclay spectrometer as far as solid angle and momentum acceptance goes. The momentum resolution is superior due to the better control of higher order aberrations, and has reached $\sim 5 \cdot 10^{-5}$. This type of split-pole spectrometer with a rather large distance from target to focal plane is not very compact; together with a bending angle of only 90° this makes background supression more difficult.

The most modern spectrometer installation, at the NIKHEF accelerator in Amsterdam, features a quadrupole-two dipole QDD system (Fig. 2). The dipoles, with a homogeneous field and curved pole faces, deflect the electrons by 75° each. The quadrupole defocusses the electrons in the dispersive plane where the spectrometer acceptance is very large anyway due to the large momentum acceptance. For the non-dispersive plane, this quadrupole creates a crossover of electron trajectories. This leads to an important decrease of dipole gap for a given solid angle. An additional multipole element between the two dipoles serves to correct imperfections. This QDD system achieves 6 msr solid angle, a momentum acceptance of 10% and an energy resolution which, at the present time, is $\sim 10^{-4}$. The degrees of freedom offered by pole face curvature and quadrupole have allowed to create a flat focal plane, a feature that is very helpful in reducing the complexity of the focal plane detector. The compactness of the QDD design together with the good possibilities for shielding is expected to lead to very low background.

In general, the field maps of high-resolution spectrometers have been studied in great detail for the verification of optical properties. Knowing the field map, the electron energy can be calculated if the absolute field is measured at one point using nuclear magnetic resonance probes. This allows to determine the energy of scattered (hence incident) electrons to an accuracy of a few times 10^{-4} in $\delta E/E$. This type of energy measurement in general is more reliable than the determination via recoil energy differences or floating wire measurements.

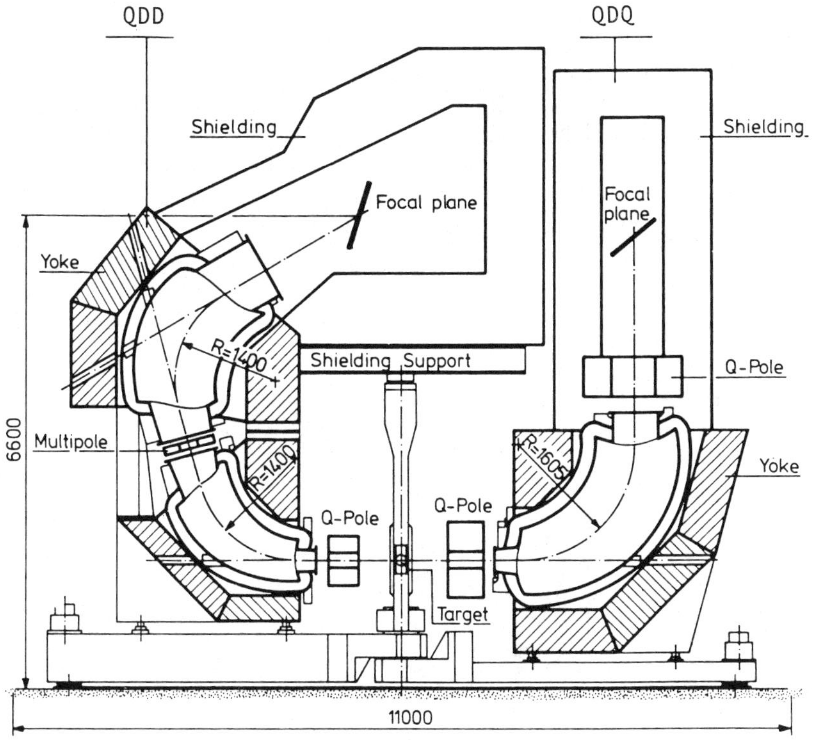

Fig. 2. NIKHEF Spectrometers

The solid angle of the spectrometer is defined by accurately machined collimators of various sizes. Unless the biggest solid angle is used (where ill-defined pieces of the vacuum chamber and alike define the solid angle) the collimators determine the solid angle with an accuracy of < 1%.

Targets

For electron scattering experiments, the targets used in general are solid ones. For the rather special techniques required for liquid or gas targets the reader is referred to the literature (Fig. 3 gives one particular example, a ^{3}He gas target working at 20° K and 10 at ; the special feature of this target are windows that are not visible to the spectrometer). The solid targets used have thicknesses of 20÷100 mg/cm^2 and size of 5÷10 cm^2. The targets are fabricated by rolling, pressing, electroplating or casting, depending on the mechanical and thermal properties of the isotope of interest.

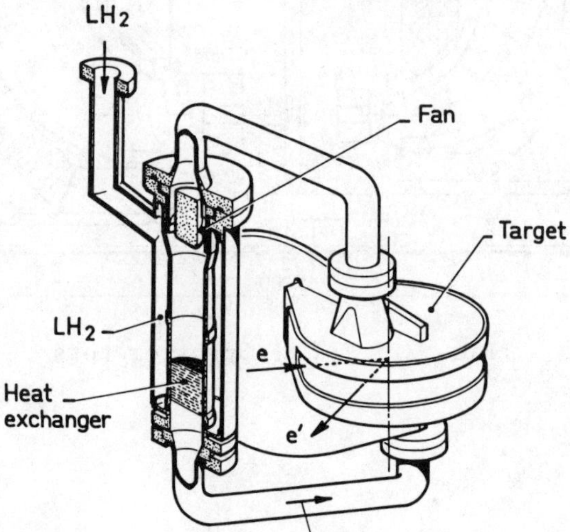

Fig. 3 ^{3}He gas target (10 at. 20 K) built at Saclay

The target thickness is determined by the energy resolution desired; energy straggling in the target gives a contribution of approximately 0.4 keV/mg/cm^2. Target homogeneity is imposed by the accuracy to be achieved, since in general the intensity distribution of the beam over the target is poorly known. In order to measure the relative density profile of the target, absorption of γ- or β rays has proven to be the best method. For a measurement of the absolute target thickness the combination of relative density profiles with the average areal density obtained from weight and surface yields target thicknesses for the area covered by the beam to an accuracy of < 1%.

For experiments at medium or large q, i.e. for small cross sections, the removal of heat deposited in the target by the high-intensity electron beam requires additional apparatus. While energy-loss systems disperse the beam vertically, additional movement of the target horizontally is often required to spread the heat. The most efficient way to remove heat is achieved by cooling the target with a jet of H_2-gas of ~1 Torr pressure. Circulating H_2 gas a high speed through nozzles pointing at the target allows a very high heat extraction rate with no disturbing effects upon the experiment.

Focal plane detectors

The detector system located in the spectrometer focal plane has a double function. It identifies the scattered particle as an electron in order to suppress the large amount of radiation present despite good shielding. It measures the coordinates and direction of the scattered electron in the focal plane for accurate energy-(loss) measurement.

In order to identify the particle detected, one generally requires a fast coincidence between the signals provided by plastic scintillators and a Čerenkov counter. The plastic scintillators, of the order of 1 cm thick, are arranged in one or two planes and are often segmented into many individual units to reduce pileup and increase directionality. The Čerenkov counter is used to identify the particles via its velocity v. The conventional Čerenkov detectors, using lucite or some liquid hydrocarbon compound, accept particles with $v/c \gtrsim 0.8$; they do not eliminate the pions that are abundant at the higher incident energies and large energy loss. Čerenkov counters using a gas at atmospheric pressure as utilized in the Saclay spectrometers allow to eliminate all particles but electrons.

The complexity of the detector used for the localization of the electron trajectory strongly depends on the type of spectrometer used and the number of aberrations to be corrected. The smallest numbers of measurements to be performed occur with double-focussing spectrometers the focal plane of which is a plane indeed. Then the localization of the track along the dispersive direction has to be measured with a resolution of a fraction of a mm. This resolution is set by the desire to define a peak of width $\Delta E/E \sim 10^{-4}$ by a number of points. In order to determine the nuclear recoil energy with equivalent resolution, a second measurement of either trajectory angle or transverse position is required. For spectrometers with a focal plane of complicated shape and with residual aberrations, two coordinates in the focal plane plus two angles have to be measured. With this information the trajectory measured at the location of the detector can be extrapolated to the focal plane, and known aberrations removed.

The type of localization detector that has found general acceptance is the drift chamber. In such a multiwire proportional chamber, the secondary electrons produced by the scattered electron drift to one or several collection wires, and the drift times are used to accurately locate the track. A spatial resolution of ~0.1mm can be obtained with a wire density of a few wires per cm. Successive measurements of several coordinates of the track are possible since this detector causes little multiple scattering of the high-energy electron. Such drift-chambers require a few hundred ns to process one track, a time that has become acceptable with modern electron accelerators of duty cycle $>10^{-2}$. If the readout electronics and interface to the data acquisition computer are fast enough, the high data rates that can occur at very low momentum transfer are acceptable.

Data reduction

In order to go from the raw data as furnished by the experimental hardware to publisheable cross sections, considerable manipulation of the data is required. Many corrections have to be applied, imperfections of the apparatus removed, target impurities eliminated, spectra decomposed into contributions of individual nuclear levels, etc.. Some of these corrections can be applied on-line during the experiment, others have to be performed after a detailed study of the results reveals some unwanted effect. These procedures, though requiring considerable effort, are of little interest to the reader who wants to obtain a general idea on the state of art of electron scattering experiments. We therefore will not discuss them here.

One particularly large "correction" should be explicitely mentioned, however. The scattering of an electron is inevitably accompanied by emission of bremsstrahlung. This leads to the fact that every peak in the energy spectrum corresponding to a well defined nuclear level has a radiative tail. The cross sections required for a theoretical understanding should have these radiative effects removed.

For the elastic scattering of main interest here, this correction is a rather simple one. For the highest-energy (elastically scattered) electrons all we need to know is the fraction of electrons that have radiated more than a maximal energy imposed by the excitation energy of the first excited state. This fraction can be calculated quite reliably. In principle these radiative corrections have been calculated using an exact theory, QED; in practice some approximations have been made. While there is little information on the uncertainty of this procedure, various checks made lead to the estimate that the radiative corrections for elastic scattering (they amount to a correction of the order of 30%) have a reliability such that the final cross section is accurate to significantly better than 1%.

NUCLEAR CHARGE DISTRIBUTIONS

The determination of ground state charge densities is the "pièce de résistance" of electron-nucleus scattering. This section tries to describe some of its development and today's state of the art.

History

Let me start by briefly listing some of the important milestones in the historical development of this field. In 1929 Sir N.F. Mott applied the brandnew Dirac equation to electron-nucleus scattering. In the electron range 0.5-3 MeV then accessible experimentally large differences between calculation (for point-like nuclei) and experiment were found. Eventually, the experiments improved to the point where they agreed with what we today refer to as the "Mott cross section"; the field lost interest, and the investigation of the nucleus by probe-nucleus scattering got its start through hadronic probes.

In 1934 E. Guth in a remarkably clear paper pointed out that the effects due to nucleon finite size should be quite large. Guth presented the low-momentum transfer expansion of the elastic form factor, and already saw the use of electron-scattering as a tool to determine the Fourier transform of the ground state charge density.

With the experiments of Lyman, Hanson and Scott, who used the 15 MeV University of Illinois Betatron, electron-nucleus scattering really got its start in 1951. Investigating nuclei between ^{12}C and ^{197}Au, LHS first measured the effect due to finite nuclear size ($F^2(q) \simeq 0.5$). This very first experiment actually already observed (tentatively, though) that nuclear radii were 20% smaller than the $1.45 \cdot A^{1/3}$ fm rule experiments using strongly interacting probes had determined.

The construction of the High Energy Physics Laboratory (HEPL) linear accelerator at Stanford, a machine that developed from an initial 200 MeV to a final 1 GeV, gave the field of electron-nucleus scattering the character it has today. Many of the features of accelerators, spectrometers and experimental techniques originated at HEPL. The determination of nucleon and nuclear structure by elastic and inelastic scattering of electrons became a thriving field. The Nobel prize awarded to Bob Hofstadter in 1961 documents the acceptance of these achievements by the nuclear physics community.

Interest

Why, even today, is it still interesting to investigate nuclei with electrons, and why are we still interested in particular in the charge density? The ground state charge density has two attractive features:

1) A measurement of size and shape of a nucleus allows us to visualize this object. For somebody thinking in terms of macroscopical, classical (non quantum-mechanical) concepts this is most important. As we will see later, electron scattering does merit the description "electron microscope" to a large degree!

2) The ground state charge density measures a basic quantity, the wave function squared of the charged nucleons (protons) integrated over all but the radial (and on occasion azimuthal) coordinates. It is very important

that any theoretical wave function agrees with these "geometrical" observables. Electron scattering is a unique tool to quantitatively extract this information from experimental data; also, electron scattering is the only tool that allows us to really study the nuclear interior, since electrons are not absorbed.

What do we expect to find if we measure the charge density $\rho(r)$ of a (spherical) nucleus? The dotted curves in fig. 4 show what the early experiments (<65) found: a proton density roughly constant in the nuclear interior, a half-density radius $c \simeq 1.2 \cdot A^{1/3}$fm and a surface thickness of 2fm. The best theoretical calculations today available (DDHF-calculations, see below) yield a much more structured density (solid curves fig. 4). This structure reflects the fact that, in the independent particle shell model nucleons occupy orbits of very distinct radial shape, as exemplified for the few lowest orbits in fig. 5. If an experiment wants to see the structure predicted by theory, then the "electron microscope" used obviously must achieve a spatial resolution $\lessdot 1$ fm, which the early experiments did not have.

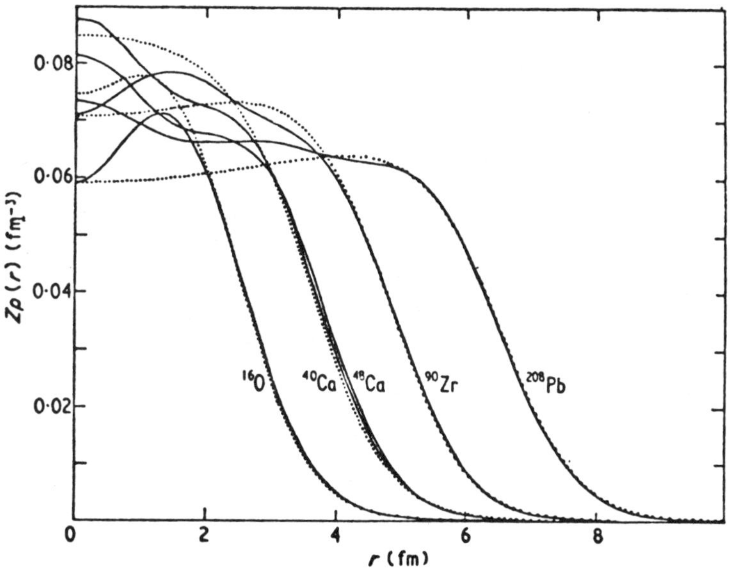

Fig. 4 Charge densities of magic nuclei given by experiment (dotted) and DDHF (solid)

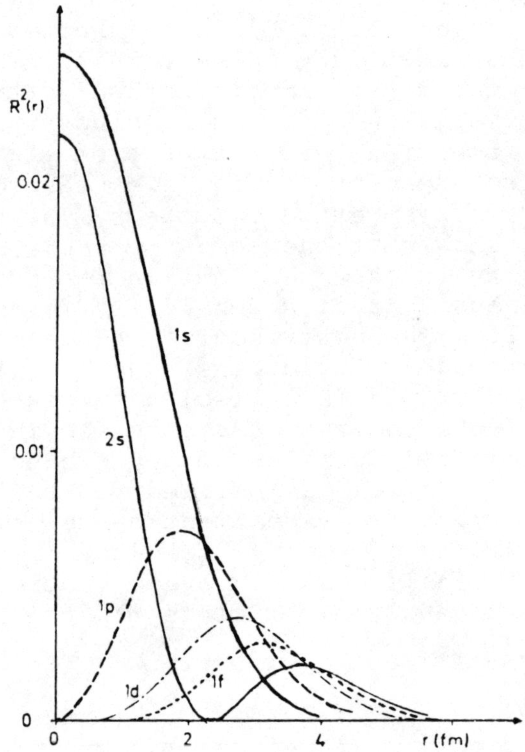

Fig. 5 DDHF radial wave functions of lowest shells

Determination of $\rho(r)$

How do we determine a charge density once we have done our elastic scattering experiment ? Assume that we deal with a spin-zero nucleus, in which case the only observable is the spherical part of $\rho(r)$ averaged over all orientations of the nucleus. The nucleus creates a distribution of the electrostatic potential $V(r)$ given by the distribution of charge $\rho(r)$. Scattering an electron off this potential allows to deduce its shape, and to determine $\rho(r)$. The relation between the scattering amplitude $f(\Theta)$ (the cross section is given by $|f(\Theta)|^2$) and $\rho(r)$ is obtained by solving the Dirac equation (see talk of T.W. Donnelly).

In order to understand what we are doing, we will first discuss the case where $Z\alpha \ll 1$. Then the plane-wave Born approximation is a valid one, and we can solve the Dirac equation in closed form

$$\frac{d\sigma}{d\Omega}(\Theta,E) = \frac{d\sigma}{d\Omega}(\Theta,E)_{Mott} \quad F^2(q) \tag{1}$$

$$\frac{d\sigma}{d\Omega_{Mott}} = \frac{Z^2\alpha^2 \cos^2\Theta/2}{4E^2 \sin^4 \Theta/2} \tag{2}$$

$$F(q) = \frac{1}{Z}\int_0^\infty \frac{\sin(qr)}{qr} \rho(r)\ 4\pi r^2 dr \tag{3}$$

$$q = 2\ E \sin \Theta/2 \tag{4}$$

$$F(q \simeq 0) = 1 - \frac{q^2 < r^2 >}{6} + \frac{q^4 < r^4 >}{120} - \ldots \tag{5}$$

The cross section factorizes into a kinematical term, σ_{Mott}, and a form factor that contains all the information on $\rho(r)$. This form factor depends on the momentum transfer q only, and represents the Fourier transform of the charge density we want to extract from the experimental data.

Above equations show that at low q we are sensitive to integral moments of $\rho(r)$, like $<r^2>$, only. At higher q we gradually see more details. The spatial resolution (the FWHM of a peak of the sin(qr)-term that samples $\rho(r)$) is roughly 1.5/q.

Unfortunately, PWBA is sufficiently precise for the very lightest nuclei ($Z \leqslant 2$) only. For the vast majority of nuclei we have to use the exact solution of the Dirac equation that describes the electron in the electrostatic potential of the nucleus. In order to determine $\rho(r)$ from the experimental data one therefore postulates a nuclear charge density (depending on a number of parameters), solves the Dirac equation for this $\rho(r)$ numerically, computes elastic cross sections and compares them to the experimental ones. Then one varies the parameters until agreement with experiment is satisfactory. This procedure is virtually exact, but entirely oblique.

Worse, one does not really know which features of $\rho(r)$ are fixed by the data, and which ones depend on the particular parametrization used. This problem is evident for the 2-parameter Fermi densities

$$\rho(r) = \text{const} \cdot (1+ e^{(r-c)/z})^{-1} \tag{6}$$

that led to the dotted curves in fig. 4. Densities at different radii are strongly coupled through model densities like the above.

In order to avoid this problem, one can expand the density in terms of a complete (orthogonal) set of functions. This looks great, but is not useful. Due to the finite maximum momentum transfer q_{max} of experiment, only a few coefficients of the expansion can be determined. The truncation of the basis brings us back to a model-density. Different radii still are strongly coupled, and a sensible error bar of $\rho(r)$ cannot be defined.

To derive a really "model-independent" density, one somehow has to account for the properties of the information one was not able to measure ($q > q_{max}$). This can be done if we have some theoretical input on the maximum amount of structure to be expected in $\rho(r)$. Such information is provided by DDHF calculations, for instance, that all predict smooth radial wave functions with structures no narrower than the peaks in the $R^2(r)$ of individual shells (fig. 5). A constraint on the finest structure to be allowed for in $\rho(r)$ can be applied directly in radial space, as done with the sum-of-Gaussians parametrization (SOG). It also can be imposed in momentum space (as done with the Fourier-Bessel parametrization) where it amounts to upper limits of $F(q > q_{max})$. Once this not-measured part of $\rho(r)$ is "replaced" by a theoretical estimate of its maximal possible effect, one can derive densities that have a realistic error bar that incorporates both the experimental uncertainties and the error due to the lack of completeness (finite q_{max}) of the data.

Most transparent, easy and without theoretical bias is the Direct Fourier Transform (DFT) I developed six years ago. Realizing that the form factor $F(q)$ as the Fourier transform of $\rho(r)$ (defined by eq. 3) is quantitatively quite similar to $(\sigma/\sigma_{Mott})^{1/2}$ even for large Z, one can define an experimental PWBA form factor by

$$F_{exp}(q) = \left(\frac{\sigma_{exp}(\Theta,E)}{\sigma_{mod}(\Theta,E)} \right)^{1/2} F_{mod}(q) \qquad (7)$$

The index "mod" refers to quantities computed using a model density ρ_{mod} that fits the experimental data. Using this model density one can connect the real world (σ_{mod}) to a PWBA-world (F_{mod}). Since σ_{exp} differs by percents only from σ_{mod}, one can quantitatively determine F_{exp} using (7). The Coulomb distortion responsible for the difference between $\sigma_{Mott} \cdot F^2_{exp}$ and σ_{exp} depends so little on ρ_{mod} (it basically depends on Z!) that the value of F_{exp} and their error bars δF_{exp} are independent of ρ_{mod}.

Once one has translated the experimental data into the PWBA-world, the determination of $\rho(r)$ proceeds via the Fourier transform of eq. (3)

$$\rho(r) = \frac{Z}{\pi^2} \int_0^\infty \frac{\sin(qr)}{qr} F_{exp}(q)\, q^2 dq \qquad (8)$$

If the data are of sufficiently good quality (and this is in the hands of the experimentalist!) a brute-force integration over the experimental form factor yields $\rho(r)$.

Of course, we still have to deal with the problem that integral (8) cannot be extended beyond $q = q_{max}$. In order to estimate the error one makes by stopping at $q = q_{max}$, one can plot the integral (8) as a function of the upper integration limit. For every radius

$$\rho(r,q_{max}) = \frac{Z}{\pi^2} \int_0^{q_{max}} \frac{\sin(qr)}{qr} F_{exp}(q)\, q^2 dq \qquad (9)$$

represents a damped oszillatory function of q_{max}. As an example, consider the place where $\rho(r)$ is most difficult to determine, r = 0, where

$$\rho(0,q_{max}) = \frac{Z}{\pi^2} \int_0^{q_{max}} F_{exp}(q)\, q^2\, dq \qquad (10)$$

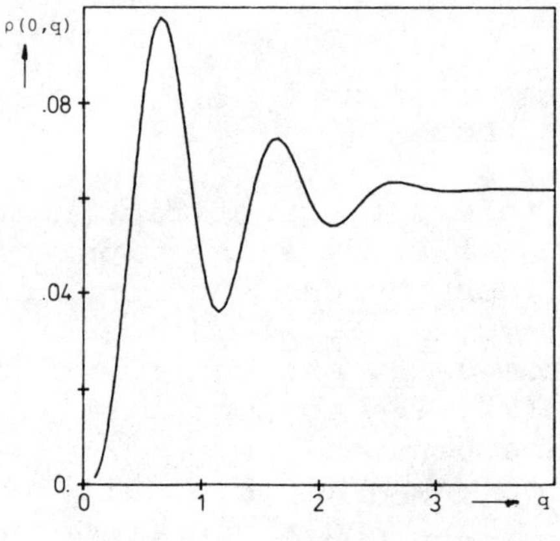

Fig. 6 $\rho(0,q)$ for ^{208}Pb

At every diffraction zero, F_{exp} changes sign and $\rho(0,q_{max})$ has a minimum or maximum (fig. 6). The difference between the last measured minimum and maximum may be taken as an honest experimental estimate of the contribution of the completeness error to $\delta\rho_{exp}(r)$. The other contributions (statistical and systematical errors of the data) are easily calculated by error propagation and eq. (8). Additional physical constraints like radial moments determined by muonic atoms experiments, the vanishing of $\rho(r)$ for $r > R$ or an exponential tail at large r can be imposed easily in momentum space before Fourier transforming.

Example: ^{208}Pb

Many experiments on Pb have been done over the years, with ever increasing range of q and improving precision. (Occasional steps backwards unfortunately have occured as well). The first experiment I know of was done in 1953 in Stanford (fig. 7), the last one we did 5 years ago in Saclay (fig. 8). Over 25 years the q-range has been expanded from $0.5 \div \text{fm}^{-1}$ to $0.2\ \text{fm}^{-1} \div 3.6\ \text{fm}^{-1}$, the dynamical range in $\sigma(\theta)$ from 3 decades to 12. Fig. 8 shows today's cross sections, all plotted for an electron energy of 500 MeV. Fig. 9 shows the same data in the form of $F^2_{exp}(q)$.

Fig. 6 displays the function $\rho(0,1)$ (eq. 10) and shows that today we have reached an excellent precision: the difference between the last maximum and minimum,

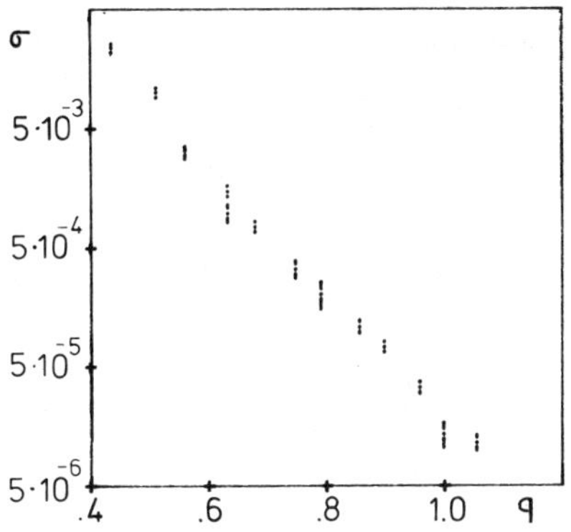

Fig. 7 Cross sections from the first electron scattering experiment on Pb

barely visible, is 1%. (For other radii it is less). The convergence with q clearly demonstrates that it was not for fun that the experiment was pushed to cross sections of $10^{-38}cm^2/sr$ (1 count/day at 20 μA beam current). To get $\rho(0)$ to converge to < 1%, one must reach $F(q_{max}) < 10^{-5}$. If the experiment stops at 2 fm^{-1}, as most experiments do, the error on $\rho(r)$ is large (fig. 6).

The result for the density is shown in fig. 10; error bars on this scale are no longer visible.

What can we learn from this experimental result? To discuss this, we have to briefly describe the most refined, DDHF, calculations for ground state wave functions today available. The shell model assumes that a picked-out nucleon moves in an average nuclear potential created by all other nucleons. It ignores the correlations that occur between nucleons once they get close enough to feel the very strong short-range repulsive nucleon-nucleon interaction. The average potential seen by one nucleon can be calculated from the wave function of all other nucleons and an effective nucleon-nucleon interaction. Once the average potential is known, the wave function can be computed by solving the Schrödinger equation. The Hartree-Fock calculations iterate the wave function of all nucleons such that the ones used to initially compute the effective potential are equal to those

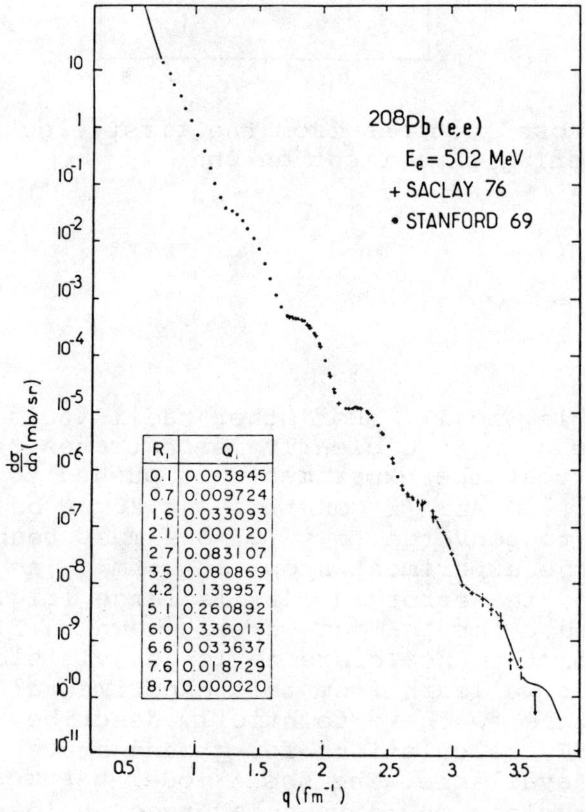

Fig. 8 Cross sections for Pb(e,e)

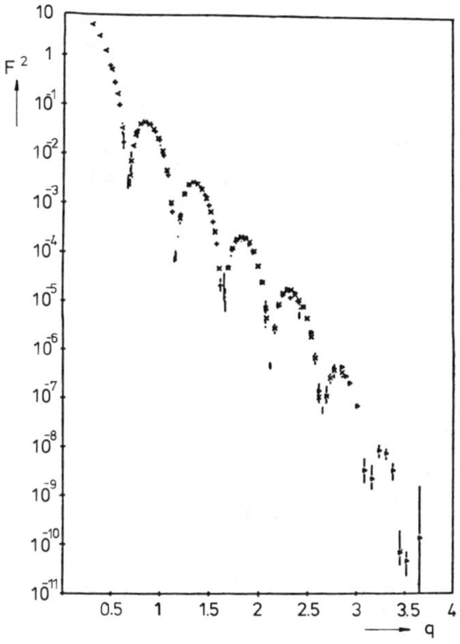

Fig. 9 Experimental form factors of ^{208}Pb derived from data of Fig. 8

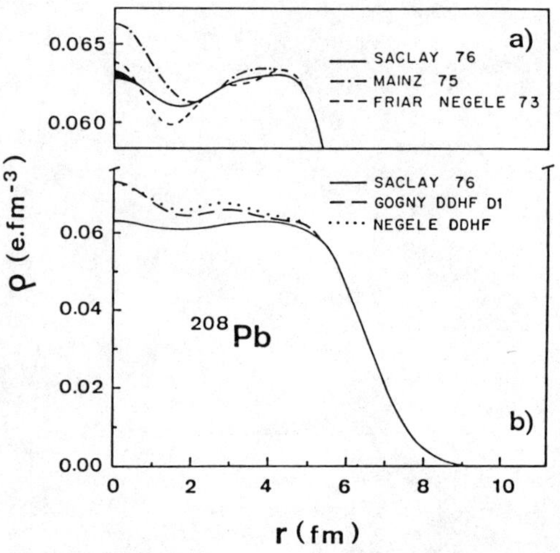

Fig. 10 Experimental charge density (solid) and DDHF predictions

obtained for the Schrödinger equation, i.e. that this procedure is selfconsistent.

The effective nucleon-nucleon interaction V_{eff} is density dependent (hence the term Density Dependent Hartree-Fock) in order to account for the very strong interaction and saturation at small internucleon distance and large density. V_{eff} can be derived from the nucleon-nucleon interaction as measured by NN scattering, or it can be parametrized in an intelligent way and fitted to nuclear observables. In either way some phenomenological input (adjustment of parameters) is inavoidable. The curve labeled Gogny in fig. 10 results from a phenomenological V_{eff}, the one labeled Negele from a V_{eff} derived to a large degree from V_{NN}. Fig. 11 shows a variety of other DDHF calculations. When comparing DDHF and experiment for ^{208}Pb (fig. 10) and other nuclei, we find that the density in the nuclear surface region is quite well explained. This should not come as a surprise since the overall radius has been fit using the phenomenology alluded to above. In the nuclear interior, the HF calculations show much more shell-structure of ρ than experiment does. It would seem that in nature the nucleons occupy less well defined orbits than HF assumes. Indeed,

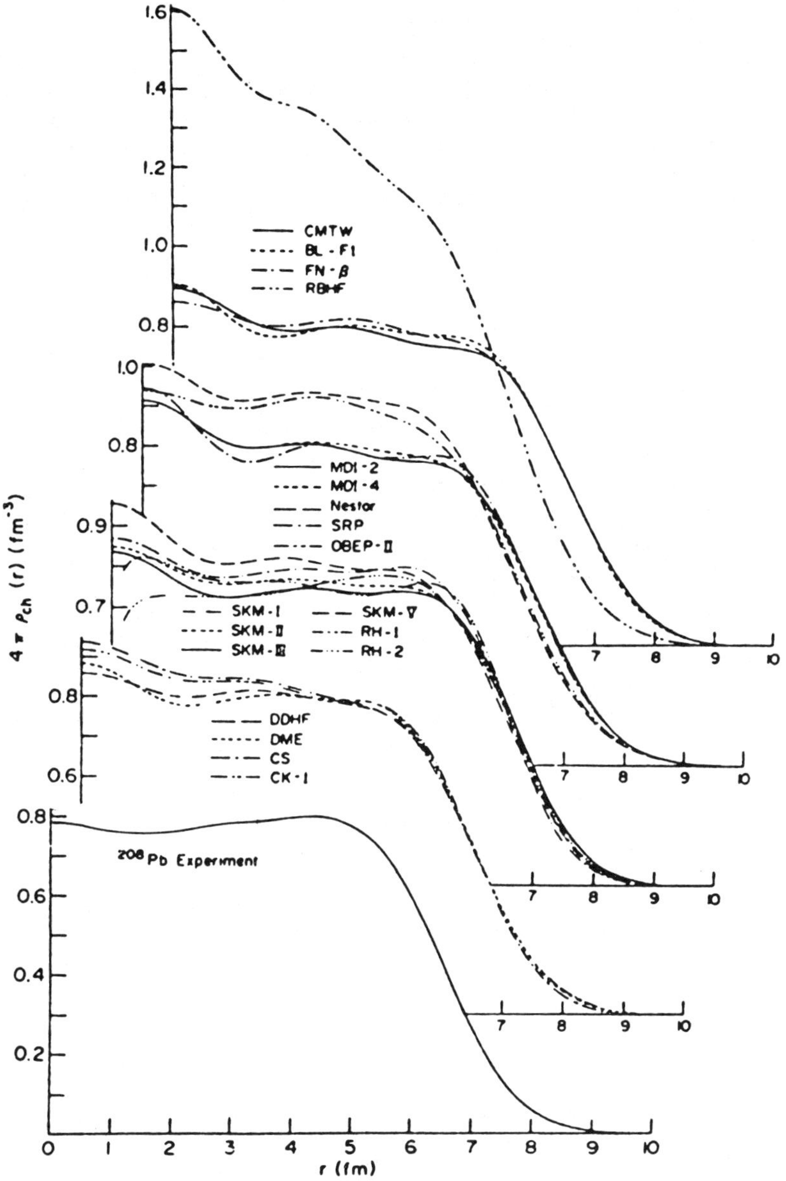

Fig. 11 Various DDHF predictions for charge density of ^{208}Pb.

the short range repulsion of V_{NN}, largely neglected in DDHF, is known to lead to a smoothing out of the shell structure of calculated densities.

This problem evokes the question whether the concept of a shell-model orbit as shown in fig. 5 makes any sense in the high-density interior of a heavy nucleus. Given the strong NN correlations at short-range and high nucleon density, this is not at all obvious. Actually, for a long time before the shell-model was successfully applied to nuclei, it was believed that this model would not make any sense exactly for this reason. Without the concept of a shell-model orbit the structure in $\rho(r)$ predicted by DDHF calculations would largely vanish.

The validity of the shell-model is of course well established from experiments with hadronic probes. These probes test only the nuclear surface region, however. Hadronic probes are strongly absorbed and hardly sensitive to the nuclear interior.

In order to see whether the concept of a shell-model orbit does make sense, one can study the charge density difference between Pb and Tl. These nuclei differ (in the shell-model) by a 3s proton, a shell that has a most distinct radial wave function (fig. 12): a peak at the nuclear center and two nodes. Such a peculiar shape should be easily detectable via electron scattering.

Looking at fig. 12 and eq. (3) we actually can see that the form factor difference $\Delta F = F_{Pb}-F_{Tl}$ should have

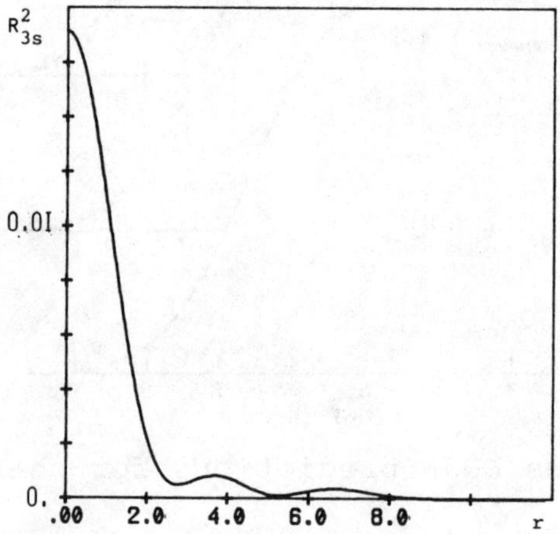

Fig. 12 3s radial wave function (folded with finite proton size).

a rather spectacular behaviour. The $\Delta\rho = \rho_{Pb} - \rho_{Tl}$ looks very similar to a $\sin(q_o r)$ function, with $q_o \simeq 2$fm. The Fourier transform of such a $\Delta\rho(r)$ should give a $\delta(q_o)$-function in q-space. The main effect of a 3s-proton thus should be a pronounced peak in $\Delta F(q)$ (or $\Delta\sigma(\Theta)$) near 2 fm^{-1}.

What does experiment say? A few years ago an experiment on ^{206}Pb and ^{205}Tl performed at Mainz measured accurate data for $q < 2$ fm^{-1}. Near 2 fm^{-1}, where the main effect of the 3s R(r) is to be expected, no significant data were taken; it obviously was not realized where a measurement would be most interesting. The experiment we performed two years ago at Saclay was designed to cover this region of momentum transfer, and fig. 13 shows the result. We indeed do find the pronounced δ-function like peak due to the 3s radial wave function of the added proton. The structure in the cross section ratios for 0.5-1.5 fm^{-1} is mainly due to core polarization, i.e. the increase in overall size of the Tl core due to the addition of one nucleon. From these cross section ratios one can extract the charge density difference, and obtain fig. 14. When comparing this $\Delta\rho(r)$ to HF, one should realize that above discussion was a bit too simpleminded.

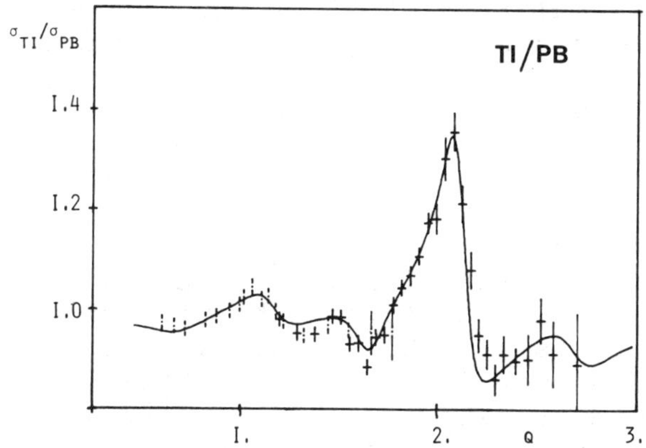

Fig. 13 Ratio of ^{205}Tl to ^{206}Pb cross sections, together with DDHF-prediction.

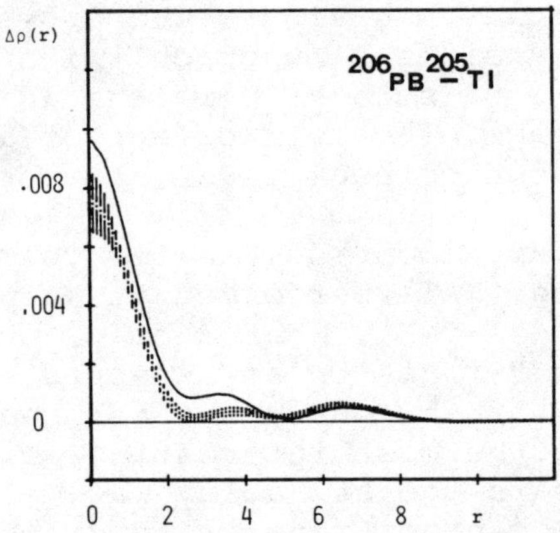

Fig. 14 Experimental Pb-Tl difference and DDHF-prediction.

Only in the strict single-particle shell-model do Pb and Tl differ by one 3s proton. In reality nuclei like ^{205}Tl, ^{206}Pb (non magic nuclei) will exhibit considerable configuration mixing; the proton hence will partly occupy other shells as well. In particular, an excitation of ^{206}Pb to a 2^+ state coupled to a $2d_{5/2}$ proton hole is one of the components that occur prominently in the ^{205}Tl ground state. Both one-proton transfer reactions and shell-model calculations tell us that the spectroscopic factor of the 3s shell is 0.7÷0.8 rather than one as assumed above.

The DDHF calculation shown as a curve in figs. 13 and 14 therefore has been performed by Campi using a 3s spectroscopic factor of 0.7, assuming that 30% of the time the additional proton occupies the $2d_{5/2}$ state. With this somewhat phenomenological adjustment, the agreement with experiment is excellent, much better than one could have hoped for. The difference between experiment and DDHF in fig. 14 is smoothly decreasing from 0fm to 6fm, and probably has nothing to do with $R^2(r)_{3s}$. Such a structureless difference can be expected if DDHF does not quite properly describe the mean increase in size of the core. Thus, overall, a real success for the independent particle shell-model!

MAGNETIC ELECTRON SCATTERING

The third part of these lectures deals with elastic electron scattering from nuclear magnetization densities of high multipole order. This type of experiment and its interpretation in terms of valence nucleon radial wave functions has very recently been described in an extensive paper to which the reader is referred.

LITERATURE

Review papers

T. de Forest, J.D. Walecka, Adv. Phys. 15 (66) 1
T.W. Donnelly, J.D. Walecka, Ann. Rev. Nucl. Sci. 25 (75) 329
R.C. Barrett, Rep. Prog. Phys. 37 (74) 1
H. Ueberall, Electronscattering from complex nuclei, Academic Press, 71
C. Ciofi degli Atti, Prog. Part. Nucl. Phys. 3 (80) 163

Experimental techniques

R. Hofstadter, Rev. Mod. Phys. 28 (56) 214
M. Chodorow etal., Rev. Sci. Instr. 26 (55) 134
P. Leconte etal., Nucl. Inst. Meth. 169 (80) 401
W. Bertozzi etal., Nucl. Inst.Meth. 162 (79) 211
C. de Vries etal., Lect. Notes in Phys., 137 (80) 258
H. Ehrenberg etal., Nucl. Instr. Meth., 105 (72) 253
W. Bertozzi etal., Nucl. Instr. Meth. 141 (77) 457

History

N.F. Mott, Proc. Roy.Soc. 124 (29) 425
E. Guth, Anz. Akad.Wiss., Wien, Math.-Naturw.Kl. 24 (34) 299
E.M. Lyman etal., Phys. Rev. 84 (51) 626
R. Hofstadter, Nuclear and Nucleon Structure, Benjamin, N.Y., 1966

Determination of charge densities

D.R. Yennie etal., Phys. Rev. 95 (54) 500
R. Hofstadter, Ann. Rev. Nucl. Sci. 7 (57) 231
I. Sick, Nucl. Phys. A218 (74) 509

B. Dreher etal., Nucl. Phys. A235 (74) 219
I. Sick, Phys. Lett. 88B (79) 245
C.W. de Jager etal., Atom Data Nucl. Data Tab. 14 (74) 479 (compilation)

(e,e)' on nuclei near ^{208}Pb

R. Hofstadter etal., Phys. Rev. 92 (53) 978
J. Heisenberg etal., Phys. Rev. Lett. 23 (69) 1402
B. Frois etal., Phys. Rev. Lett. 38 (77) 152
H. Euteneuer etal., Nucl. Phys. A298 (78) 452
J.M. Cavedon etal., Phys. Rev. Lett., to be publ.
I. Sick, Nucl. Phys. A208 (73) 557

Hartree-Fock calculations

H.A. Bethe, Ann. Rev. Nucl. Sci. 21 (71) 93
J. Negele, Phys. Rev. C1 (70) 1260
D. Gogny etal., Phys. Rev. C21 (80) 1568
M.R. Strayer etal., Phys. Rev. C8 (73) 1269
J.P. Svenne, Adv. Nucl. Phys. 11 171

Magnetic elastic scattering

S. Platchkov etal., Phys. Rev. C25 (82) 2318
I. Sick, Com Nucl. Part. Phys. 9 (80) 55

Recent work in conference proceedings

W. Bertozzi (Santa Fé) AIP Conf. Proc. 26, 409
I. Sick (Santa Fé) AIP Conf. Proc. 26, 388
J. Negele, (Zurich 77) Proc. ICOHEPANS, p.17
D. Drechsel (Vancouver 79) Nucl. Phys. 335 (80) 17
I. Sick (Vancouver 79) Nucl. Phys. 335 (80) 555
I. Sick (Berkeley 80) Nucl. Phys. A354 (81) 37
W. Bertozzi (Versailles 81) Nucl. Phys. A374 (82) 109

Δ DYNAMICS*

E.J. Moniz

Center for Theoretical Physics
Laboratory for Nuclear Science and Department of Physics
Massachusetts Institute of Technology
Cambridge, Massachusetts 02139

I. INTRODUCTION AND OUTLINE

We shall discuss in these lectures the properties and interactions with nuclei of Δ's. To do so, we must establish first a common idea about what a Δ is. This will be the topic of discussion in Chapter II, wherein an isobar model is developed which allows us to discuss precisely the propagation and dynamics of the Δ as an "unstable particle." The need for the discussion is grounded in the apparently very different descriptions of the Δ offered by conventional meson-nucleon scattering theory and by the quark model. In the first approach, the Δ is associated with the broad resonance seen in the L = 1, J = 3/2, T = 3/2 πN scattering channel. In the quark approach, the Δ is viewed spectroscopically as the first excited state of the nucleon and is usually treated implicitly as if it were a stable particle. We shall see that the phenomenological isobar model is compatible with both languages, provides a dynamically consistent approach to the many-body problem, and leads to dynamical modifications of quark model predictions of Δ properties.

In Chapter III, we discuss the extraction of Δ-nucleus interaction parameters from an analysis of intermediate energy pion- and photon-initiated nuclear reactions. The central result is the strength of the Δ-nucleus optical potential, which is found to have a large imaginary part associated with intermediate coupling to purely nuclear states (i.e., pion annihilation channels). The isospin signature provided by the pion annihilation dynamics both lends support to this interpretation and leads to new predictions for inelastic pion

*This work is supported in part through funds provided by the U.S. DEPARTMENT OF ENERGY (DOE) under contract DE-AC02-76ER03069.

reactions.

Finally, we discuss briefly in Chapter IV, the role of Δ's in low energy nuclear properties. This subject has received considerable attention because of the special role played by the Δ in nuclear spin-isospin excitations (the Δ is a static spin-flip nucleon excitation in the quark model). Our focus will be on the ambiguities inherent in such discussions both because of the dynamical nature of the resonance discussed in Chapter II and because the quarks are the "true" nucleon internal degrees of freedom.

Let me stress at the outset that these lectures are not a review of the field. This will be clear from the short list of references. However, these references will lead the interested reader to a wider literature.

II. THE ISOBAR MODEL

Our first task is to develop a dynamically consistent, quantitatively accurate, simple-to-use model for the $\pi N \rightleftarrows \Delta$ system. We use the (bare) isobar model. This is easily phrased in terms of a phenomenological Lagrangian

$$\mathcal{L}_{int} = \frac{g_{\pi N\Delta}(k^2)}{m_\Delta} \psi_\Delta^+ \underset{\sim}{T}^+ \vec{S}^+ \psi_N \cdot \vec{\nabla}\underset{\sim}{\phi}_\pi + \text{h.c.} \tag{1}$$

The model is that we have elementary pions, nucleons and Δ's, with an interaction consisting of direct $\Delta \rightleftarrows \pi N$ coupling. The form of the coupling is determined by the channel quantum numbers for the Δ; namely, the gradient reflects the P-wave nature of the interaction while $S^+(T^+)$ are spin (isospin) $1/2 \rightarrow 3/2$ transition operators. The spin and isospin operators will be discussed below more fully. Finally, the coupling strength is given by a phenomenological vertex function, $g_{\pi N\Delta}(k^2)$, where k is the πN c. m. momentum. We stress that the Δ appearing in Equation (1) is a stable object; its propagator is described in the usual way

$$G_\Delta^o(E) = (E^+ - M_\Delta - T_\Delta)^{-1} \tag{2}$$

where M_Δ is the bare mass and T_Δ is the kinetic energy operator. (Note that we shall consistently treat nucleon and Δ motion nonrelativistically; relativistic kinematics will be used for the pion for any quantitative results). However, the "physical" Δ propagation is modified by coupling to the πN channel

$$G_\Delta(E) = G_\Delta^o(E) + G_\Delta^o(E)\Sigma_{\pi N}(E)G_\Delta(E) \tag{3}$$

$$\Sigma_{\pi N}(E) = V^{+}_{\pi N\Delta}\, G^{o}_{\pi N}(E)\, V_{\pi N\Delta} \tag{4}$$

$$G^{o}_{\pi N}(E) = (E^{+} - (m_{\pi} + T_{\pi}) - (m_{N} + T_{N}))^{-1} \tag{5}$$

where $V^{+}_{\pi N\Delta}$ is the $\pi N \to \Delta$ vertex operator given by Equation (1). Here, $\Sigma_{\pi N}$ is the Δ self-energy generated by Δ decay into the πN channel, propagation of the free πN system, and reformation of the Δ. Since the πN channel is open, the self energy $\Sigma_{\pi N}$ has an imaginary part, and this corresponds to the Δ width. In this model, πN scattering in the resonant channel is generated simply by coupling to and decay from the physical Δ:

$$t_{\pi N}(E) = V_{\pi N\Delta}\, G_{\Delta}(E)\, V^{+}_{\pi N\Delta} \tag{6}$$

We demonstrate explicitly that $t_{\pi N}$ satisfies the elastic unitarity relation:

$$\begin{aligned} t^{+}_{\pi N} - t_{\pi N} &= V_{\pi N\Delta}(G^{+}_{\Delta} - G_{\Delta})V^{+}_{\pi N\Delta} \\ &= V_{\pi N\Delta}\, G^{+}_{\Delta}(\Sigma^{+}_{\pi N} - \Sigma_{\pi N})G_{\Delta}\, V_{\pi N\Delta} \\ &= V_{\pi N\Delta}\, G^{+}_{\Delta}\, V^{+}_{\pi N\Delta}(G^{o+}_{\pi N} - G^{o}_{\pi N})(V^{+}_{\pi N\Delta}\, G_{\Delta}\, V_{\pi N\Delta}) \\ &= t^{+}_{\pi N}\, 2\pi i\, \delta(E - (m_{\pi} + T_{\pi}) - (m_{N} + T_{N}))t_{\pi N} \end{aligned} \tag{7}$$

The last line in Equation (7) is the standard unitarity relation; the expectation value of this operator relation for forward scattering yields the optical theorem. For future use, we note that such unitarity relations will play a significant role in the discussion of Chapter III and that an important relation, used in passing from the first to the second line in Equation (7), is

$$T = V + VGT \tag{8}$$

$$T^{+} - T = T^{+}(G^{+} - G)T + \Omega^{+}(V^{+} - V)\Omega \tag{9}$$

$$\Omega = 1 + GT \tag{10}$$

Note that this relation, Equation (9), holds for any linear equation of the type given by Equations (8) or (3).

Since the isobar model defined by Equations (1)-(6) is central to much of what follows, we shall be rather explicit in writing out the Δ-propagator and πN t-matrix. First, the vertex operator can be written as

$$V_{\pi N\Delta}(k^2) = gh(k^2)\ \vec{k}\cdot\vec{S}^{+}T^{+}_{\alpha} \tag{11}$$

where α is the pion isospin label and we define the "form factor" normalization as h(0) = 1. Therefore, for scattering the pion from momentum $\vec{q}$ to momentum $\vec{p}$ in the πN c. m. system, we have

$$\langle\vec{p},\ \beta\ |\ t_{\pi N}(E)|\vec{q},\alpha\rangle = gh(p^2)\vec{p}\cdot\vec{S}\ T_{\beta}\ \frac{1}{D(E)}\ gh(q^2)\vec{q}\cdot\vec{S}^{+}T^{+}_{\alpha} \tag{12}$$

where $D(E)^{-1}$ is the full propagator. The transition spin operator $\vec{S}^{+}$ describes the angular momentum coupling $[L_{\pi N} = 1 \otimes S_N = 1/2]^{3/2}$ and so can be constructed explicitly as

$$\langle 3/2\ m_{\Delta}|\vec{S}^{+}|1/2\ m_N\rangle \equiv \sum_{m'} \langle 1m',\ 1/2\ m_N|3/2\ m_{\Delta}\rangle\hat{\varepsilon}_{m'} \tag{13}$$

where $\hat{\varepsilon}_m$ are the unit vectors

$$\hat{\varepsilon}_0 = \hat{z},\ \hat{\varepsilon}_{\pm 1} = \mp\frac{\hat{x}\pm i\hat{y}}{\sqrt{2}} \tag{14}$$

Note that $\vec{S}^{+}$ connects two-component nucleon spinors with four-component Δ spinors; for example, we have

$$S^{+}_{Z} = \sqrt{\frac{2}{3}}\begin{pmatrix}0 & 0\\ 1 & 0\\ 0 & 1\\ 0 & 0\end{pmatrix} \tag{15}$$

A very useful relation is given by the product

$$S_i S^{+}_j = \frac{2}{3}\delta_{ij} - \frac{i}{3}\ \varepsilon_{ijk}\ \sigma_k = (P_{3/2})_{ij} \tag{16}$$

where i, j are Cartesian indices. This is precisely the combination which enters in $t_{\pi N}$ and is nothing more than the J = 3/2 projection operator for the L = 1 πN system. This discussion goes through unmodified for the transition isospin operator, since the unit isospin of the pion now plays the role of L = 1 in the angular momentum discussion. The inverse Δ propagator D(E) is obtained by solving Equation (3) with the vertex function Equation (11). Since SS^{+} is just a projection operator, the solution is straightforward:

$$D(E) = E - M_{\Delta} - \Sigma_{\pi N}(E) \tag{17}$$

$$\Sigma_{\pi N}(E) = \frac{1}{3} g^2 \int \frac{d\vec{k}}{(2\pi)^3} \frac{k^2 h^2(k^2)}{E^+ - (m_\pi + m_N) - k^2/2\mu} \tag{18}$$

Clearly, the πN scattering phase shift (see Equation (12)) is carried by the propagator:

$$D(E) = |D(E)| e^{-i\delta_{33}(E)} \tag{19}$$

This provides the connection with data. We choose the simple monopole form factor

$$h(k^2) = (1 + k^2/\alpha^2)^{-1} \tag{20}$$

and fit the three parameters M_Δ, g, and α so as to best fit the experimental phase shift $\delta_{33}(E)$. At resonance, defined by $\delta_{33}(E_R) = \pi/2$, the real part of D(E) vanishes, so that

$$E_R = m_\Delta + \text{Re}\, \Sigma_{\pi N}(E_R) \tag{21}$$

The imaginary part of $\Sigma_{\pi N}$ gives the resonance width in a Breit-Wigner form for D(E), i.e., $\text{Im}\Sigma_{\pi N}(E) = -\Gamma(E)/2$.

In actually fitting the πN phase shifts, relativistic kinematics must be used for the pion. A method for doing this consistent with translation invariance for nonrelativistic Δ's is described in Reference 1. We simply quote here the resulting modification of Equation (18):

$$\begin{gathered} \Sigma_{\pi N}(E) = \frac{1}{3} g^2 \int \frac{d\vec{k}}{(2\pi)^3} \frac{k^2 h^2(k^2)}{(E-\varepsilon_k)^2 - \omega_k^2 + i\eta} \\ \varepsilon_k = m_N + k^2/2m_N\,, \quad \omega_k = (m_\pi^2 + k^2)^{\frac{1}{2}} \end{gathered} \tag{22}$$

The πN resonant phase shift is then described very well with the parameters

$$\begin{aligned} m_\Delta &= 1415 \text{ MeV} \\ g &= 2.5/m_\pi \\ \alpha &= 1.9 \text{ fm}^{-1} \end{aligned} \tag{23}$$

Note in particular that the bare mass is considerably larger than the resonance position $E_R \approx 1235$ MeV, although the precise value is not terribly important in this context (different relativistic extension for the self-energy would change M_Δ somewhat; for example, a Blankenbecler-Sugar form gives $M_\Delta = 1370$ MeV[2]).

We repeat that the main attraction of the isobar model is that it is simple while still admitting the proper dynamical treatment of the coupled Δ and πN channels. Nevertheless, one may legitimately ask about the relationship of the model to more basic physics, such as the quark model. We assert that the bare isobar model is quite compatible[3] with the quark model calculations, particularly with so-called chiral bag models, although this assertion is based only on an association and not on a formal demonstration. The quark bag calculations treat the Δ as if it were a stable three-quark system confined in a cavity. The mass, magnetic moment, etc. are computed on this basis. Clearly, this is an approximation which ignores the dynamical effects arising from coupling to the decay channel. The chiral bag models go beyond the simple bag model by explicitly treating pions outside the bag, with a pion-bag coupling defined at the bag surface (the details of the coupling differ with different groups,[4,5] but this is not important for our argument). Clearly, we have in mind the association of our bare isobar with the static three quark bag. The $\pi N\Delta$ vertex function then represents the pion coupling to the bag, which we treat completely phenomenologically. We can be somewhat more quantitative in considering the bare mass parameter. We argue that the static three-quark bag energy should not be fitted to the resonance energy but rather to the bare mass. Jaffe and Low[6] have essentially proposed a method for determining precisely this quantity in the scattering of two "elementary" particles. They suggest that the "bare masses" coincide with the poles of the P-matrix, defined for the L-th partial wave by

$$P_L(k)/k = \frac{\hat{j}_L'(kb) - \tan\,\delta_L(k)\hat{n}_L'(kb)}{\hat{j}_L(kb) - \tan\,\delta_L(k)\hat{n}_L(kb)} \tag{24}$$

where k is the c. m. momentum, b is the matching radius, $\hat{j}_L$, $(\hat{n}_L)$ are the Riccati-Bessel (-Hankel) functions, and $\delta_L(k)$ is the experimental phase shift. The poles of P occur for

$$\tan\,\delta_L = \frac{\hat{j}_L(kb)}{\hat{n}_L(kb)} \tag{25}$$

This has a simple physical interpretation since, outside the interaction region, the scattering wavefunction has the form

$$\psi_L^{(+)}(k,r) \sim \hat{j}_L(kr) - \tan\,\delta_L(k)\hat{n}_L(kr) \tag{26}$$

Therefore, the poles correspond to a vanishing wavefunction at the matching radius b. Physically, this corresponds to the imposition of confining boundary conditions on the internal quark wavefunction at the matching radius. Given the experimental phase shift δ_{33}, we can easily find the P-matrix poles for any given matching radius b. For b $\sim$ 1 fm, the pole position is $\sim$1350 MeV, in crude agreement with the Δ bare mass in the isobar model. Thus, we conclude there is no obvious conflict between the isobar model and the commonly practiced phenomenological quark model exercises.

It is amusing to note that the dynamical effects which must be present because of the $\Delta \rightleftarrows \pi N$ coupling influence significantly the calculation and extraction from data of Δ properties such as the magnetic moment. The basic point is simple. We have no difficulty defining the magnetic moment of the bare Δ. If we persist in the association with the quark model outlined above, we might expect that the bare Δ^{++} magnetic moment has the SU(6) value $\mu(\Delta^{++})/\mu(p) = 2$. However, the dressed Δ "magnetic moment" is more problematical. We really must calculate the convection and magnetization currents for the interacting $\pi N \rightleftarrows \Delta$ system and compare with measurements of pion-nucleon bremsstrahlung $\pi^+ p \to \pi^+ p\gamma$. If we define an effective magnetic moment for the dressed Δ as the long wavelength limit of the photon coupling to the dressed Δ, then we must include terms corresponding, for example, to photon coupling to internal pions (basically, pions contributing to the free space Δ self-energy). This is shown schematically in Figure 1. In the long wavelength limit, it is easy to see that this will contribute to μ_Δ^{eff} a term like

$$(\mu_\Delta^{eff})_\pi \underset{k\to 0}{\sim} \frac{m_\Delta}{m_\pi} \frac{\partial \Sigma_{\pi N}(E)}{\partial E} \tag{27}$$

The mass ratio enters because we are coupling to the pion convection current; the energy derivative of $\Sigma_{\pi N}(E)$ enters because the photon couples inside the πN bubble which generates $\Sigma_{\pi N}$. This is strictly a dynamical effect. The effective moment is now both energy dependent

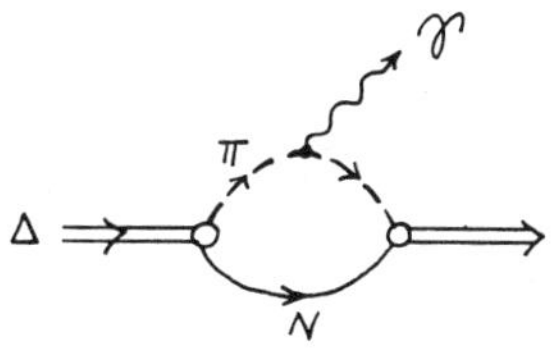

Fig. 1. Internal pion contribution to the Δ electromagnetic current.

and complex. This term, Equation (27), is large, giving a model dependent result of order one magneton in the imaginary part near resonance. We show in Figure 2 the result of fitting the π^+p bremsstrahlung cross section including a full dynamical treatment[3] of Δ propagation and keeping only the bare Δ magnetic moment a free parameter. From a fit to many spectra, we find

$$\mu(\Delta^{++})/\mu(p) \sim 2.5 \text{ to } 3.3 \tag{28}$$

Furthermore, we find that polarization assymetry, theoretical results for which are shown in Figure 3, is very sensitive to this quantity. Nevertheless, the main point is just that dynamical treatment of the Δ is crucial for understanding its propatation and free-space properties. It is also crucial for constructing a consistent theory of its propagation in the nucleus, to which we now turn.

III. Δ-NUCLEUS INTERACTIONS

Our main task in this chapter is development and application of the pion optical potential within the Δ isobar framework developed above. The theoretical development will be carried through as though πN scattering proceeds only through the Δ. Clearly, there are small background partial waves and these will be included in all results compared with data. In the formal development, we shall stress the constraints placed upon the optical potential by our knowledge of the experimental reaction cross section. The vehicle for imposing these constraints is the extension of the unitarity relation discussed above to the π-nucleus case. Of course, this implies that we must know something about pion-nucleus reactions as a guide to the theoretical development. Therefore, we start the chapter with a brief, selective overview of experimental information. An important point will be that pion annihilation is a very important reaction channel in competition with direct nucleon knockout. This will be followed by the formal development and by application of the theory to pion elastic scattering data. The basic picture that will emerge of pion-nucleus interactions is of pion multiple scattering but with strongly modified intermediate Δ propagation. The modification is essentially a spreading of the resonance due to coupling to the annihilation channels $\Delta N \rightleftarrows NN$. The strength of this coupling will lead to an expectation that a new reaction mechanism, namely, nucleon knockout by the Δ, should be important. The evidence for this will be discussed, together with implications for certain nuclear structure studies with pions. Finally, we shall turn briefly to intermediate energy photonuclear reactions and find that the limited data available on coherent photoreactions is consistent with the dynamical picture outlined above.

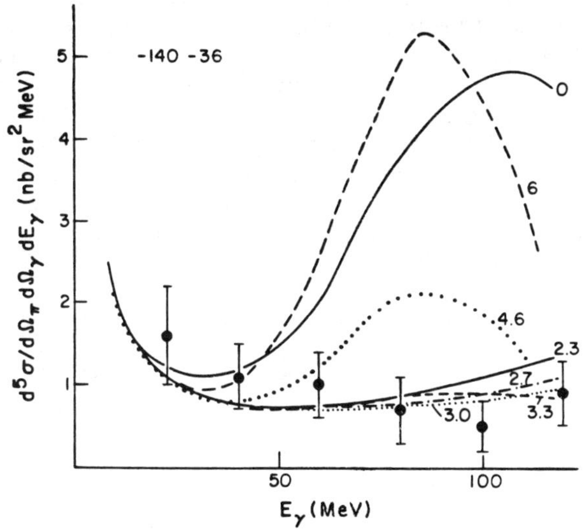

Fig. 2. Cross section for $\pi^+p \rightarrow \pi^+p\gamma$ for incoming pion kinetic energy T_π = 290 MeV, outgoing pion angle $\Theta_\pi = 50.5°$, and outgoing photon angles $\alpha = -140°$, $\beta = -36°$. The photon angles are related to the conventional polar and azimuthal angles by $\cos\Theta = \cos\alpha \cos\beta$ and $\tan\phi = \tan\beta \cos\alpha$. Data from D.I. Sober, et.al., Phys. Rev. D11 (1975) 1017. Theoretical curves, taken from Reference 3, are labelled according to the value of the parameter $\mu(\Delta^{++})/\mu(p)$.

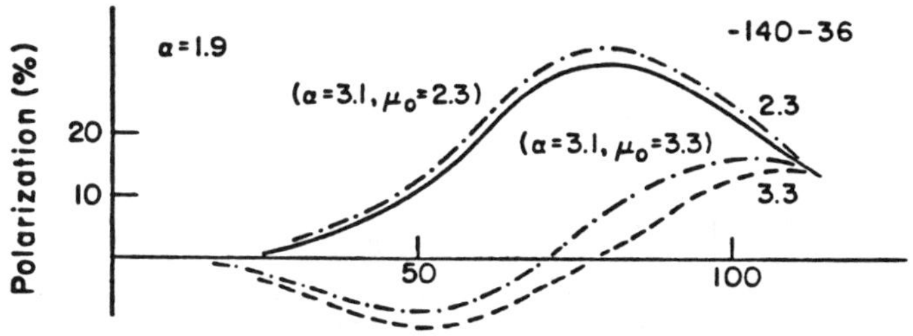

Fig. 3. Calculated polarization asymmetry for the kinematics of Figure 2. Theoretical results are shown for two vertex range parameters, $\alpha = 1.9\ \text{fm}^{-1}$ and $\alpha = 3.1\ \text{fm}^{-1}$.

A. An Experimental Primer

Before building a theory of pion-nucleus interactions based upon the Δ, we should establish Δ-dominance of pion scattering, $\pi N \to \pi N$, and pion annihilation, $\pi NN \to NN$. Figure 4 shows the total cross section for πN scattering in various charge states. The cross section is totally dominated by the J = 3/2, T = 3/2 Δ resonance, which occurs for a pion laboratory kinetic energy $T_\pi \simeq 190$ MeV ($k_{lab} \simeq 290$ MeV/c). The solid line in Figure 4 corresponds to the contribution coming from the resonant partial wave alone. There is little background. The cross section is very large: $\sigma_{\pi^- p} = \sigma_{\pi^+ n} \simeq 200$ mb at resonance. The Δ width is $\Gamma \simeq 115$ MeV, leading to an average propagation distance in free space $d_\Delta \simeq 1$ fm. Resonance dominance leads to a strong isospin dependence, as is evident from Figure 4: the $\pi^+ p \to \pi^+ p$ cross section is nine times that for $\pi^+ n \to \pi^+ n$. Finally the large width of the Δ implies that the relevant nuclear degrees of freedom are those associated with single particle motion.

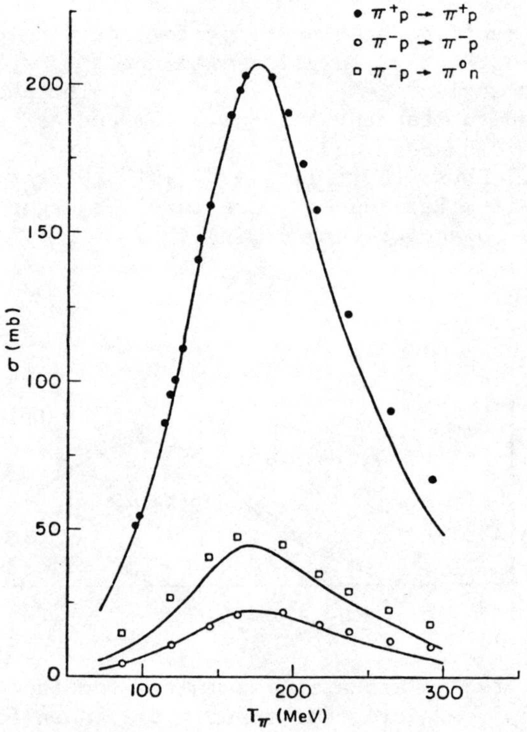

Fig. 4. Pion-nucleon total cross section as a function of pion kinetic energy. The solid lines give the resonant J = 3/2, T = 3/2 contributions.

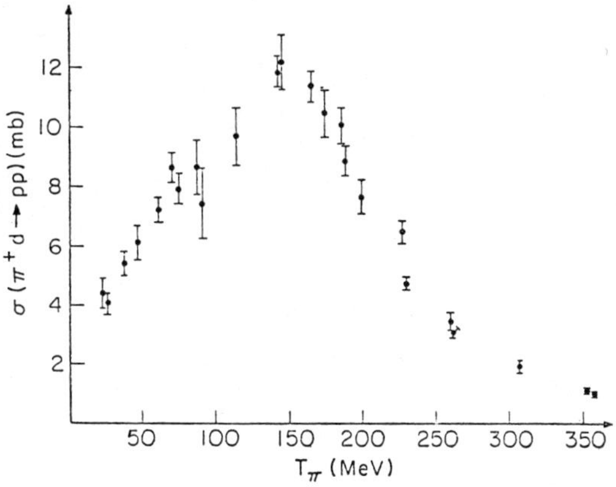

Fig. 5. Total cross section for π^+ annihilation on the deuteron. Data from C. Richard-Serre, et.al., Nucl. Phys. B20 (1970) 413.

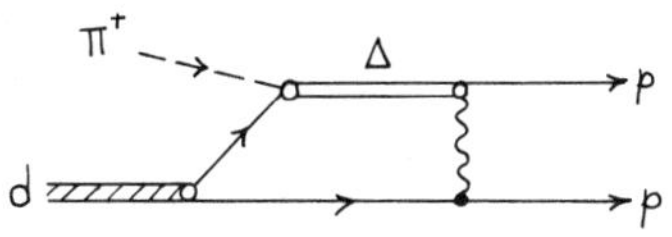

Fig. 6. Pion absorption through the Δ.

The cross section for π^+ annihilation on the deuteron is shown in Figure 5. The peak in the cross section shows clearly that the the Δ acts as a doorway for pion annihilation. This is indicated schematically in Figure 6. Later, we shall use the fact that pion annihilation appears to take place primarily on isoscalar nucleon pairs. For example, measurements of pion annihilation on ^{4}He [7] show that absorption on isoscalar pairs is rougly fifty times stronger than on isovector pairs. A simple argument leads us to expect this. Assume, following Figure 6, that absorption occurs on nucleon pairs with relative angular momentum $L^i_{NN} = 0$ (presumably, the process is reasonably short-ranged) and that the intermediate ΔN state has relative angular momentum $L_{\Delta N} = 0$. The latter assumption should be reasonable since the relative ΔN momentum up to resonance $P_{\Delta N} \lesssim .75\ \mathrm{km}^{-1}$, implying a strong preference for small $L_{\Delta N}$. However, $L_{\Delta N} = 0$ implies that overall parity is positive and that total angular momentum $J = S_{\Delta N} = 1$ or 2. The $\Delta N \to NN$ coupling enforces the

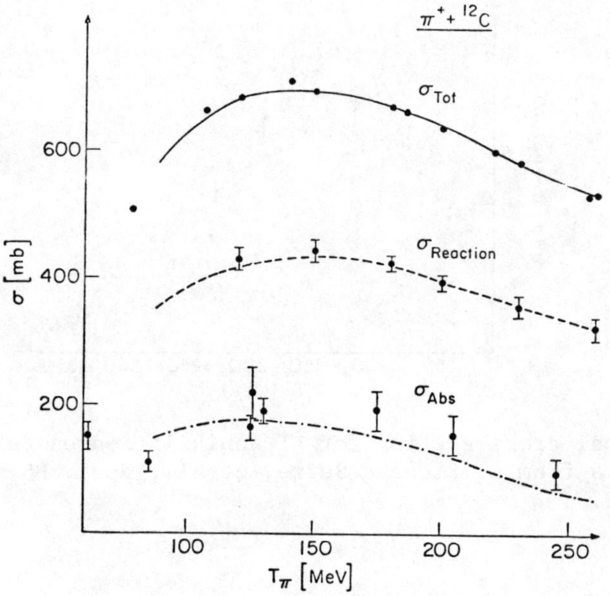

Fig. 7. Total, total reaction, and total π annihilation cross sections for $\pi^+ - {}^{12}C$ scattering. Theoretical curves are results of Δ-hole calculations. See References 14 and 8.

constraint that the overall isospin is T = 1, so that, together with even parity, the total spin of the final nucleons is $S^f_{NN} = 0$. Even parity then implies that $J = L^f_{NN} = 2$. Now, even parity implies that the pion angular momentum L_π is odd, requiring that the initial NN angular momentum $S^i_{NN} = 1$. Finally, this, together with $L^i_{NN} = 0$, implies that the isospin of the absorbing pair $T^i_{NN} = 0$. Therefore, for the s-wave pion annihilation mechanism, we must start with a pn pair. This isospin constraint will play an important role in various consideration below. We note that pion annihilation is a very important reaction channel for π-nucleus interactions. For example, the total cross section for π-${}^{12}C$ scattering, shown in Figure 7, consists of roughly equal contributions from elastic scattering, inelastic scattering, and annihilation.

Pion-nucleus scattering displays features of strong absorption scattering. This is not too surprising, since the distance over which the elastic channel wavefunction is damped is estimated naively

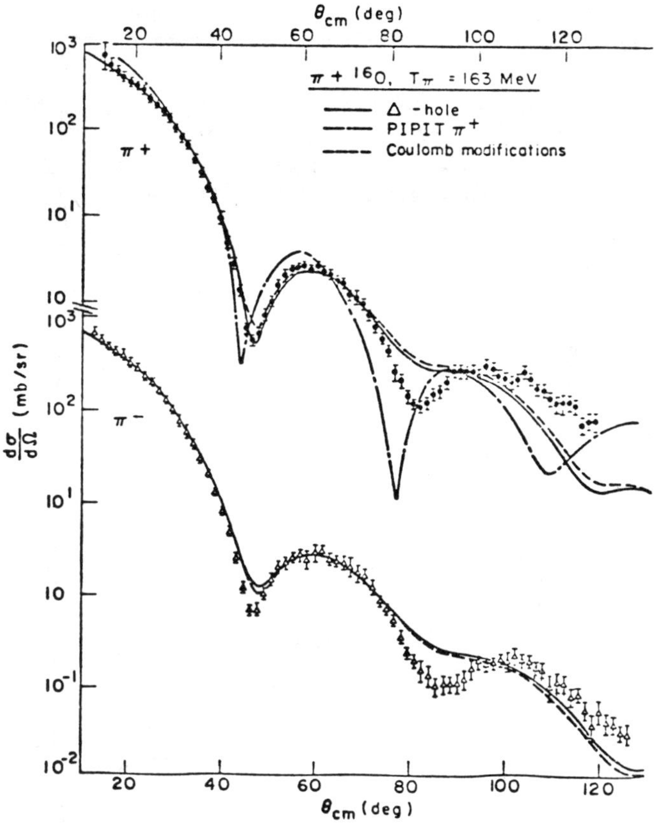

Fig. 8(a). Pion elastic scattering differential cross sections (a) π^+ - ^{16}O at 163 MeV[13]. Data from J. Jansen et al., Phys. Lett. 77B (1978) 359.

to be $d_\pi = 2/\lambda\bar{\sigma} \approx 1$ fm at resonance, and this is much smaller than the nuclear size. The diffractive-like structure in the elastic scattering differential cross section is shown for ^{16}O, ^{40}Ca and ^{208}Pb in Fig. 8. However, strong absorption does not explain pion reactions quantitatively. For example, the partial wave amplitudes corresponding to the π^+-^{16}O cross section shown in Fig. 8(a) are shown in the Argand plot of Figure 9. The amplitudes shown are $T_L = (S_L - 1)/2i$, where S_L is the partial wave S-matrix. Clearly, S_L is small for the central partial waves only, and even here it may be surprising that these central partial waves have an elastic cross section larger than the reaction cross section. The opposite is true for the peripheral partial waves $L \gtrsim 5$.

We shall not enter into a detailed discussion of pion inelastic scattering. However, we will make use of the isospin structure of nuclear particle-hole excitation. We shall consider only T = 0

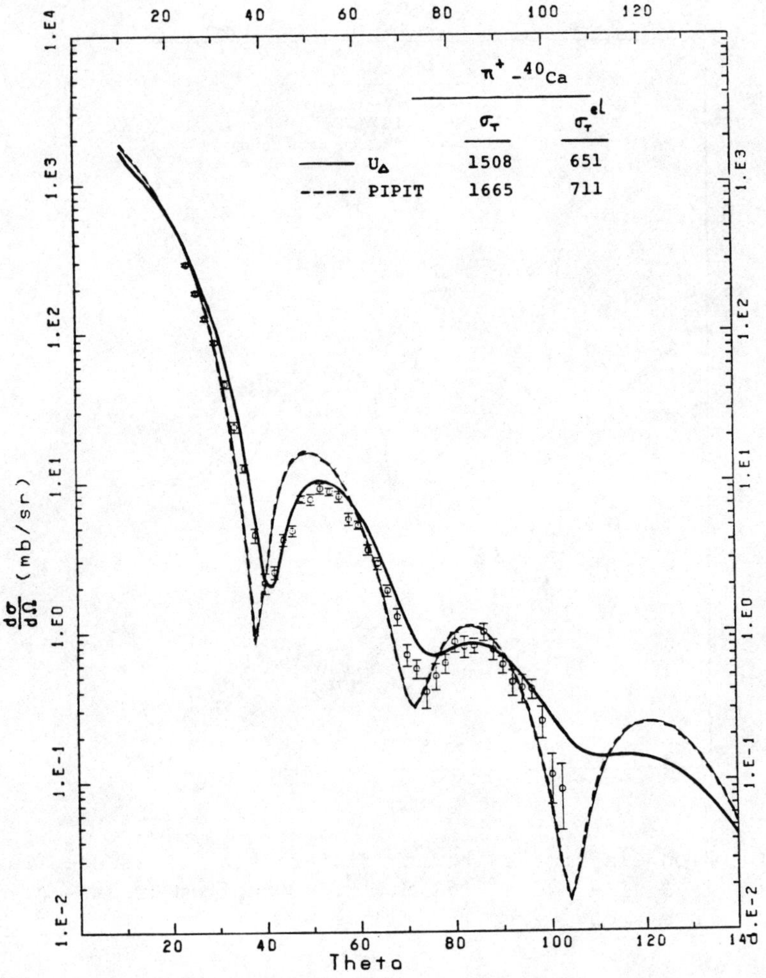

Fig. 8(b). Pion elastic scattering differential cross sections (b) $\pi^+ - {}^{40}Ca$ at 130 MeV[16]. Data from P. Gretillat et al., Nucl. Phys. A364 (1981) 270.

nuclear ground states. First, if the final p-h state is a state of pure isospin T = 0,1 then the ratio of π^+ and π^- inelastic cross sections is the same. However, the amplitudes for $\pi^\pm$ excitation of T = 1 states have a relative sign difference, so that a final state of mixed isospin will lead to unequal cross sections. Clearly, proton-enhanced transitions stand out more strongly in π^+ scattering, and vice versa for neutron-enhanced transitions. Further, if the excitation mechanism is scattering through the Δ, as indicated in Figure 10(a), the cross section for isoscalar p-h states is four times that for the isovector state with the same configuration.

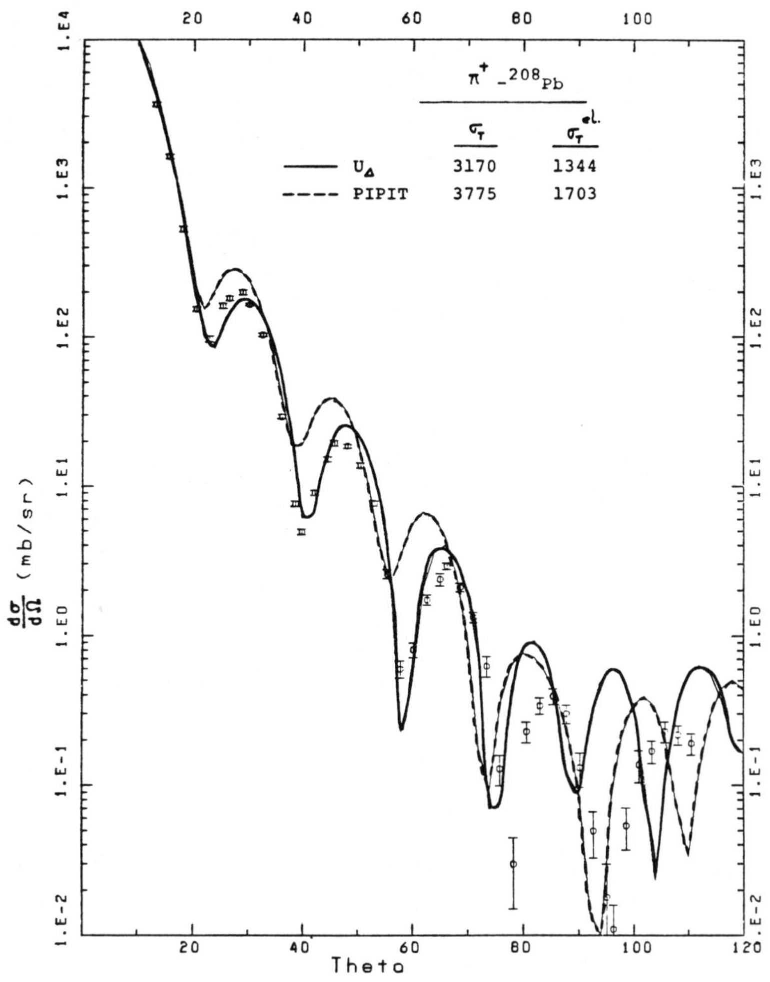

Fig. 8(c). Pion elastic scattering differential cross sections (c) $\pi^+ - {}^{208}Pb$ at 162 MeV[16]. Data from C. Olmer et al., Phys. Rev. C21 (1980) 254.

The study of isospin mixing in pion scattering has proven quite fruitful. We shall see that such studies may lead to information on Δ-nucleus interactions.

Finally, we show in Figure 11 an inclusive "deep inelastic" pion spectrum for scattering from ^{16}O. The central point here is that the energy dependence for large angles is qualitatively the same as that found with electron scattering at comparable momentum transfer. In the latter case, the reaction mechanism is simply quasifree nucleon knockout. For the pion case, we must keep in

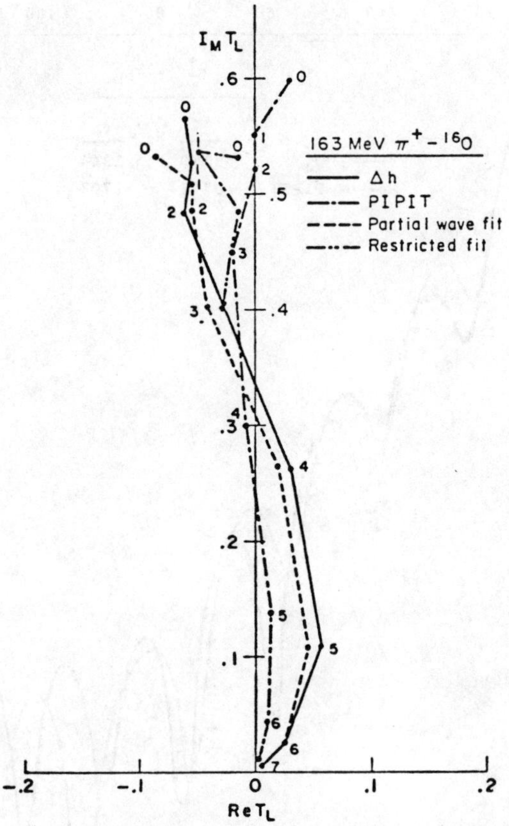

Fig. 9. Argand diagram for the partial wave amplitudes[13] associated with Figure 8(a). Also shown are the partial wave amplitudes obtained from a fit to the data and those produced by the PIPIT optical potential.[15]

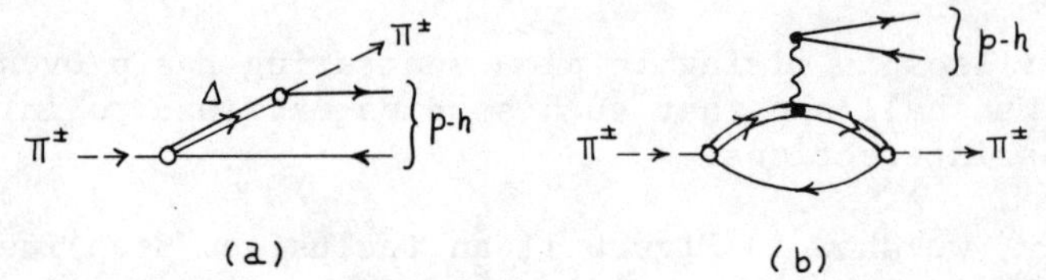

Fig. 10. Particle-hole excitation (a) the (standard) impulse approximation, (b) Δ knockout mechanism.

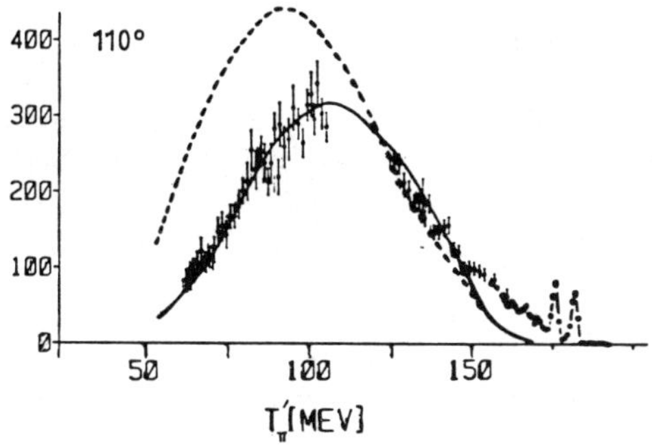

Fig. 11. Large angle $\pi^+ - {}^{16}O$ inelastic scattering, with incident pion kinetic energy 163 MeV and final pion angle $\Theta = 110^\circ$. T'_π is the final pion kinetic energy. Solid and dashed curves give the results of the Δ-hole and static calculations, respectively. See Reference 9.

mind that annihilation is a strongly competing reaction mechanism. We shall see that this is important for quantitative understanding of the inclusive pion spectrum.[9]

B. The Pion Optical Potential and Unitarity

We start the formal development with a brief general discussion of the optical potential and the role of unitarity. This discussion follows one to be presented in a review written with Lenz.[8] We assume that the Hamiltonian has the form

$$H = H_A + T_\pi + V_{\pi A} \tag{29}$$

where H_A is the nuclear Hamiltonian and V_π gives the projectile interaction with each of the target nucleons

$$V_{\pi A} = \sum_{i=1}^{A} V_i \tag{30}$$

The transition operator obeys the standard equation

$$T_{\pi A}(E) = V_{\pi A} + V_{\pi A}\, G(E)\, T_{\pi A}(E) \tag{31}$$

$$G(E) = (E^+ - T_\pi - H_A)^{-1} \tag{32}$$

The optical potential is <u>defined</u> by a Lippmann-Schwinger equation for $T_{\pi A}$ with iteration only in the nuclear ground state

$$T_{\pi A}(E) \equiv U(E) + U(E)\, P_0 G(E)\, T_{\pi A}(E) \tag{33}$$

$$U(E) = V_{\pi A} + V_{\pi A}\, Q_0\, G(E)\, U(E) \tag{34}$$

where P_0 is the ground state projection operator

$$P_0 = |\Psi_0\rangle \langle\Psi_0| \tag{35}$$

$$P_0 + Q_0 = 1 \tag{36}$$

It follows from Equations (32) and (34) that the optical potential is energy dependent and complex above inelastic threshold. For elastic scattering, the optical potential is the expectation value of U(E) in the nuclear ground state wavefunction

$$U_0(E) = \langle\Psi_0|U(E)|\Psi_0\rangle \tag{37}$$

The unitarity relation will provide important guidance in construction of the optical potential. Using Equations (8)-(10), we get from Equations (33) and (34)

$$\begin{aligned} T^+_{\pi A} - T_{\pi A} &= T^+_{\pi A}\, P_0\, 2\pi i\ \delta(E-T_\pi-H_A) P_0 T_{\pi A} \\ &\quad + \Omega(E)^+ U^+(E) Q_0\ 2\pi i\ \delta(E-T_\pi-H_A) Q_0 U(E)\Omega(E) \end{aligned} \tag{38}$$

$$\Omega(E) = P_0(1+G(E)T_{\pi A}(E)) \tag{39}$$

The optical theorem then reads

$$\begin{aligned} \sigma_{Total} &= -i\,\frac{E_k}{k}\,\langle\vec{k},\Psi_0|T^+_{\pi A}-T_{\pi A}|\vec{k},\Psi_0\rangle \\ &= \sigma_{Elastic} + \sigma_{Reaction} \end{aligned} \tag{40}$$

$$\sigma_{Elastic} = 2\pi\,\frac{E_k}{k}\int \frac{d\vec{k}'}{(2\pi)^3}\,\delta(E_k-E_{k'})\,|\langle\vec{k}',\Psi_0|T_{\pi A}|\vec{k},\Psi_0\rangle|^2 \tag{41}$$

$$\sigma_{\text{Reaction}} = 2\pi \frac{E_k}{k} \sum_{\alpha\neq 0} \int \frac{dk'}{(2\pi)^3} \delta(E_k - E_{k'} - \varepsilon_\alpha) |\langle \vec{k}', \Psi_\alpha | U(E) | \psi_{\vec{k}}^{(+)} \Psi_0 \rangle|^2 \tag{42}$$

In writing Equation (42), we have used the fact that the Moller operator $\Omega(E)$ generates the elastic channel distorted wave

$$\Omega(E_k)|\vec{k}\rangle = |\psi_{\vec{k}}^{(+)}\rangle \tag{43}$$

The important result in Equations (38)-(42) is that the imaginary part of the optical potential reflects directly the reaction channels. Conversely, it is the "reactive content" of the optical potential which guides our approximations in theoretically evaluating the optical potential. We will use this in the following section, where we explicitly construct the optical potential within the isobar framework of Chapter II.

C. The Δ-Hole Approach

We assume in the formal development that the pion interacts only through the $\pi N \rightleftarrows \Delta$ coupling. For the Δ dynamics, we take both a binding potential U_Δ, assumed to be similar to the nucleon shell model potential, and a $\Delta N \rightleftarrows NN$ interaction to describe the important pion annihilation reaction. For the nuclear model, we simply use the noninteracting shell model. The optical potential is then written fairly simply in the projection operator formalism. We define a set of projection operators

$$P_0 + D + P_1 + Q_1 + Q_2 = 1 \tag{44}$$

where P_0 projects into states of the pion plus nuclear ground state, D projects onto Δ-hole states, P_1 projects onto π-particle-hole states, Q_1 projects onto states with the π or Δ plus at least two nucleon holes, and Q_2 projects onto states with no π or Δ present (the important states here are those with at least two p-h pairs). The nonzero interactions are then V_{P_0D}, V_{DP_1}, V_{DQ_2}, V_{DD}, $V_{P_1Q_2}$, $V_{Q_1Q_2}$ and $V_{Q_1Q_1}$. Projecting Equation (31) onto the nuclear ground state, we have the elastic scattering amplitude

$$T_{P_0P_0} = V_{P_0D} \mathcal{G}_D V_{DP_0} \tag{45}$$

$$\mathcal{G}_D^{-1} = G_D^{0\,-1} - V_{DP_1} G_{P_1}^0 V_{P_1D} - V_{DD} - V_{DP_0} G_{P_0}^0 V_{P_0D} - \tilde{V}_{DD} \tag{46}$$

where V_{DP_0} is the $\pi N \to \Delta$ coupling and $\mathcal{G}_D$ is the full Δ-hole propagator. The first two terms in $\mathcal{G}_D^{-1}$ give the dressed Δ propagator, Equation (17), together with kinetic energy of Δ propagation and Pauli restriction of the Δ decay in the nucleus; we denote these terms by $(G_D^{0\,-1} - V_{DP_1} G_{P_1}^0 V_{P_1 D}) = (E - E_R + i\Gamma/2 - T_\Delta - \delta W)$, where δW is the Pauli blocking term. The next term in Equation (46) takes care of the Δ binding and hole energy $V_{DD} = U_\Delta + H_{A-1}$. The next term describes intermediate propagation in the elastic channel. Therefore, if we remove this term, we have the pion optical potential

$$U_0(E) = V_{P_0 D} \tilde{\mathcal{G}}_D(E) V_{DP_0} \tag{47}$$

$$\tilde{\mathcal{G}}_D^{-1}(E) = E - E_R + i\Gamma/2 - H_\Delta - \delta W - V_{DD} \tag{48}$$

$$H_\Delta = T_\Delta + V_\Delta + H_{A-1} \tag{49}$$

The last term, $\tilde{V}_{DD}$, is extremely complicated and contains all the Q_1 and Q_2 terms. If this term is dropped, we have all terms containing only intermediate one-hole states. This defines the first order optical potential $U_0^{(1)}(E)$. This approximation is unitary[8]

$$U_0^{(1)+} - U_0^{(1)} = U_0^{(1)+} P_1 (G_{P_1}^{0+} - G_{P_1}^0) P_1 U_0^{(1)} \tag{50}$$

The reactive content is nucleon knockout, with the πN t-matrix modified by binding effects and Pauli blocking. This reactive content is certainly reasonable. The one-hole approximation gives the correct low density limit, so that $U_0^{(1)}$ provides guidance for a phenomenological optical potential. Nevertheless, we have seen that knockout accounts for only half the reaction cross section. The important annihilation contribution is contained in $\tilde{V}_{DD}$, through terms such as $V_{DQ_2} G_{Q_2} V_{Q_2 D}$. Guided by the fact that the Δ acts as doorway for annihilation, we simply take $\tilde{V}_{DD}$ phenomenologically as a spreading potential with a strength determined by a fit to data. Before discussing this further, we make a slight digression to emphasize once more the importance of the reaction content in determining the optical potential.

Kuo et al.[10] pointed out that one could construct an energy-independent optical potential which could reproduce the elastic channel scattering wavefunction at all energies:

$$U_0(E)|\psi_k^{(+)}> \equiv \bar{U}|\psi_k^{(+)}> \tag{51}$$

where $U_0(E)$ is the standard energy-dependent optical potential and $\bar{U}$ is the energy-independent optical potential. To determine $\bar{U}$ microscopically,[11] it is convenient to introduce dual states $<\tilde{\psi}_k|$ such that

$$<\tilde{\psi}_{\vec{k}'}|\psi_{\vec{k}}^{(+)}> = (2\pi)^3\delta(\vec{k} - \vec{k}') \tag{52}$$

An integral equation for these dual states is obtained easily after removing explicitly the noninteracting pieces:

$$<\vec{q}|\psi_{\vec{k}}^{(+)}> \equiv (2\pi)^3\delta(\vec{q} - \vec{k}) + \phi_{\vec{k}}^{(+)}(\vec{q}) \tag{53}$$

$$<\tilde{\psi}_{\vec{k}'}|\vec{q}> = (2\pi)^3\delta(\vec{q} - \vec{k}') - \tilde{\phi}_{\vec{k}'}(\vec{q}) \tag{54}$$

$$\tilde{\phi}_{\vec{k}'}(\vec{k}) = \phi_{\vec{k}}^{(+)}(\vec{k}') - \int \frac{d\vec{q}}{(2\pi)^3}\,\phi_{\vec{k}}^{(+)}(\vec{q})\,\tilde{\phi}_{\vec{k}'}(\vec{q}) \tag{55}$$

With these dual states, Equation (51) is clearly satisfied if

$$\bar{U} = \int \frac{d\vec{k}'}{(2\pi)^3}\,U_0(E_{k'})\,|\psi_{\vec{k}'}^{(+)}><\tilde{\psi}_{\vec{k}'}| \tag{56}$$

Since we know (at least formally) $U_0(E)$ and $|\psi_k^{(+)}>$ in a multiple scattering expansion, we can evaluate $\bar{U}$ in a multiple scattering expansion. This is where the physics comes in. For $U_0(E)$, we found that truncation at the lowest order (i.e., the first order optical potential $U_0^{(1)}$) gave a unitary approximation (Equation (50)) with a reasonable reactive content (quasifree nucleon knockout). Indeed, this is the reason that a multiple scattering expansion in the (medium-modified) projectile-nucleon t-matrix makes sense and provides guidance for a reasonable phenomenology. Unfortunately, $\bar{U}$ does not have these properties. For example, it follows immediately from Equation (53), (54) and (56) that the "first-order" $\bar{U}$ (i.e., the term lowest order in the density) is just

$$\begin{aligned} <\vec{p}|\bar{U}^{(1)}|\vec{q}> &= <\vec{p}|U_0^{(1)}(E_q)|\vec{q}> \\ &\approx \rho(\vec{p}-\vec{q})\,t_{\pi N}(\vec{p},\vec{q};E_q) \end{aligned} \tag{57}$$

where $\rho(q)$ is the nuclear ground state form factor and, for illustrative reasons, the static approximation has been used to obtain the second line in Equation (57). The diseases here are evident in that the energy variable in $t_{\pi N}$ is associated with the momentum (note that $\bar{U}$ is here non-symmetric and is defined specifically so as to give $|\psi_K^{(+)}>$ only). This truncation does not lead to a reasonable reactive content; basically, at any energy it gives quasi-free knockout averaged over all energies. The nonlocality of the optical potential is associated not with interaction parameters of the Hamiltonian but with the energy variation of the elementary amplitude; for example, a resonance in $t_{\pi N}$ leads through a strong energy dependence to a long-range nonlocality. Thus, while $\bar{U}$ is mathematically definable, it is not physical in that it does not generally lead to a truncation scheme which respects the nature of the physical reaction channels. Again, we stress the main point: construction of the optical potential, which necessarily contains phenomenological elements when we go beyond first order, must be guided by experimental information on the reaction channels, with the associated constraints imposed via unitarity.

We return now to the Δ-hole model. Our basic result is summarized by Equations (47)-(49). All terms but $\tilde{V}_{DD}$ are evaluated microscopically [12,13]. The coupling to multi-hole states, $\tilde{V}_{DD}$, will be evaluated phenomenologically as a local Δ spreading potential

$$\tilde{V}_{DD} \rightarrow V_{sp}(r) = V_c \frac{\rho(r)}{\rho(0)} + V_{LS}\, f(r) 2\vec{L}_\Delta \cdot \vec{\Sigma}_\Delta \qquad (58)$$

where V_C and V_{LS} are complex parameters. The central term is taken proportional to the nuclear density (this will be refined somewhat below). The spin-orbit part is surface peaked; for light nuclei, the form $f(r) = \mu r^2 \exp(-\mu r^2)$ has been used.[14] For p-shell nuclei ^{12}C and ^{16}O, the strength parameters are roughly energy independent and equal to

$$V_C \simeq (20 - 45i)\mathrm{MeV} \qquad (59)$$

$$V_{LS} \simeq (-10-4i)\mathrm{MeV} \qquad (60)$$

These parameters are extracted from an overall fit to elastic scattering differential cross sections and to the total cross section. The theoretical results are shown in Figures 7, 8(a) and 9. On the Argand plot, we also show the results obtained with a standard first order optical potential,[15] essentially given by Equation (57) but with $E_q \rightarrow E$ (i.e., the static approximation to $U_0^{(1)}(E)$). Clearly, the medium effects play a substantial role, with central and peripheral partial waves affected quite differently. For a fuller discussion of the role of each of the dynamical ingredients in Equation (48), see Reference (13). For heavier nuclei, a local density approximation for the Δ propagator in Equation (48) has been made.[16] This greatly shortens the calculation while retaining the essential physics. A slightly stronger central potential $\mathrm{Im}V_c \approx -50$ MeV produces the results shown in Figures 8(b) and (c) for ^{40}Ca and ^{208}Pb, respectively. For comparison, the static results[15] are also shown.

The most striking feature of the Δ-nucleus spreading potential is the very large imaginary part of the central potential needed to fit the data. As stated above, we expect that this is associated largely with strong coupling to the pion annihilation channels via the Δ as a doorway. We shall use this to test the model further. Playing once again the unitarity game, we have from Equations (45)-(49)

$$T^{+}_{P_0P_0} - T_{P_0P_0} = V_{P_0D}\,\mathcal{G}^{+}_{D}\,\{V_{DP_1}(G^{0\,+}_{P_1} - G^{0}_{P_1})V_{P_1D}$$
$$+ V_{DP_0}(G^{0\,+}_{P_0} - G^{0}_{P_0})V_{P_0D} + (V^{+}_{sp} - V_{sp})\}\,\mathcal{G}_{D}\,V_{DP_0} \tag{61}$$

The three terms inside the curly brackets then correspond to the partial cross sections for nucleon knockout, elastic scattering and pion annihilation, respectively. Actually, the decomposition of the reaction cross section is not expected to be precise. Clearly, the "spreading potential" $\tilde{V}_{DD}$ in Equation (46) must contain, for example, intermediate coupling to the P_1 space (even though each term must contain at least two holes at some point), so that we do not really have a clean separation of the annihilation cross section. Nevertheless, we perform the separation indicated in Equation (61) as a rough guide and show the results[14] in Figure 7. The results are certainly of the appropriate scale. Actually, this could be guessed more simply, since we have $-\mathrm{Im}V_{sp} \approx \Gamma/2$. That is, the Δ has a roughly equal chance for decaying via the free-space process $\Delta \rightarrow \pi N$ or via the absorption process $\Delta \rightarrow NNN^{-1}$ in nuclear metter. This leads to the rough equality of the inelastic and annihilation cross sections. Further, Lee and Ohta[17] have evaluated the spreading potential microscopically using an absorption model fit to NN inelastic scattering. They find, for nuclear matter, $\mathrm{Im}V_{sp} \approx -33$ MeV, with over half the contribution coming from the ΔN relative s-wave. Most of the rest comes from relative P-waves, which would allow for absorption on isovector pairs.

The isospin dependence of the spreading interaction leads to further tests and predictions. We shall stick for now to the simple model that pion annihilation occurs on isoscalar (therefore, pn) pairs via the Δ. For π^+ scattering, the favored scattering process is on protons, forming a Δ^{++}. In this case, the spreading potential in Equation (51) should be proportional to the neutron density, not to the nuclear density. This leads immediately to an isotope dependence[18] of the spreading effect. For example, with V_c fit to scattering on ^{16}O, the spreading potential for Δ^{++} should be 25% stronger in ^{18}O (i.e., the ratio of neutrons, assuming that the shapes of the neutron densities in ^{16}O and ^{18}O are nearly the same). Since the spreading effect reduces the effective strength of the π^+p scattering amplitude at resonance, the cross section for proton knockout by π^+ mesons should be smaller in ^{18}O than in ^{16}O. A simple local density calculation based upon this idea gives a ratio[18] for backward inclusive π^+ scattering of 0.87. The experimental result is ∿0.84,[19] lending further support to the picture developed above. Indeed, Thies[9] has found that the spreading potential affects strongly the energy spectrum of inelastically scattered pions. His results are shown in Figure 11 both with and without the medium corrections. The medium modification of the πN scattering amplitude is large and leads to excellent agreement[9] with the data.

The need for strong medium corrections can be seen also in medium energy photoreactions. Coherent π^0 photoproduction from nuclei provides a good example. Photoexcitation of the Δ is described by a vertex operator

$$V_{\gamma N\Delta} = \frac{g_{\gamma N\Delta}}{M_\Delta} \hat{\varepsilon}_{\vec{k}\lambda} \times \vec{k} \cdot \vec{S}^+ T_3^+ \tag{62}$$

In contrast to the pion coupling, Equation (11), the $\gamma N\Delta$ coupling is transverse, reflecting the M1 nature of the process. Δ -excitation is the dominant mechanism for π^0 production in the resonance region. The nuclear coherent π^0 photoproduction amplitude then has a form similar to that in Equation (45);

$$T_{\gamma\pi 0} = V_{P_0 D} \mathscr{G}_D V_{D\gamma} \tag{63}$$

the only difference being that the Δ-hole states are excited by the photon. This can be rewritten in a distorted wave form[20]

$$\langle\vec{p};\Psi_0|T_{\gamma\pi 0}|\vec{k}\lambda;\Psi_0\rangle = \langle\psi^{(-)}_{\vec{p}};\Psi_0|V_{P_0 D}\tilde{\mathscr{G}}_D V_{D\gamma}|\vec{k}\lambda;\Psi_0\rangle \tag{64}$$

$$\langle\psi^{(-)}_{\vec{p}}| = \langle\vec{p}|[V_{P_0 D}\mathscr{G}_D V_{DP_0} G^0_{P_0} + 1] \tag{65}$$

Note that, although the full distorted wave $\psi^{(-)}_{\vec{p}}$ is written in Equation (64), the production operator $V_{P_0 D}\tilde{\mathscr{G}}_D V_{D\gamma}$ must still be medium-modified (see Equations (48) and (49)). This is important since, for example, the spreading potential weakens the production operator at resonance but simultaneously decreases optical absorption of the pion wavefunction. The latter effect is shown explicitly in

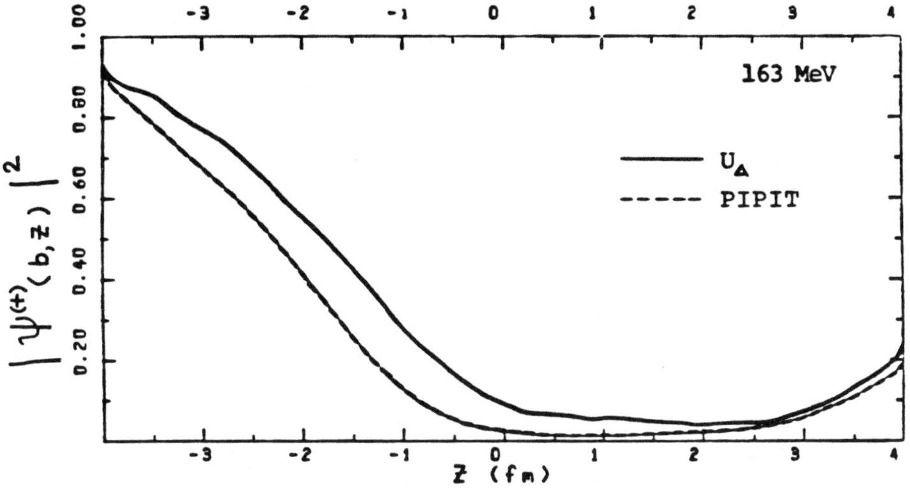

Fig. 12. Probability density of the pion elastic channel wavefunction at impact parameter b = 1 fm for π^+ - ^{16}O scattering with incident pion kinetic energy 163 MeV. Solid line includes Δ medium corrections; dashed line is the PIPIT result. See Reference 16.

Figure 12, where we plot the probability density of the pion wavefunction at impact parameter b = 1 fm for 163 MeV π^+ - ^{16}O scattering[16]; the increase for large Z is due to the fact that the real part of the optical potential is somewhat attractive at this energy. Clearly, there is a substantially larger probability for finding the elastic channel pion inside the nucleus when non-static effects are included, thereby compensating somewhat for the weaker production amplitude. In Figure 13, a Δ-hole calculation[20] of the $^{12}C(\gamma,\pi^0)^{12}C$ coherent cross section, for k_γ = 240 MeV, is compared with recent Bonn data.[21] We note that the data includes contributions from particle-stable excited states and has a poor angular resolution; the latter effect "spreads out" the angular distribution. Clearly, the Δ spreading effect is important for a quantitative understanding of the data.

We have seen that the description of many intermediate energy reactions can be unified rather simply in terms of a strong phenomenological Δ spreading potential associated with pion annihilation. Nevertheless, there are still many open questions, both with regard to refinements of the phenomenological Δ-nucleus interaction and with regard to understanding its microscopic origins in the nuclear many-body physics. Inelastic pion scattering data may play an important role in these investigations. We indicate this with some brief comments on the isospin structure of p-h excitation.

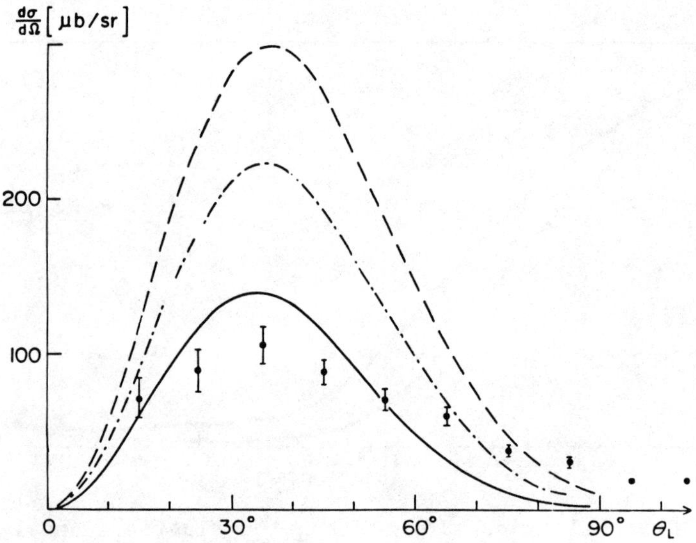

Fig. 13. Angular distrubution for coherent $^{12}C(\gamma,\pi^0)^{12}C$ with $E\gamma \approx 235$ MeV. Theoretical results from Reference 20: solid and dashed results are those of full and impulse calculations, respectively; dot-dashed curve gives the result with no Δ spreading potential. Data from Reference 21

Consider $\pi^\pm$ inelastic scattering from a $T_i = 0$ ground state to a final p-h state of isospin T_F (= 0,1). Let $f_{\pi^\pm;T_f}$ denote the amplitude for this process. Isospin conservation then implies that

$$f_{\pi^+;0} = f_{\pi^-;0} \,, \quad f_{\pi^+;1} = -f_{\pi^-;1} \tag{66}$$

so that, as noted earlier, $\pi^\pm$ cross sections to a state of pure isospin are the same. The standard mechanism for exciting nuclear p-h states is indicated in Figure 10(a). This amplitude we denote as $f^0_{\pi^\pm;T_f}$; assuming Δ-dominance, we have the isospin relations

$$f^0_{\pi^\pm;1} = \pm \tfrac{1}{2} f^0_{\pi^\pm;0} \tag{67}$$

where we assume that the final states have the same space-spin configurations but different isospin T_f. Thus, if the final nuclear state is impure in isospin

$$|\psi_f\rangle = \alpha|p\text{-}h(T_f=0)\rangle + \beta|p\text{-}h(T_f=1)\rangle$$
$$\alpha^2 + \beta^2 = 1 \tag{68}$$

then the impulse approximation for the ratio of $\pi^{\pm}$ amplitudes is

$$\frac{f^0_{\pi^+}}{f^0_{\pi^-}} = \frac{\alpha + \beta/2}{\alpha - \beta/2} \tag{69}$$

Note that the π-nucleus dynamics (given the underlying assumption of Δ-dominance) factors out of the ratio, allowing a clean determination of the isospin admixture. However, the strong Δ-nucleus interaction discussed above removes the "model independence." In particular, the strong effective Δ-nucleon interaction implies that we should consider an additional reaction mechanism,[8] indicated in Figure 10(b). Here, the Δ-N interaction excites the p-h configuration directly. Assuming that the Δ-N interaction has purely $T_{\Delta N} = 1$ (recall that this follows from our association of the spreading potential with pion annihilation), then the isospin recoupling coefficients give[22]

$$f^{\Delta}_{\pi^+;0} = f^{\Delta}_{\pi^-;0}\ , \quad f^{\Delta}_{\pi^+;1} = -f^{\Delta}_{\pi^-;1}$$
$$f^{\Delta}_{\pi^{\pm};1} = \mp\frac{5}{6}\, f^{\Delta}_{\pi^{\pm};0} \tag{70}$$

where $f^{\Delta}_{\pi^{\pm};\,T_f}$ is the amplitude for exciting the p-h state with isospin T_f via the reaction mechanism of Figure 10(b). Note that these isospin considerations apply equally well to continuum knockout of nucleons. That is, proton (neutron) knockout leads to a final state with $\alpha = 1/\sqrt{2}$, $\beta = 1/\sqrt{2}$ $(-1/\sqrt{2})$ in Equation (68). Thus, the ratio of amplitudes for proton knockout by $\pi^{\pm}$ mesons is

$$\frac{F(\pi^+,\pi^+p)}{F(\pi^-,\pi^-p)} = 3\,\frac{1-\xi/9}{1-11\xi/3} \xrightarrow[\xi\to 0]{} 3(1 + \frac{32}{9}\xi + \ldots) \tag{71}$$

$$\xi \equiv -f^{\Delta}_{\pi^+;0}\Big/ f^{0}_{\pi^+;0} \tag{72}$$

For the general case of an isospin mixed state, defined by Equation (68), the $\pi^{\pm}$ ratio is

$$\frac{F(\pi^+,\pi^{+\prime})}{F(\pi^-,\pi^{-\prime})} = \frac{(\alpha+\beta/2) - \xi(\alpha-5\beta/6)}{(\alpha-\beta/2) - \xi(\alpha+5\beta/6)} \tag{73}$$

$$\xrightarrow[\beta<<1]{} 1 + \beta\frac{1+5\xi/3}{1-\xi} + \ldots \tag{74}$$

$$\xrightarrow[\alpha<<1]{} -(1+4\alpha\frac{1-\xi}{1+5\xi/3} + \ldots) \tag{75}$$

Clearly, the Δ-nucleus dynamics summarized by the parameter ξ modifies the isospin admixture coefficient extracted from $\pi^{\pm}$ cross section ratios.

The magnitude of the corrections in Equation (71) and (73) depend sensitively on the details of the Δ-N effective interaction (which generates the spreading potential). For example, in the knockout process, the Δ now transfers a considerable momentum to the ejected nucleon. This will certainly test the spatial dependence of the interaction. Similarly, excitation of isospin-mixed p-h states of different spin-parity can be used to refine the model. In the latter case, one would have to rely upon other measurements of the isospin admixture coefficients for the specific states involved. Karapiperis[22] has evaluated these effects for the specific Δ-N S-wave model alluded to above, with a finite range for the effective force. Just to get an order-of-magnitude estimate, we note that recent S.I.N. coincidence data[23] for the knockout process (see Equation (71)) imply that $\xi \sim 0.05$ to 0.1. The sign agrees with a static limit evaluation of ξ in the resonance region. Using this in the isospin mixing expressions, we find that the effect of small isovector (isoscalar) admixtures is amplified (reduced) by about 20%. Thus, the Δ-nucleus interaction plays a significant role even in "qualitative" nuclear structure studies with pions. Such effects must now be explored systematically in a variety of reactions, including single and double charge exchange, using a reasonable model for the ΔN interaction.

IV. Δ'S AND NUCLEAR BOUND STATES

So far, we have discussed Δ propagation and dynamics above the pion scattering threshold. Clearly, we would like to extend our knowledge of the Δ-nucleus system to lower energy, where "virtual excitation" of the Δ plays a role in mediating the nuclear force and modifying nuclear properties, such as electromagnetic moments. The reasons for doing so are reinforced by our conclusion that the Δ couples strongly to the pion annihilation channels, since this is basically the same physics which leads us to think of Δ-excitation as a mediator of the nuclear force. Considerable effort has been expended on including Δ's in calculations of low energy nuclear properties. For example, perturbative estimates have been made of the Δ contribution to nuclear isovector magnetic moments via two-body exchange currents; Δ-hole components have been included nonperturbatively in nuclear wavefunctions to address the nuclear spin-isospin response function. In the quark language, the Δ may influence long-wavelength nuclear properties in the $\sigma\tau$ channel because $N \to \Delta$ excitation involves no change of the quark orbital. However, quantitative results are all rather controversial, depending sensitively on the short range structure of the Δ-N interaction. We shall make no attempt to review the situation. Rather, we will focus on the narrow question of defining a "Δ probability." We shall see that a dynamical model of the Δ, as presented in Chapter II, implies a strong model dependence in describing Δ's as wavefunction components.

A first remark is that, given a Δ component in a nuclear wavefunction, "measurement" of this component is basically a theoretical problem.[24] The basic difficulty is simply that such components, involving large energy virtual excitations, are extremely short-ranged.

$$\psi_{\Delta N}(r) \underset{r\to\infty}{\sim} e^{-r/d}$$

$$d = [2\mu_{\Delta N}(m_\Delta - m_N)]^{-\frac{1}{2}} \approx \frac{1}{3}\ \text{fm} \tag{76}$$

This is well inside the interaction region of the strong force. In the first place, our understanding of strong interaction dynamics at this distance is close to zero. Secondly, since there is no asymptotic region in which to sample the ΔN wavefunction, one cannot escape the complications of final state interactions in the analysis of any experiment (e.g., a knockout reaction). Consequently, a consistent analysis requires a model of all the short-range interactions, presumably including the complications of relativity. This is possible, although it has not been done. It is also probably irrelevant, since the π, N, Δ degrees of freedom are presumably not a very efficient way to describe hadron physics at a scale of ∿1/3 fm.

Amado[25] has stressed the difficulty with defining a probability within a dynamical model of the $\Delta \rightleftarrows \pi N$ system, such as the isobar model discussed in Chapter II. The only "model-independent" way of defining the Δ wavefunction component entails going to the complex-energy resonance pole (where the Δ is "on-shell"), but the resulting amplitude is not constrained by normalization. Recall that the bound state wavefunction for a standard two-body problem can be written in momentum representation as

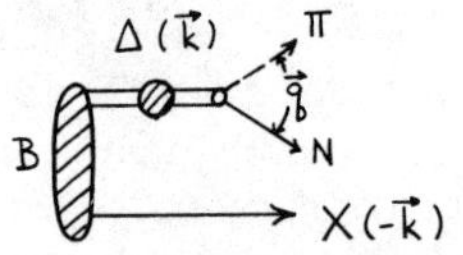

Fig. 14. Schematic representation of Δ component in bound state wavefunction.

$$\psi_B(k) = -\frac{\langle k|V|\psi_B\rangle}{B+k^2/2\mu} \tag{77}$$

showing explicitly the pole at $k^2 = -2\mu B$. The numerator gives the amplitude for separating the constituents with relative momentum k. For the Δ-nucleus system, we wish to perform a similar separation. This is indicated in Figure 14, where X is the residual nucleus. Writing the Hamiltonian as

$$\begin{aligned} H &= T_\pi + T_N + T_X + V_{\pi N} + V_X \\ V_X &\equiv V_{\pi X} + V_{NX} \end{aligned} \tag{78}$$

we have[25] from the Schrödinger equation

$$\psi_B[X(-\vec{k});\pi N(\vec{k},\vec{q})^+] = -\frac{\langle X(-\vec{k});\pi N(\vec{k},\vec{q})^+|V_X|\psi_B\rangle}{B+k^2/2M_X+k^2/2(M_N+M_\pi)+E_q} \tag{79}$$

where $\vec{k}$ is the ΔX relative momentum and $\vec{q}$ is the πN relative momentum. The plus in the state vector indicates an interacting wavefunction

$$\begin{aligned} \langle\pi N(\vec{k},\vec{q})^+| &= \langle\pi N(\vec{k},\vec{q})^0| \\ &+ \int \frac{d\vec{q}\,'}{(2\pi)^3}\frac{\langle\vec{q}|t_{\pi N}(E_q)|\vec{q}\,'\rangle}{E_q - E_q{}'}\langle\pi N(\vec{k},\vec{q})^0| \end{aligned} \tag{80}$$

where the state vectors on the right hand side of Equation (80) are just plane waves. The resonance pole is contained in the transition matrix (see Chapter II)

$$\langle\vec{q}|t_{\pi N}(E)|\vec{q}\,'\rangle = \frac{V(\vec{q})V(\vec{q}\,')}{E-\varepsilon_R} \tag{81}$$

where $V(\vec{q})$ is the vertex function and ε_R is the (complex) pole position. Therefore, we can write near the pole

$$\psi_B[X(-\vec{k});\pi N(\vec{k},\vec{q})^+] \approx -\frac{V(\vec{q})}{\varepsilon_q-\varepsilon_R}\left\{\frac{f(k)}{B+k^2/2\mu_{X,\Delta}+\varepsilon_R}\right\} \tag{82}$$

$$f(k) = \int \frac{d\vec{q}\,'}{(2\pi)^3}\frac{V(\vec{q}\,')}{\varepsilon_R-\varepsilon_{q'}} \langle X(-k);\pi N(\vec{k},\vec{q})^0|V_X|\psi_B\rangle \tag{83}$$

The term in curly brackets plays the role of the Δ-x wavefunction but is defined only at the (dressed) Δ pole. The residue at this complex energy pole is not part of the bound state wavefunction normalization integral.[25] Thus the "Δ probability" in the wavefunction is a highly model dependent quantity. Clearly, the difficulty comes about because of the large width of the Δ generated by coupling to the πN channel. In this context, we recall from Chapter II that the Δ self-energy led to a large mass shift as well as a width. At the energies relevant to low-lying nuclear states, this mass shift may be considerably modified; for example, Pauli blocking of the self-energy would give the Δ a considerably larger, rather model dependent mass. This effect should be included even in perturbative estimates of Δ effects on nuclear properties.

The difficulties mentioned above enter already at the $\Delta \rightleftarrows \pi N$ level. However, an additional level of ambiguity is revealed upon introducing quarks as the true nucleon internal degrees of freedom. For example, imagine that we have a two-nucleon bound state wavefunction in terms of the coordinates of the six quarks $\psi_B(\vec{r}_1 \ldots \vec{r}_6)$, where we have suppressed spin-flavor-color labels. However, it is important to recognize that ψ_B must be antisymmetric under the exchange of any two quarks. We may now ask for the probability that ψ_B contains two nucleons, one nucleon plus a Δ, etc. The problem is that this question is not defined. There are many ways to provide a probability interpretation in terms of three-quark cluster wavefunctions. One prescription would be to define a channel wavefunction as

$$\psi_{nm}(\vec{r}) = \int (d\vec{\rho}\,d\vec{\xi})_{123}(d\vec{\rho}\,d\vec{\xi})_{456}\,\phi_n^*(\vec{r}_1\ldots\vec{r}_3)\,\phi_m^*(\vec{r}_4\ldots\vec{r}_6)\,\psi_B(\vec{r}_1\ldots\vec{r}_6) \tag{84}$$

where we have assumed that the center-of-mass is at the origin, $\vec{r}$ is the relative separation of the (123) and (456) clusters, and $\vec{\rho}$ and $\vec{\xi}$ are internal coordinates for the three-quark cluster. The integral of the square of ψ_{nm} gives the probability for finding the clusters in the states n and m. Clearly, this prescription is reasonable when the clusters are well separated. However, where they are overlapping, the prescription is completely arbitrary and presumably exaggerates considerably the probability for excited state clusters.

For example, a wavefunction of the resonating group type built upon the cluster ground states

$$\psi_{RGM} \equiv \mathcal{A}[\phi_0(\vec{r}_1 \ldots \vec{r}_3)\phi_0(\vec{r}_4 \ldots \vec{r}_6)f(\vec{R}_{123}-\vec{R}_{456})] \tag{85}$$

where $\mathcal{A}$ is the antisymmetrization operator, would lead in Equation (84) to substantial probabilities for excited state clusters. It is perhaps worth noting that there is considerable room for model-dependence in such definitions; for example, the probability for the two nucleons in the deuteron to be outside 1.5 fm (where the "bags" are reasonably separated) is only about 50%. The real point is that definitions, such as that in Equation (84), of cluster probabilities are basically meaningless. Meaningful comparisons should be made at the level of expectation values of one-, two- and possibly higher-body operators within a specific dynamical model of the many-body system.

REFERENCES

1. E. J. Moniz and A. Sevgen, Phys. Rev. C24 (1981) 224.
2. R. M. Woloshyn, E. J. Moniz and R. Aaron, Phys. Rev. C13 (1976) 286.
3. L. Heller, J. C. Martinez and E. J. Moniz, to be published.
4. G. E. Brown and M. Rho, Phys. Lett. 82B (1979) 177.
5. S. Théberge, A. W. Thomas and G. A. Miller, Phys. Rev. D22 (1980) 2838.
6. R. L. Jaffe and F. E. Low, Phys. Rev. D19 (1979) 2105.
7. D. Ashery et al., Phys. Rev. Lett. 47 (1981) 895.
8. F. Lenz and E. J. Moniz, to be published in Advances in Nuclear Physics.
9. M. Thies, Nucl. Phys. A382 (1982) 434.
10. T. T. S. Kuo, F. Osterfeld and S. Y. Lee, Phys. Rev. Lett. 45 (1980) 7861.
11. M. S. Hussein and E. J. Moniz, to be published.
12. M. Hirata, F. Lenz and K. Yazaki, Ann. Phys. (N.Y.) 108 (1977) 116.
13. M. Hirata, J. H. Koch, F. Lenz and E. J. Moniz, Ann. Phys. (N.Y.) 120 (1979) 205.
14. Y. Horikawa, M. Thies and F. Lenz, Nucl. Phys. A345 (1980) 386.
15. R. A. Eisenstein and F. Tabakin, Comp. Phys. Commun. 12 (1976) 237.
16. B. Karaoglu, MIT thesis (1982).
17. T. S.-H. Lee and K. Ohta, Phys. Rev. C25 (1982) 3043.
18. B. Karaoglu, T. Karapiperis and E. J. Moniz, Phys. Rev. C22 (1980) 1806.
19. I. Navon et al., Phys. Lett. 95B (1980) 365.
20. J. H. Koch and E. J. Moniz, Phys. Rev. C20 (1979) 235 and to be published.
21. J. Arends et al., Bonn preprint (1982).
22. T. Karapiperis, MIT thesis (1982).

23. Th. S. Bauger et al., SIN Newsletter (1982); Q. Ingram, private communication.
24. R. D. Amado, Phys. Rev. C19 (1979) 1473.
25. R. D. Amado, Phys. Rev. C19 (1979) 1095.

PARITY NON-CONSERVATION IN MUONIC HELIUM ATOMS

John Bailey

NIKHEF-K
Postbus 4395
1009 AJ Amsterdam

I shall describe a difficult experiment, which may succeed if several unknown factors are favourable.

The reason for doing the experiment is as follows. The electroweak interaction Hamiltonian H between a lepton (electron or muon) and a nucleus is:

$$H = NC_{1n}H_{1n} + ZC_{1p}H_{1p} + S_nC_{2n}H_{2n} + S_pC_{2p}H_{2p}$$

where

$$C_{2p} = -C_{2n} = 1{\cdot}2\,\varepsilon\ ; \quad C_{1p} = \varepsilon = \tfrac{1}{2}(1-4\sin^2\theta_W) \simeq 0\ ; \quad C_{1n} = -\tfrac{1}{2}\ ;$$

θ_W is the Glashow angle, so C_{1n} is much bigger than the other C's. This Hamiltonian contains 4 constants C, and different experiments test different combinations of these 4 constants, most measurements until now having been confined to electrons. In the standard model, the same 4 constants are also valid for muons, but this should be tested. The present experiment attempts to do this by making two measurements, on the muonic atoms of the two He isotopes. It is based on a suggestion be Bernabéu et al. (1974). Muonic atoms are smaller than normal atoms by a factor $m_\mu/m_e = 200$, so the muon is very close to the nucleus, thus enhancing weak-interaction effects (which are short-range).

As with most experiments looking for PNC (= Parity Non-Conservation) effects, we get an enhancement by using two quantum states which have opposite parities, but nearly the same energy. For muonic atoms with an idealized point nucleus, we know that the Dirac equation predicts that the $2s_{\frac{1}{2}}$ and $2p_{\frac{1}{2}}$ levels have identical energies. When we go to a quantized theory with creation and annihilation, there are several effects which change the energy of the 2s state;

the biggest is an increase of the 2s binding energy by the "vacuum polarization"† diagram where the photon from the μ^- temporarily becomes an e^+e^- bubble. (There is a second vacuum polarization diagram where the photon from the μ^- becomes a $\mu^+\mu^-$ bubble, but this is much smaller. The Lamb shift diagram, which decreases the 2s binding, is between the 2 vacuum polarization terms in magnitude).
However a real nucleus has a finite size, which decreases the 2s binding energy, as you can see from Isaac Newton's famous theorem on the field of a spherical shell of matter which exerts an inverse-square force -- outside the shell, its field is the same as if all the matter in the shell were concentrated at its centre, while inside the shell the field is zero. (Remember that the μ^- in a 2s state spends a lot of its time inside the nucleus, whereas the 2p states are nearly completely outside it). The finite-size effect is smaller than the vacuum polarization effect for Z=1, but much larger than it for Z>4. Clearly between these values of Z, we may find an isotope in whose muonic atom the 2 effects nearly balance, and the 2s state has nearly the same energy as one of the two 2p states. Some atomic effects further restrict our choice - we wish our 2s metastable state to have a long lifetime, so to reduce collission rates we will work in a gas; Li is perhaps possible, but out first preference is clearly He, which has given very encouraging results in the experiments of Zavattini and his colleagues.

The electro-weak theory has two important consequences. First, the true energy eigenstate is not $|2s\rangle$ but $(|2s\rangle + \eta|2p\rangle)$ where $\eta \approx 4\times10^{-9}$ for ^{4}He. As a result, the "2s" state can decay to the 1s state not only by the normal means of emitting two photons, whose energies sum to 8.2 keV, but occasionally ($B = 5\times10^{-6}$ probability) by emitting one photon of 8.2 keV instead. Second, these mono-energetic photons are not isotropic, but have a slight tendency to go near the muon spin direction, with angular distribution $(1+A\cos\theta)$, where A=1%(^{3}He) or 2%(^{4}He). (Remember that C_{1n} is multiplied by neutron number N, which accounts for the difference between our 2 experiments on the 2 He isotopes).

---

† To confuse the student, the word "polarization" is used with several meanings in this talk. You may express the Coulomb force between 2 charged bodies as a sum of virtual photons emitted by one charge and absorbed by the other. Each photon may briefly turn into an e^+e^- pair which then turns back into the photon. So the vacuum behaves like a dielectric between the charged bodies; plus and minus charges appear on the surfaces of the dielectric, and the force between the bodies is modified by this "polarization" of the dielectric. Analogously we speak of "vacuum polarization"; it increases the Coulomb force at very small distances.

Now we can measure this value of "A" easily - for example, if we continuously rotate the muon spin with a fixed transverse magnetic field, we can pick out the small signal at the known frequency with good selectivity, as you know from tuning your TV receivers. Selecting the mono-energetic photons at the top end of the huge background spectrum of single photons (from 0 to 8.2 keV, fortunately falling to zero at 8.2 keV) can be done in the same way you can use a glass of water with some dissolved detergent to detect a few ultra violet photons in the presence of an intense background of visible light.

Manufacturers add some fluorescent stuff to detergent to make your shirts seem extra brilliant and "clean". Visible light does not affect a molecule of this stuff, but an ultra-violet photon with energy above a certain sharp threshold will raise one deep-bound electron to a level from which the de-excitation is by emitting a visible photon, and then emitting the remaining energy in another form - an infra-red photon which warms up the surroundings of the molecule, for example. Then a glass of water containing some of this fluorescer is a very selective ("low-pass filter") detector, since it is nearly completely transparent to photons of energy below the threshold. Now nature provides some analogous selective low-pass transmission filters in the X-ray region, and they are called atoms; each atom has several thresholds ("absorption edges") which are the energies needed to remove an electron from its various shells, to infinity (or at least to the Fermi surface in a solid). The vacancy thus produced in a shell is filled by an electron falling from a higher shell, but NOT from infinity (or the Fermi surface), hence the fluorescence photon has lower energy than the absorbed photon. We are particularly interested in the elements Co (K-edge 7.7 keV) and Ho (L3-edge 8.1 keV), in thin slices in the form of the semi-conductor CoP or the pink transparent scintillating crystal HoF_3, respectively. The fluorescent photons from either of these can be caught in gas proportional counters, cheap and covering large solid angle, whose energy resolution is not remarkable but is good enough to distinguish K X-rays from L X-rays in Co (or L from M in Ho). The energy which remains in the semi-conductor or crystal produces a pulse in coincidence with the proportional-counter pulse, with pulse-height discrimination on both, and this coincidence has the 10^4 background rejection factor which we need. (This factor comes chiefly from the sharpness of the edge, and the fact that a huge flux of photons with energy just below the edge pass harmlessly through our primary detector; the proportional counter secondary detectors are sheltered from them, of course.). I do not yet know many electron-hole pairs I can collect from CoP, or how many photons from HoF_3, in a few microseconds, but it is likely that these or some similar materials will work.

You have doubtless thought of two other sources of photons whose direction is correlated with the spin. The commonest decay mode of the μ^- is to emit an e^- (and 2 neutrinos); these e^- have a distribution in energy (up to 53 MeV maximum) and in angle: $(1+\alpha \cos\theta)$ where α depends on the e^- energy. There is also a rarer decay mode where the μ^- emits an e^- and a low-energy γ (and 2 neutrinos) simultaneously - the "radiative" decay. Now both the normal decay with the e^- producing one or more bremsstrahlung photons as it goes through matter, and the radiative decay, will give us a background of photons correlated with the μ^- spin direction, even if we minimize the bremsstrahlung by making our apparatus of low-Z materials (and perhaps using magnetic fields to confine the e^-). But these photons are not peaked at 8.2 keV, and by measuring their spectral shape above 8.33 keV (K-edge of Ni) we can determine accurately what background to subtract from under the 8.2 keV peak.

Finally I take a beam of μ^- from an accelerator and shoot it into a container of He. Each μ^- slows down, captures into obit around a He nucleus (unless it has stopped in the walls or gone clean through the container) and cascades down through the levels of the μ^- He atom (or ion), emitting a photon or an Auger electron at each transition, until it reaches the 2s metastable state or the 1s ground state, all within a few ns. The 1s μ^- then sit around for 2μs until they decay (except that 2% of them are absorbed by the nucleus), whereas of the 2s μ^-, 40% decay but 60% make a transition to 1s, emitting <u>delayed</u> X-rays, which are what we study in this experiment. We can use a beam of 4×10^7 μ^-/sec, which is two μ^- arriving every 50 ns (due to the accelerator RF structure, a cluster of about 2 arrives at the same moment, then nothing till the next cluster 50 ns later), provided we gate out the "prompt" X-rays during about 5 ns after each beam pulse arrives, so that at any moment there are about 80 μ^- inside our apparatus. The initial beam has ~100% polarization (meaning that all the μ^- spins are oriented parallel to each other, along the beam direction). Unfortunately, during the cascade the μ^- lose most of this polarization; apart from some tricks (polarizing ^{3}He nuclei, or mixing ^{4}He in some polarized H gas) they will certainly end up with $P < 16\%$, and indeed perhaps only $P \approx 4\%$ (not yet measured in 2s state). Moreover most of them end up in the ground state; the yield of the metastable state which we want is only about $Y = 4\%$ (measured by Zavattini).

Now if you multiply together all the small numbers (B,A,P,Y) which I have told you about, and some others which I have not told you about, you get an extremely small number; assuming reasonably pessimistic values for numbers which have not yet been measured, you need to run for 40 weeks at the SIN accelerator this year to see a signal of 2 standard deviations. Buth within a few years SIN will have a higher intensity, and then the experiment should succeed within one year of real time.

PARTICIPANTS

ABRAHAMS, K.	E.C.N., Westerduinweg 3, Postbus 1, 1755 ZG Petten, The Netherlands
ADELBERGER, E.G.	Nuclear Physics lab. G1-10, University of Washington, Seattle WA-98195, U.S.A.
ALONSO, C.E., Miss	Dep. Fisica Atomica y N., Fac. de Fisica, Sevilla, Spain
ALLAART, K.	Natuurkundig Laboratorium, Vrije Universiteit, De Boelelaan 1081, 1081 HV Amsterdam, The Netherlands
ARIAS, J.M.	Dep. Fisica At. y Nuclear, Fac. de Fisica, Sevilla, Spain
ARIMA, A.	University of Tokyo, Dept. of Physics, 3-1, Hongo 7 Chome, Bunkyo-Ku, Tokyo 113, Japan
ATKINSON, D.	Instituut voor Theoretische Natuurkunde, Universiteitscomplex "Paddepoel", P.O. Box 800, 9700 AV Groningen, The Netherlands
AVILA-AGUIRRE, O.L., Mrs.	Nuclear Physics Lab., Keble Road, Oxford OX1 3RH, Great Britain
AZGUI, F., Miss	G.S.I., Darmstadt 11-6100, West Germany
BAILEY, J.	NIKHEF, Sectie K, Postbus 4395, 1009 AJ Amsterdam, The Netherlands
BAKKUM, E.A.	R. v.d. Graaff lab., Princetonplein 4, Utrecht, The Netherlands

BAKKUM, E.L.	R.J. v.d. Graaff lab., Princetonplein 4, Utrecht, The Netherlands
BARFIELD, A.F., Mrs.	Dept. of Physics, University of Arizona, Building 81, Tucson, Arizona 85721, USA
BAYRAKTAR, B.N.	Nükleer Fizik Lab., Haceteppe Univ. Fizik, Ens. Beytepe, Ankara, Turkey
BERKHOUT, J.B.R.	Natuurkundig Laboratorium der Vrije Universiteit, De Boelelaan 1081, 1081 HV Amsterdam, The Netherlands
BERLIJN, J.J.H.	CNC-11, Mail Stop H824, P.O.B. 1663, Los Alamos, NM 87545, U.S.A.
BLASI, N., Miss	K.V.I., Zernikelaan 25, 9747 AA Groningen, The Netherlands
BLOENNIGEN, F.	Institut-Laue-Langevin, 156x Centre de Tri, F-38042 Grenoble, France
BOEGLIN, W.	Institute of Physics, Klingelbergstr. 82, 4056 Basel, Switzerland
BRANDENBURG, S.	K.V.I., Zernikelaan 25, 9747 AA Groningen, The Netherlands
BRUSSAARD, P.J.	Fysisch Laboratorium, Rijksuniversiteit Utrecht, Postbus 80000, 3508 TA Utrecht, The Netherlands
BURGHARDT, A.J.C.	NIKHEF-K, Oosterringdijk 18, Amsterdam, The Netherlands
BURNS, D.T.	Kelvin Lab., N.E.L., East Kilbride, Glasgow, Scotland, Great Britain
CAVEDON, J.M.	DPh-N/HE, CEN Saclay, 91191 Gif-sur-Yvette, France
CHANFRAY, G.	Institut Physique Nucléaire, Université Claude Bernard, 43 Avenue du 11 Novembre, 69622 Villeurbanne Cedex, France
CORNELISSENS, T., Miss	L.U.C., Universitaire Campus, B-3610 Diepenbeek, Belgium

DALEY, H.J.	The University of Arizona, College of Liberal Arts, Dept. of Physics, Bldg. 81, Tucson, Arizona 85721, U.S.A.
DEAN, G.W.R.	Dept. of Natural Philosophy, The University, Glasgow G12 8QQ, Great Britain
DELFINI, M.G., Miss	R. v.d. Graaff Lab., Princetonplein 4, Utrecht, The Netherlands
DEJBAKHSH, H., Miss	Brookhaven National Laboratory, Ass. Universities Inc., Dept. of Physics, Upton, Long Island, NY 11973, U.S.A.
DIEPERINK, A.E.L.	K.V.I. der Rijksuniversiteit, Universiteitscomplex "De Paddepoel", Zernikelaan 25, 9447 AA Groningen, The Netherlands
DINCKLAGE, R.D. von	CERN/EP Div., CH-1211 Geneve 23, Switzerland
DIOSZEGI, I.	Institute of Isotopes, 1525 Budapest, P.O. Box 77, Hungary
DODIG-CRNKOVIC, G., Mrs.	Theoretical Physics Dept., Inst. R. Boskovic, Zagreb 41000, Bijenicka 54, Yugoslavia
DOGAN, N., Miss	Haceteppe University, Dept. of Physics, Ankara, Turkey
DONNE, A.J.H.	Vrije Universiteit, De Boelelaan 1081, 1081 HV Amsterdam, The Netherlands
DONNELLY, T.W.	Massachusetts Inst. of Technology, Center for Theoretical Physics, Cambridge, Mass. 02139, U.S.A.
DORNINGER, Chr.	Atominstitut, A-1020 Wien, Schüttelstrasse 115, Austria
DRUNEN, J.J.M. van	Bureau Congresses of the Minister's Secretariat, Min. of Education and Science, Nieuwe Uitleg 1, 2514 BP The Hague, The Netherlands
ESCUDERO, J.I.	Dep. Fisica At. y Nuclear, Fac. de Fisica, Sevilla, Spain

FAROOQ, A.	University of Birmingham, Dept. of Physics, Birmingham B15 2TT, Great Britain
FIELDS, Chr. A.	Nuclear Physics Lab., Box 446, University of Colorado, Boulder, CO 80309, U.S.A.
FRENCH, J.B.	Dep. of Physics and Astronomy, University of Rochester, Rochester, 14627 NY, U.S.A.
GARG, U.	University of Notre Dame, College of Science, Dept. of Physics, Notre Dame, Ind. 46556, U.S.A.
GAZIS EVANGELOS, N.	NTU, Physics Lab. B.', Zografou T.T.624, Athens, Greece
GLAUDEMANS, P.W.M.	Fysisch Laboratorium, Rijksuniversiteit Utrecht, P.O. Box 80000, 3508 TA Utrecht, The Netherlands
GOUDOEVER, J. van	NIKHEF-K, Oosterringdijk 18, Amsterdam, The Netherlands
GRASDIJK, P.	K.V.I., Zernikelaan 25, 9747 AA Groningen, The Netherlands
GUERRA, Miss M.F.	Centro de Fisica Nuclear, Av. Gama Pinto 2, 1699 Lisbon, Portugal
HAIDENBAUER, J.	Inst. für Theor. Physik, Universitätsplatz 5, A-8010 Graz, Austria
HAMILL, J.J.	K.V.I., Zernikelaan 25, 9747 AA Groningen, The Netherlands
HAERTING, A.W.	Institut für Theoret. Physik, Universität Regensburg, D-8400 Regensburg, West Germany
HAQUE, A.M.I.	Institut für Kernphysik der Universität zu Köln, Zülpicherstrasse 77, D-5000 Köln 77, West Germany
HARTER, H.	Institut für Kernphysik der Universität zu Köln, Zülpicherstrasse 77, D-5000 Köln 77, West Germany

HENGEVELD, W.	Nat. Lab. der Vrije Universiteit, De Boelelaan 1081, 1081 HV Amsterdam, The Netherlands
HERDER, J.W.A. den	NIKHEF-K, P.O. Box 4395, 1009 AJ Amsterdam, The Netherlands
HERATH-BANDA, M.A.	M.P.I. für Kernphysik, Postfach 103980, Heidelberg, West Germany
HOOFT, G. 't	Instituut voor Theoretische Fysica, Rijksuniversiteit Utrecht, Princetonplein 5, Postbus 80006, 3508 TA Utrecht, The Netherlands
KISS, A.Z.	Atomki H-4001, Debrecen, Pf. 51, Hungary
KLERK, B.A.S. de, Miss	Ministry of Education and Science, Bureau Congresses, Nieuwe Uitleg 1, 2514 BP The Hague, The Netherlands
KOCZON, P.	Ins. of Nucl. Res., Dept. IA, 05400 SWIERK, Poland
KOK, P.J.J.	E.C.N., P.O. Box 1, 1755 ZG Petten, The Netherlands
KONIJN, J.	NIKHEF-K, P.O. Box 4395, 1009 AJ Amsterdam, The Netherlands
KOUTROULOS, G.C.	Dept. of Theoretical Physics, Univ. of Thessaloniki, Thessaloniki, Greece
LAAT, C.T.A.M. de	Technische Hogeschool Delft, Afd. der Technische Natuurkunde, Vakgroep Experimentele Kernfysica, Postbus 5042, 2600 GA Delft, The Netherlands
LANE, S.M.	Nuclear Science Building, 1st floor, University of Bradford, Bradford, Great Britain
LÜTZENKIRCHEN, K.	Institut für Kernchemie, Johannes-Gutenberg Universität, 6500 Mainz, W. Germany
MAHAUX, C.	Lab. de Physique Théorique, Institut de Physique de l'Université de l'Etat, Sart Tilman, 4000 Liège I, Belgium

MALDEGHEM, J.L.D. van	Institute for Nuclear Physics, Proeftuinstraat 86, B-9000 Gent, Belgium
MEZIANI, Z.	DPH-N/HE, CEN Saclay, 91191 Gif-sur-Yvette, France
MILLER, G.J.	Kelvin Lab., N.E.L., East Kilbride, Great Britain
MONIZ, E.	Massachusetts Institute of Technology, Center for Theoretical Physics, Cambridge, Mass. 02139, U.S.A.
MOONEN, W.H.L.	Technische Hogeschool Eindhoven, afd. Natuurkunde, cycl. 02.13, P.O.B. 513, 5600 MB Eindhoven, The Netherlands
NAGAOKA, R.	c/o Prof. A.Arima, Univ. of Tokyo, Dept. of Physics, 3-1. Hongo 7 Chome, Bunkyo-Ku, Tokyo 113, Japan
NATH, S.	Cyclotron Inst. Texas A & M Univ., College Station, Tx. 77843, U.S.A.
OVERVELD, C.W.A.M. van	Technische Hogeschool Eindhoven, afd. Natuurkunde, Cycl. 02.13, Postbus 513, 5600 MB Eindhoven, The Netherlands
PEDROCCHI, V.G.	Physics Dept. Univers. of Texas, Austin, Tx. 78712, U.S.A.
PENNINGA, J.	Natuurkundig Lab., Vrije Universiteit, De Boelelaan 1081, 1081 HV Amsterdam, The Netherlands
PEREGO, A.	Instituto di Fisica A., Garbano, Italy
PETIT, R.M.A.L., Mrs.	Technische Hogeschool Eindhoven, Afd. Natuurkunde, Cycl. 02.14, Postbus 513, 5600 MB Eindhoven, The Netherlands
PROVOOST, D.M.R.	L.U.C., Universitaire campus, 3610 Diepenbeek, Belgium
RAVENSWAAY, R.O. van	Bureau Congresses of the Minister's Secretariat, Ministry of Education and Science, Nieuwe Uitleg 1, 2514 BP The Hague The Netherlands

RAWAS, E.M.E.	Dept. of Physics, Univ. of Birmingham, P.O. Box 363, Birmingham B15 2TT, Great Britain
ROUSSEV, R.P.	IRNE, Boul. Lenin 72, Sofia 1184, Bulgaria
RUYL, J.F.A.G.	E.C.N., P.O. Box 1, 1755 ZG Petten, The Netherlands
SAGHAI, B.	DPHN-HE, CEN Saclay, 91191 Gif-sur-Yvette, France
SALVERDA, A.G.	Bureau Congresses of the Minister's Secretariat, Ministry of Education and Science, Nieuwe Uitleg 1, 2514 BP The Hague, The Netherlands
SAN JOSE VILLACORTA, C.	c/o Joan Alcover 16,3^{o}, Palma de Mallorca, Baleares, Spain
SCHAD, L.	MPI für Kernphysik Heidelberg, West Germany
SCHWALM, D.	Physikalisches Institut, Universität Heidelberg, Filosofenweg 2, D-69 Heidelberg, B.R.D.
SCHMIDT, H.R.	Max-Planck-Institut für Kernphysik, Box 103980, 6900 Heidelberg, West Germany
SICK, I.	Physikalisches Institut der Universität Basel, Klingelbergstrasse 82, 4056 Basel, Switzerland
SRINIVASA RAO, K.	Matscience, The Institute of Mathematical Sciences, Madras 600113, India
STWERTKA, P.M.	Nuclear Structure Research, University of Rochester, Rochester NY 14627, U.S.A.
THIEU, P. van	Institute de Physique Nucléaire, BP.1, 91406 Orsay, France
TIELENS, T.A.A.	E.C.N., P.O. Box 1, 1755 ZG Petten, The Netherlands

VAZQUES, A.D.J.	Electron Accelerator Lab., Yale Univ., New Haven, Ct. 06520, U.S.A.
VEERMAN, H.P.J.	NIKHEF-K. Oosterringdijk 18, Amsterdam, The Netherlands
VOGEL, P.	Caltech, N.B. Laboratory of Physics, 161-33 Pasadena, California, U.S.A.
WALET, N.R.	Fysisch Laboratorium, Princetonplein 5, Postbus 80000, 3508 TA Utrecht, The Netherlands
WAN, S.L.	Kelvin Lab., N.E.L., East Kilbride, Scotland, Great Britain
WILSCHUT, H.W.E.M.	Zernikelaan 25, 9747 AA Groningen, The Netherlands
WINTER, L.C. de	Fysisch Laboratorium, Princetonplein 5, Postbus 80000, 3508 TA Utrecht, The Netherlands
WITT HUBERTS, P.K.A. de	NIKHEF-K, Postbus 4395, 1009 AJ Amsterdam, The Netherlands
YAZICI, F.	Haceteppe Univ., Dept. of Physics (83), Ankara, Turkey
ZACHAROV, I.	NIKHEF-K, Oosterringdijk 18, Amsterdam, The Netherlands
ZALMSTRA, J.J.A.	Vrije Universiteit, De Boelelaan 1081, 1081 HV Amsterdam, The Netherlands

INDEX